The NEW AGE of CERAMICS

GINGKO PRESS

The New Age of Ceramics

First Published in the United States of America, September, 2016

First Edition

Gingko Press, Inc.

1321 Fifth Street

Berkeley, CA 94710, USA

www.gingkopress.com

ISBN: 978-1-58423-624-5

Printed in China

Cover by Laurent Craste
Back cover by Zemer Peled

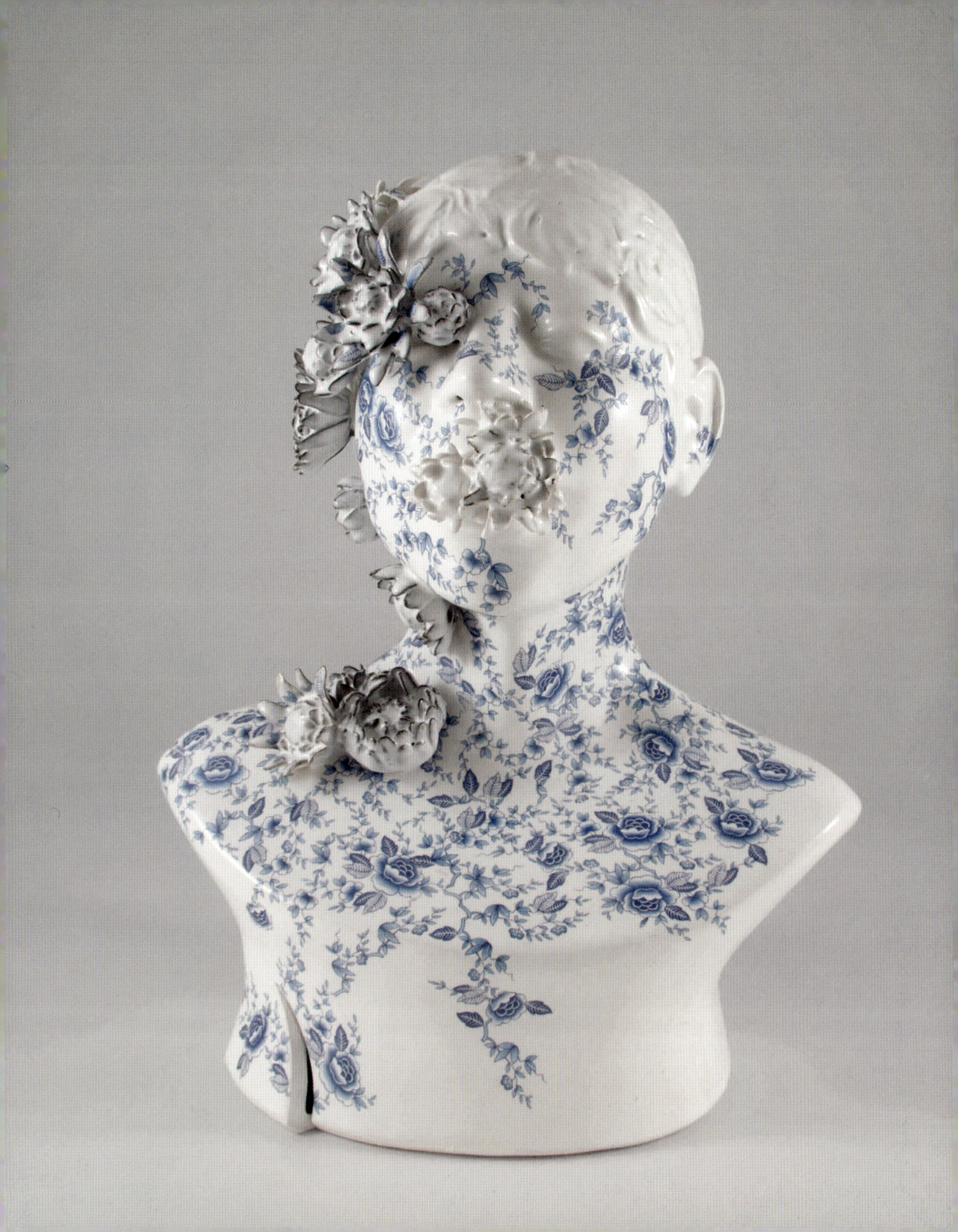

CONTENTS

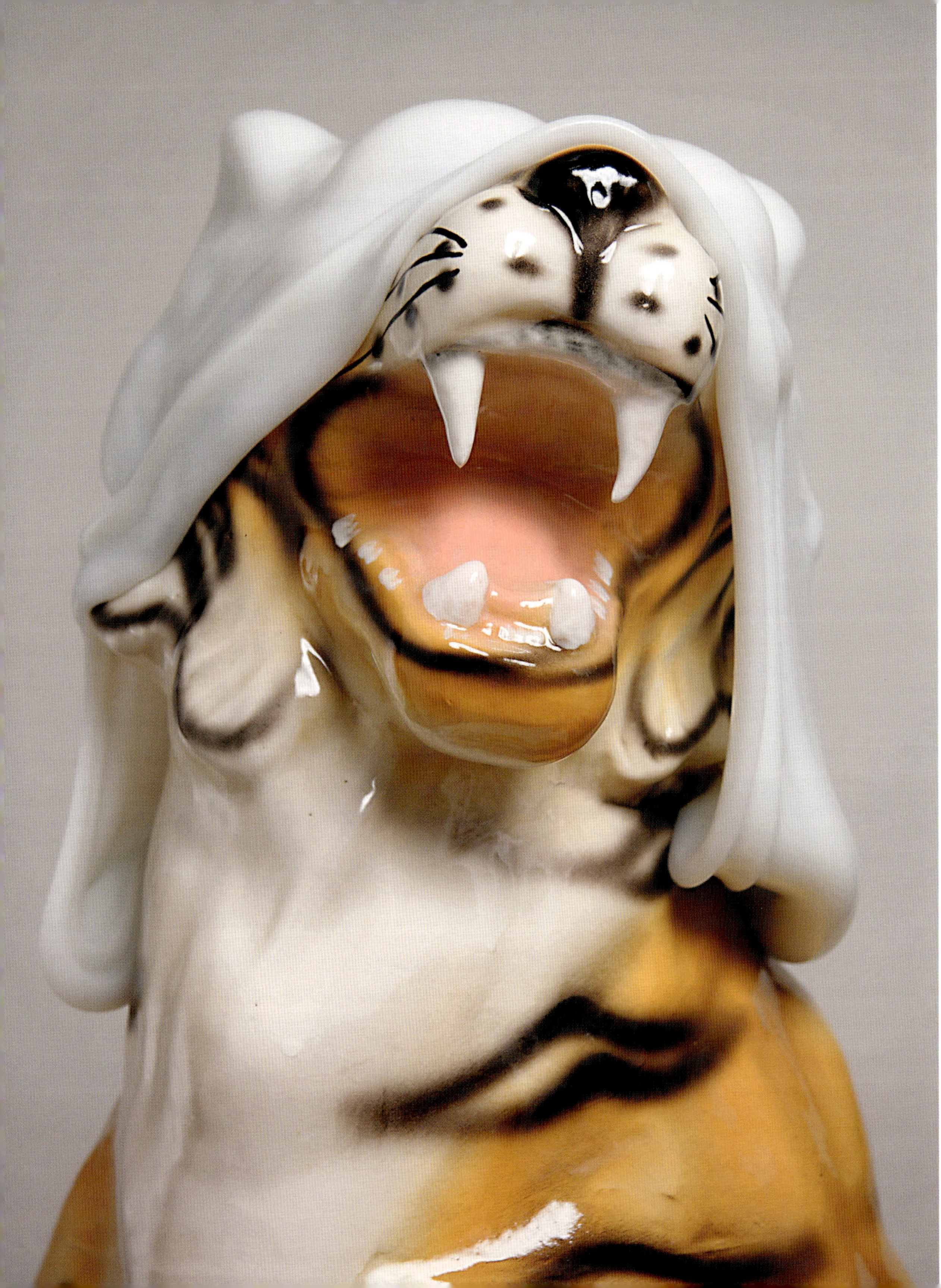

Work by Paul Wood

INTRODUCTION

By Hannah Stouffer

As we stray further from the tangible world of traditional handcraft we have gained a heightened desire for corporeal authenticity. Attaining a tangible connection with materials has become increasingly important. We want to see the imperfections; to feel something. Though the ceramic medium never left, there has been a resurgence in the appreciation of clay due to its tactile and intimate nature. Ceramics exist in a wide range of uses and appearances, from functional to abstract, utilitarian and sculptural, and in the galleries and homes of appreciators. The versatility of the material has kept it relevant as traditional methods merge with current practices.

The ceramic medium has occupied a physical space in our daily routines since the dawn of man. With an ancient history that is neither restricted to geographic location or class, clay has proven that it can withstand the passing of time beyond era or genre. The history of ceramic works exist in almost every developed culture, from usable homeware to decorative art objects, regal gold gilding and bone china, dating back nearly 20,000 years. Similar methods of production are still being used today in contemporary practice, as ceramicists push the boundaries and capabilities of clay. Artists working within the medium today enjoy a staggering variety of methods and techniques, resulting in endless experimentations in shape, form and function.

Clay is as malleable as it is timeless and the appreciation and attention it generates will continue to expand as long as there are passionate creators pushing the capabilities of the medium. The cultural rebirth we are experiencing with ceramics shines a light on new artists embracing its revival and pays homage to the contemporary masters of our time, with a deep appreciation for its historical context. Our drive to feel physical achievements, as embodied by ceramics, grounds us with their subtle reminders of tradition and tactile intimacy.

FUNCTIONAL FORMS

BEN MEDANSKY

CODY HOYT

GRIT CERAMICS

CHRIS HAWKINS

BODIL MANZ

HEATH CERAMICS

BEN MEDANSKY

LOCATION Los Angeles, CA
YEAR OF BIRTH 1988
MATERIALS Ceramic, stoneware
EXPERIENCE 15 years

Ben Medansky's desire to make forms was forged by the bright blue skies and soaring mineral formations of his home state of Arizona and informed by his studies at his alma mater, the School of the Art Institute of Chicago. It was then honed through his work in the studios of his favorite LA contemporary artists. The inspiration for Ben's work is a culmination of his experience in these three landscapes, particularly Los Angeles, with its current reconstruction and varied architectural styles. In the summer of 2012, at age 24, Ben created Ben Medansky Ceramics in a small Los Angeles studio. Since then, he has become an internationally known artist, working from a 3,000 sq. foot space of his own. The result: a meditation on minimalism, on the handmade and manmade, on mechanics re-mastered in earthen and eternal material.

Photos courtesy of the artist

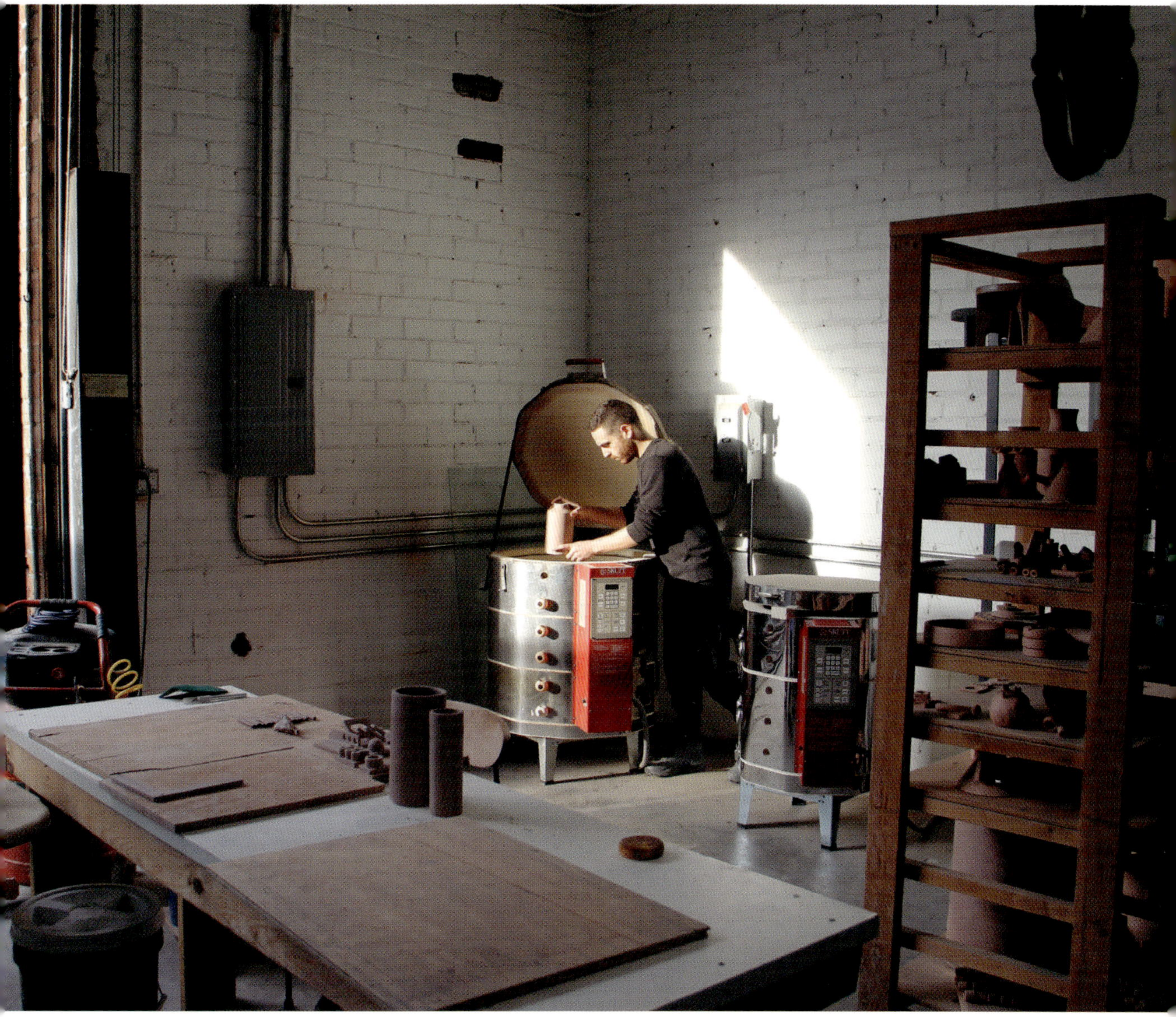

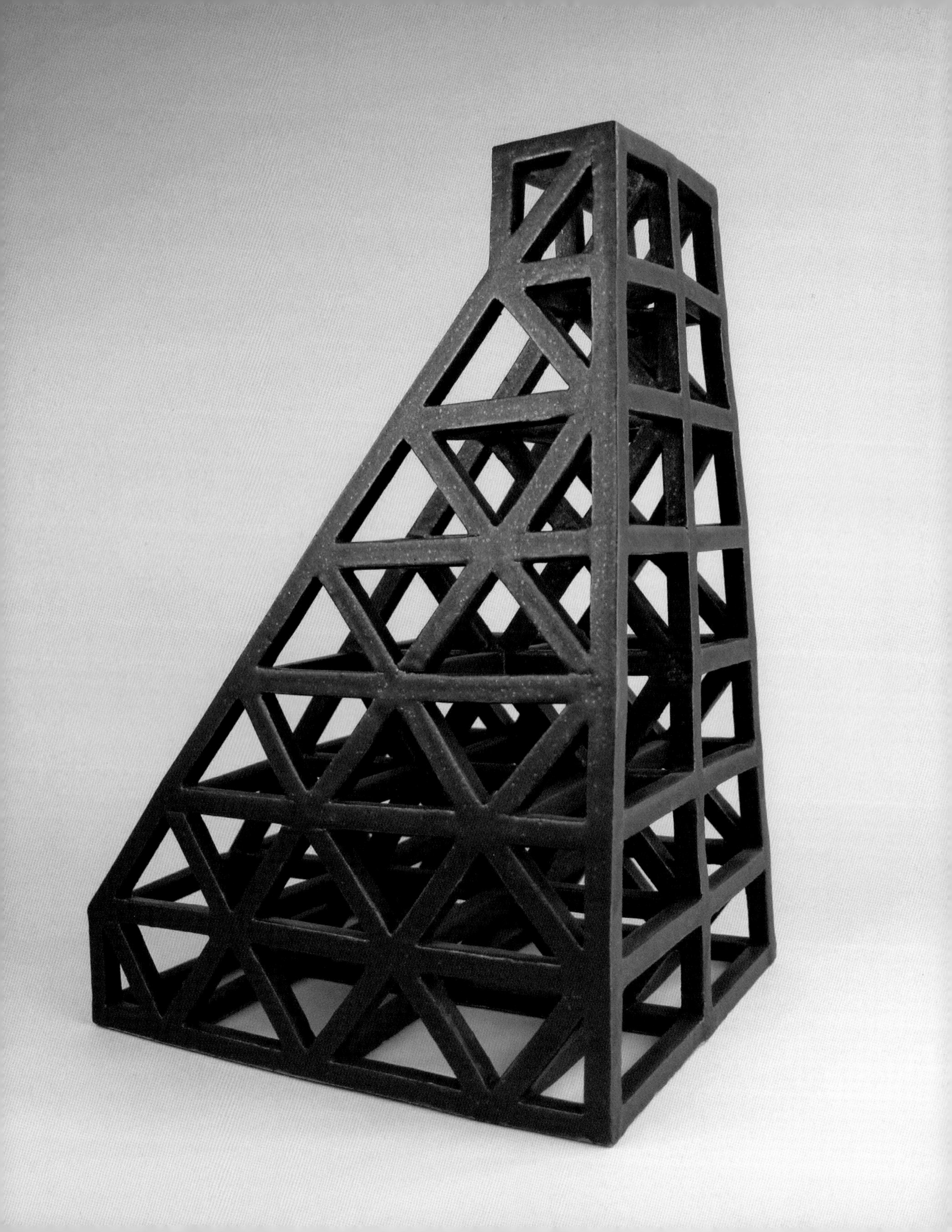

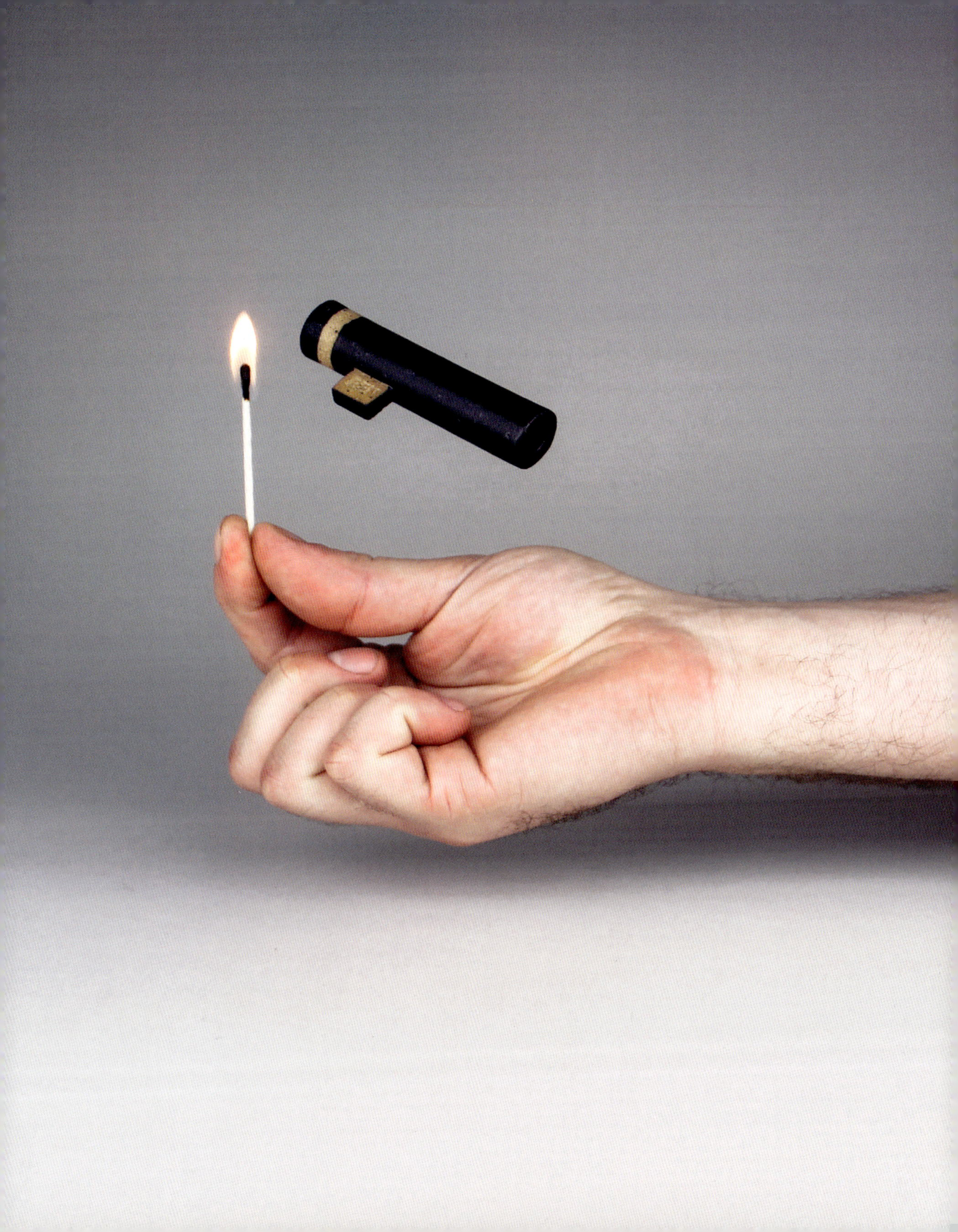

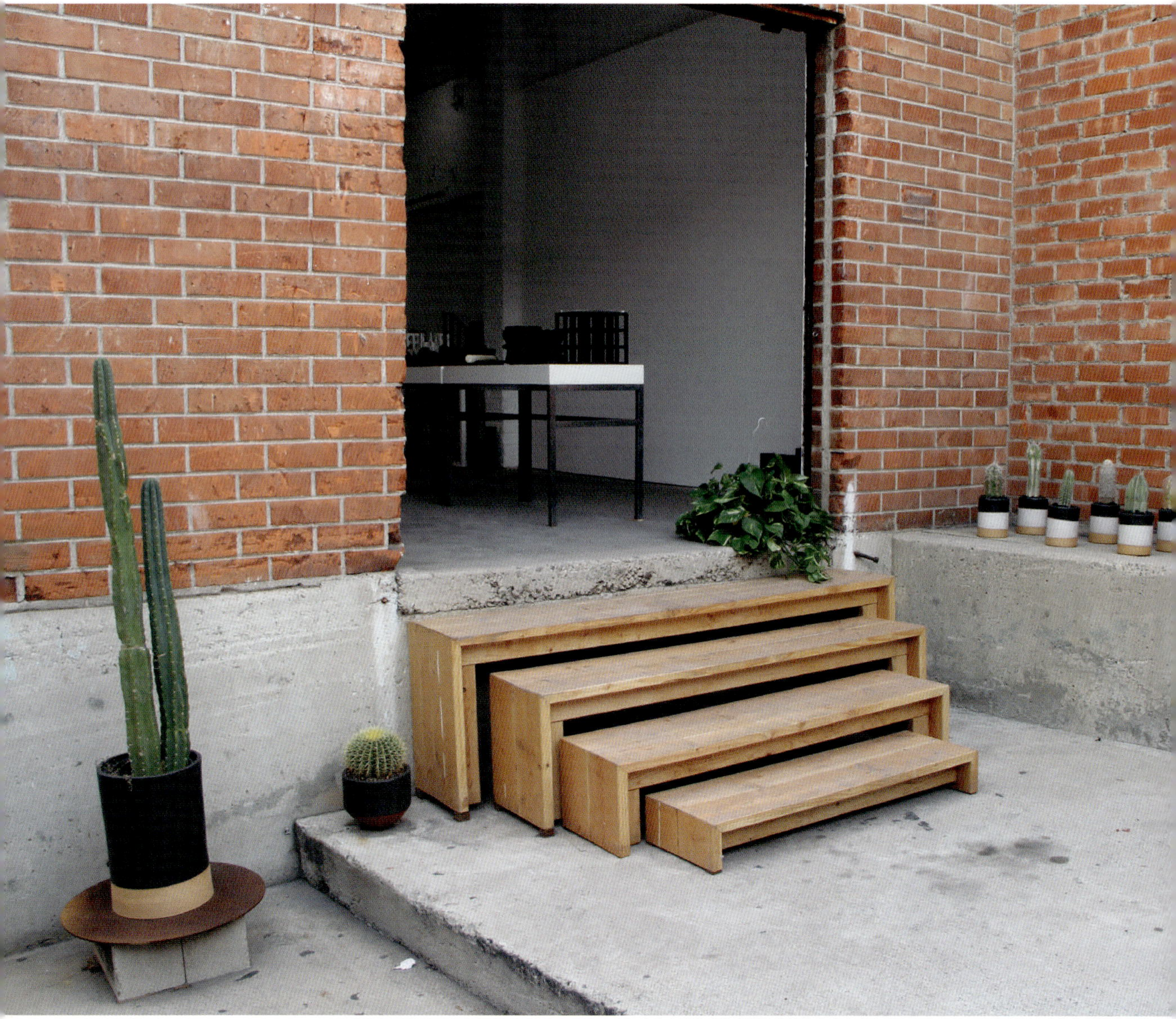

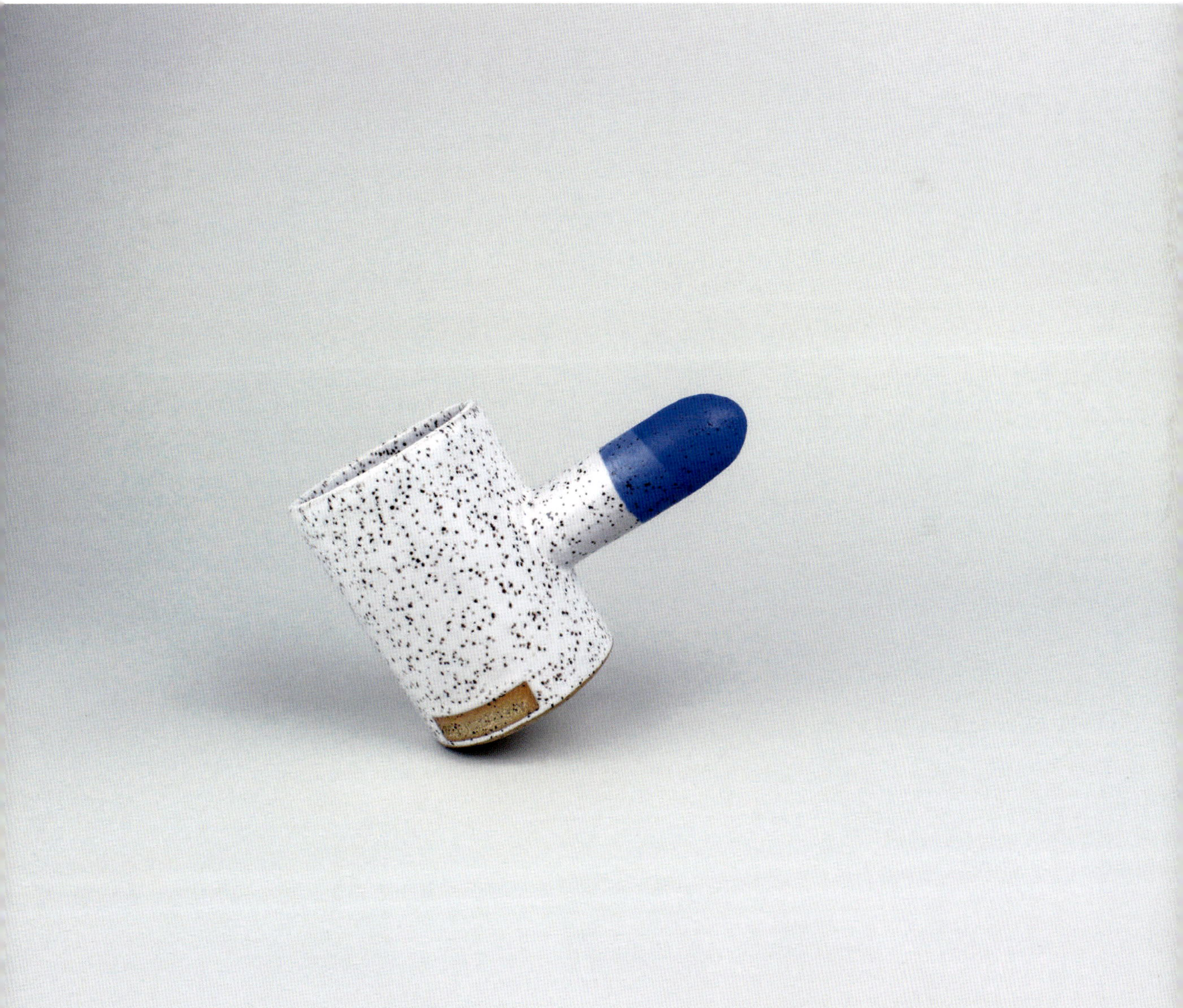

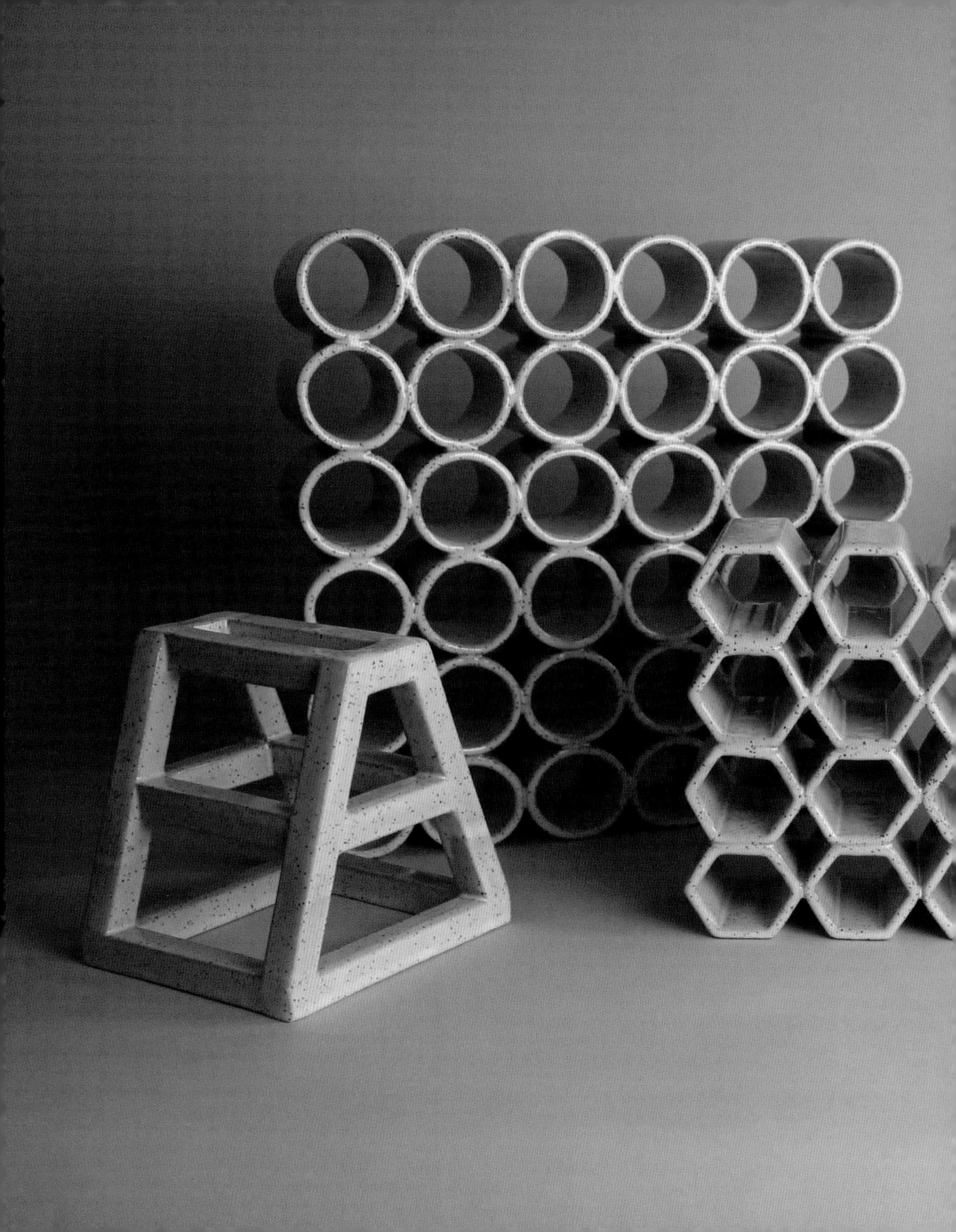

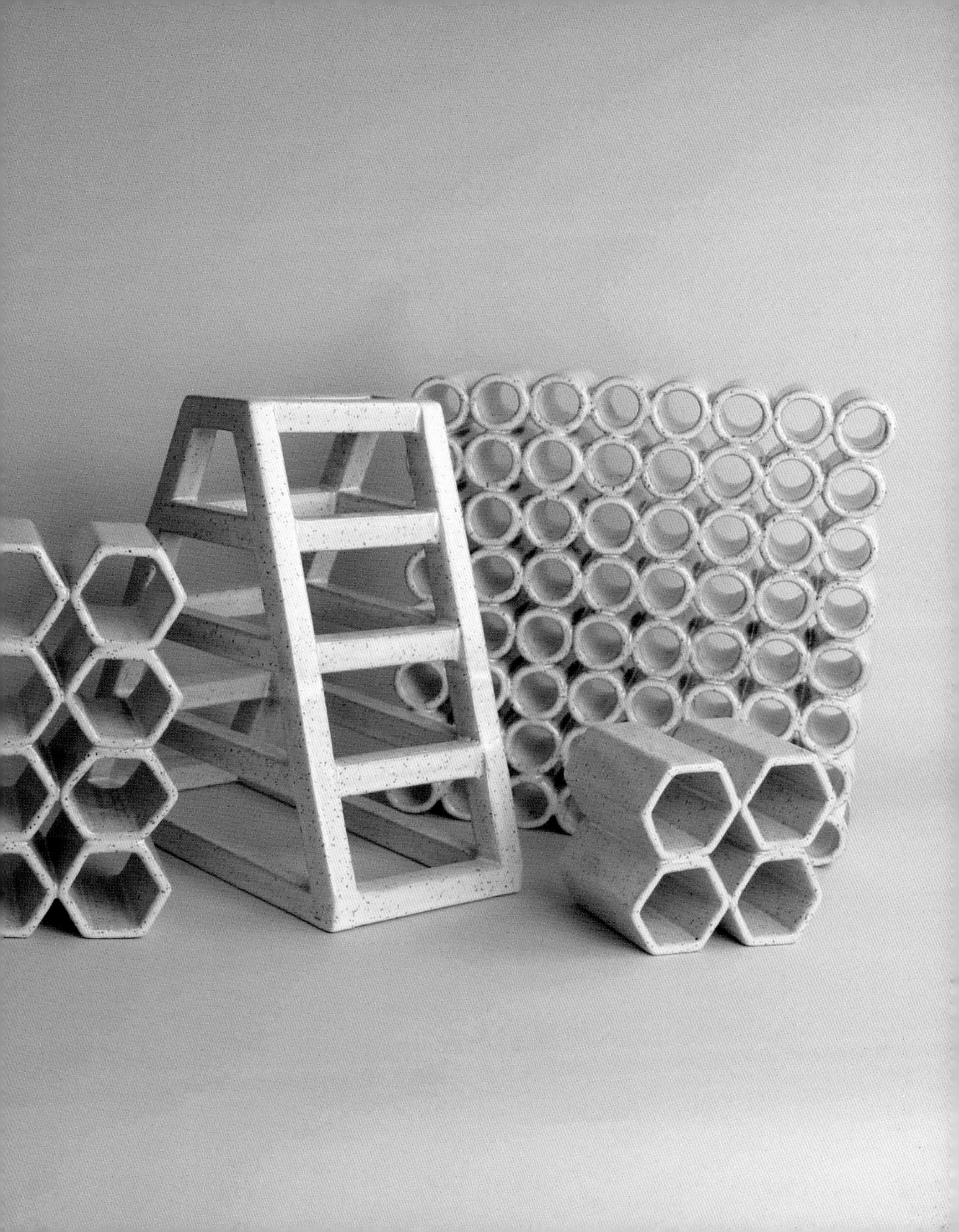

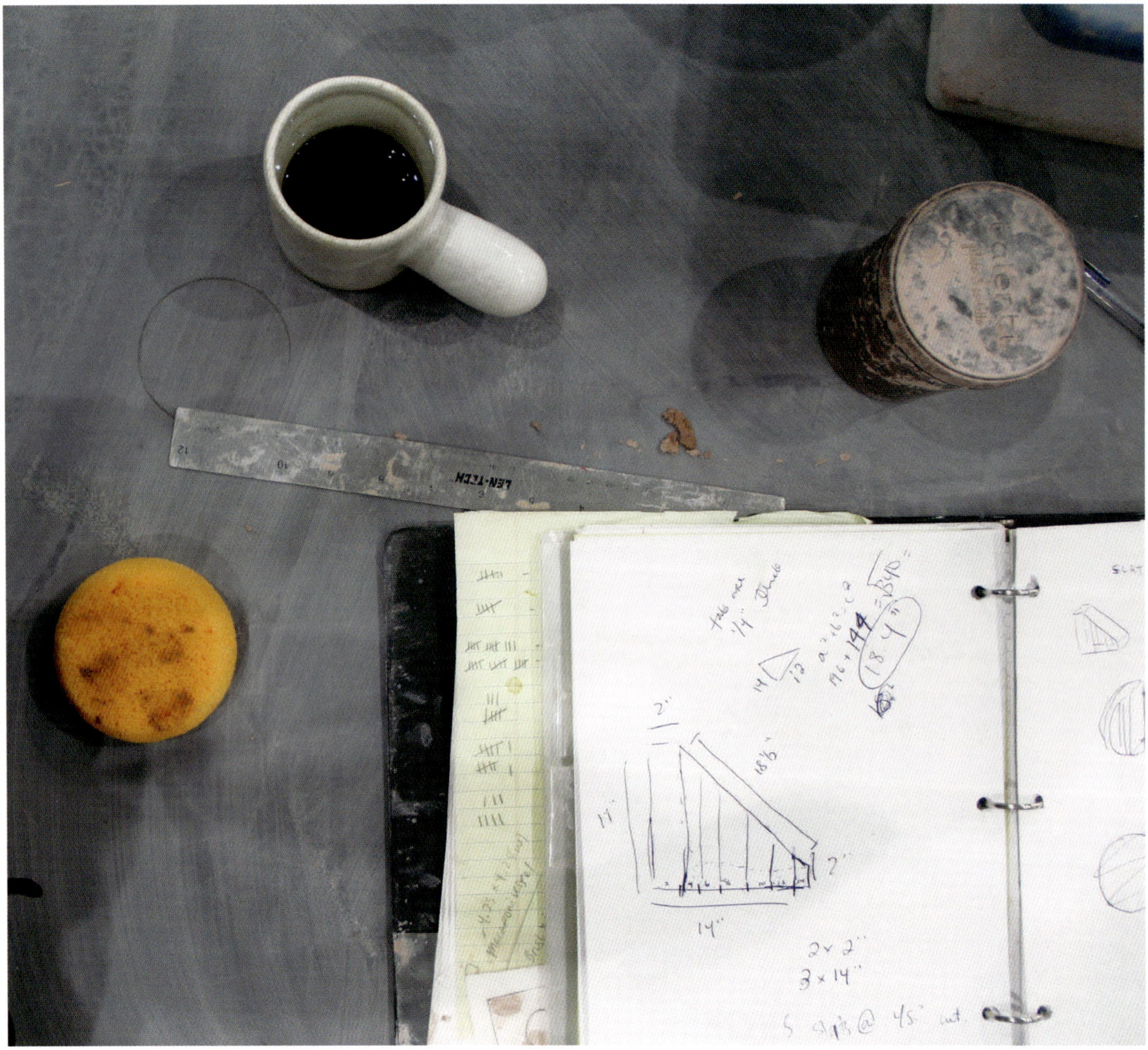
2"
18½"
14"
2"
14"
2 x 2"
3 x 14"

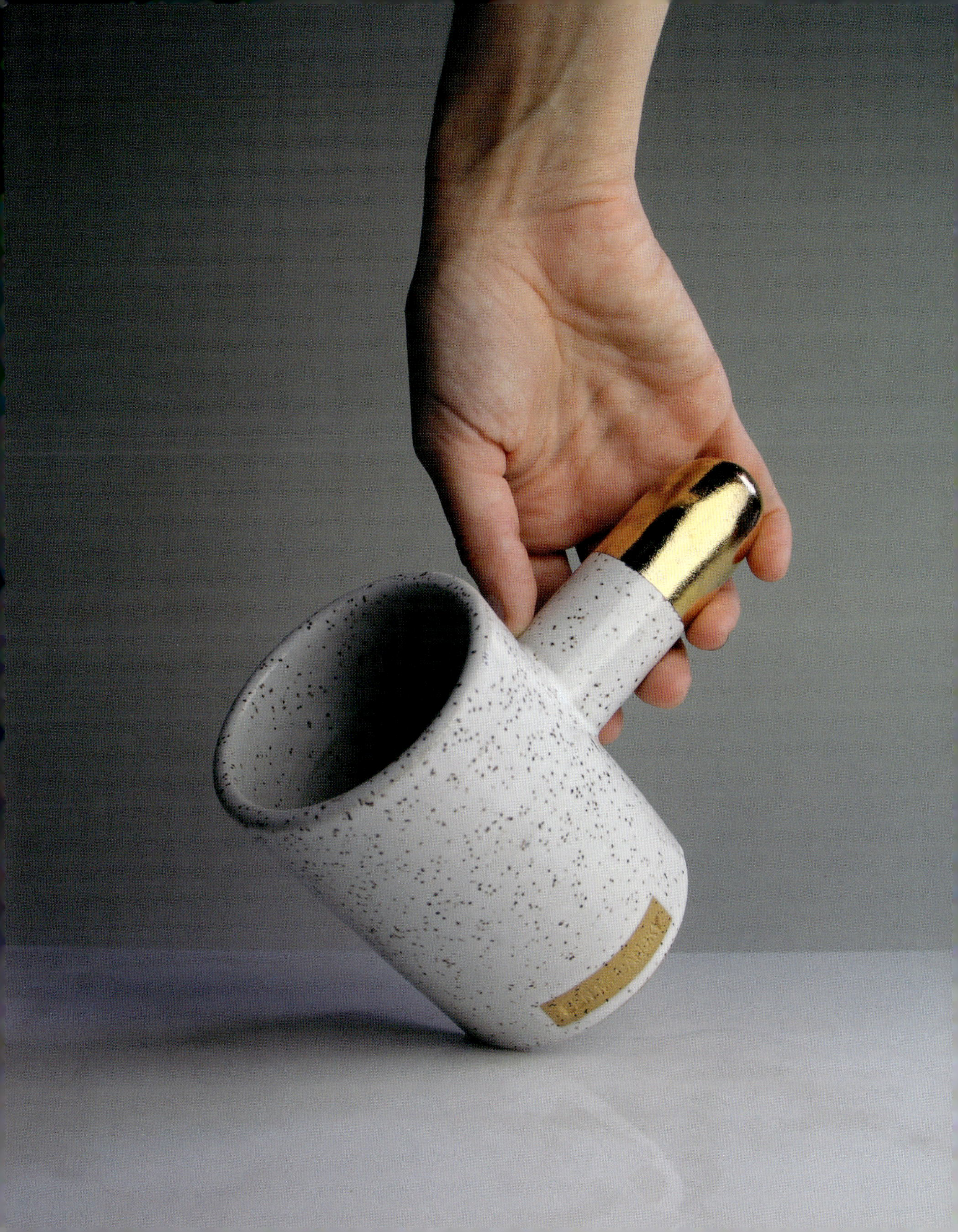

CODY HOYT

LOCATION Brooklyn, NY
YEAR OF BIRTH 1980
MATERIALS Hand built ceramics
EXPERIENCE 4 years

'Most of what I do consists of hand building vessels from clay slabs. While the slab is still flat I decorate the surface with inlay work created from other types and colors of wet clay, like parquetry. In some cases the slab is cut from a cross section of different clays, which results in pattern extending from the surface all the way through the walls of the vessel. I once fire the pieces to cone 5. I finish the surfaces with sand-paper before and after firing.

Whether it's poured in place concrete or moulded and vitrified ceramic — the slab is a symbol of stolid foundational stability in our transient, fleeting, pixelat-ed present. It's an undeniable rigid mass. A slab can be folded upon itself to create an object and retain those attributes of tangible dependability. I feel that formal complexity in the work grants it a longevity and stability that echoes this. An augmented surface further exemplifies the object-ness and engages the viewer. The experience of the object is its reason for being.'

Born and raised in Florida, Cody Hoyt received his B.F.A. from the Massachusetts College of Art in 2007. He has had solo exhibitions in New York at Patrick Parrish Gallery, Jeff Bailey Gallery, and the Ace Hotel, and at Andrew Rafacz Gallery, Chicago. His work has been shown in Dwell, Interview Magazine, Elle Decor, Architectural Digest, American Craft, and the New York Times Magazine. He lives and works in Brooklyn.

Photos courtesy of Clemens Kois and the artist

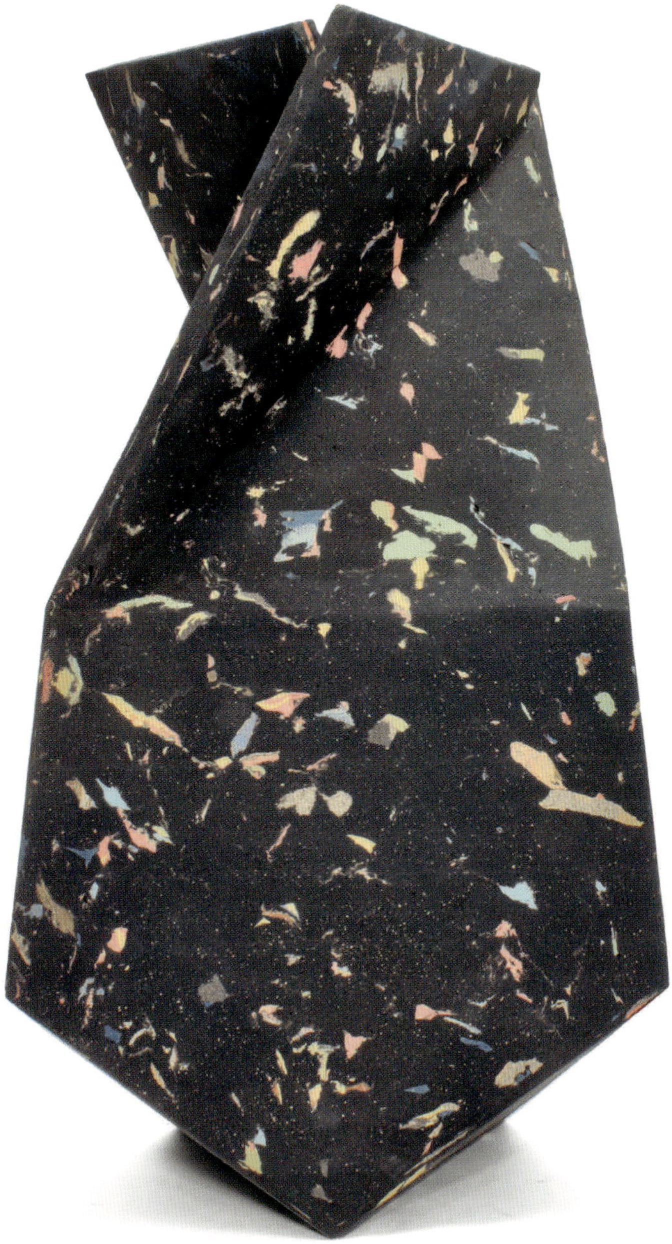

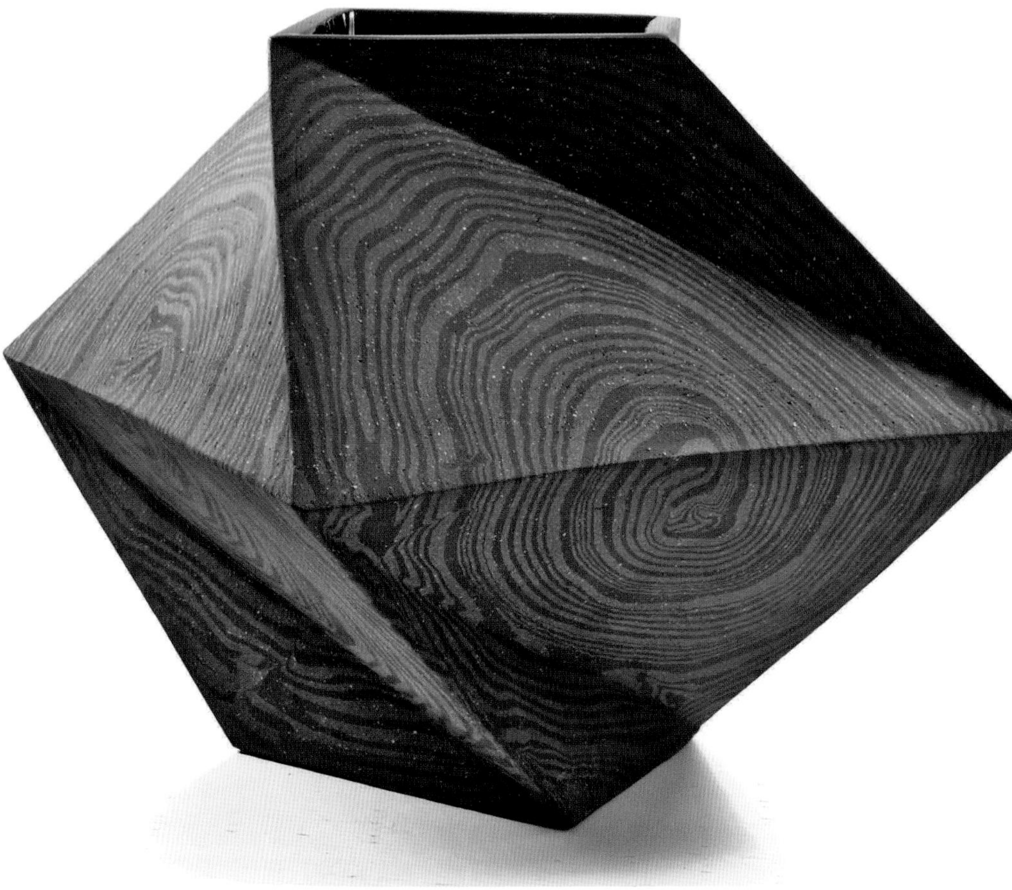

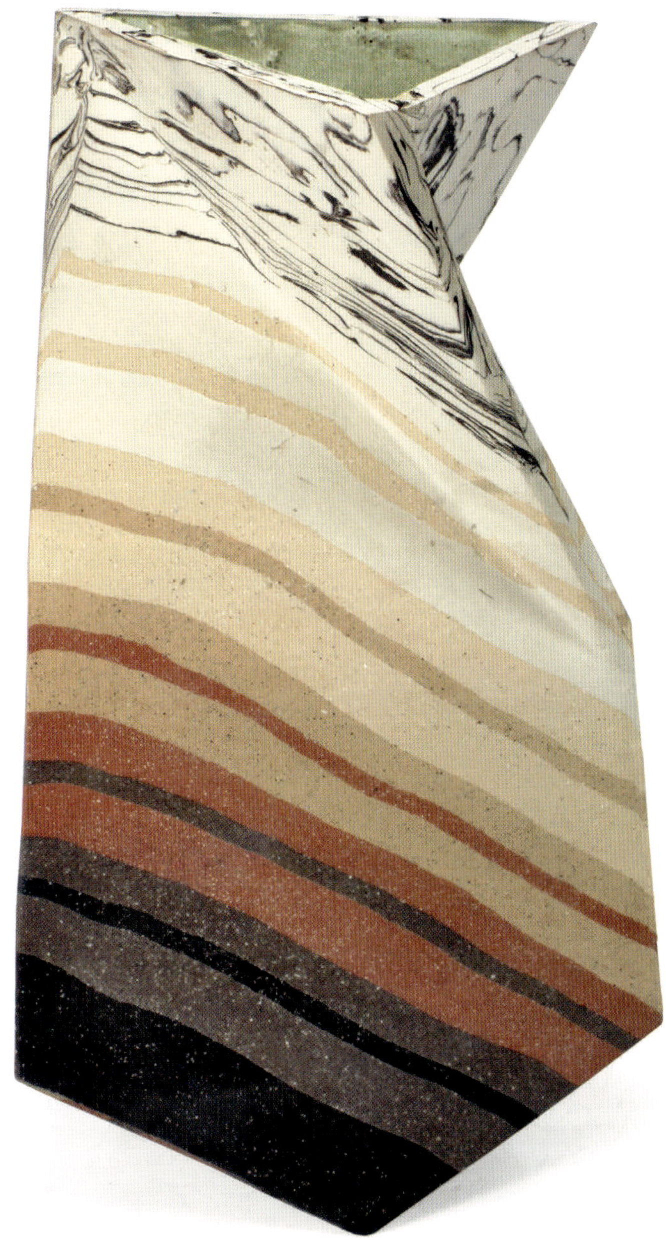

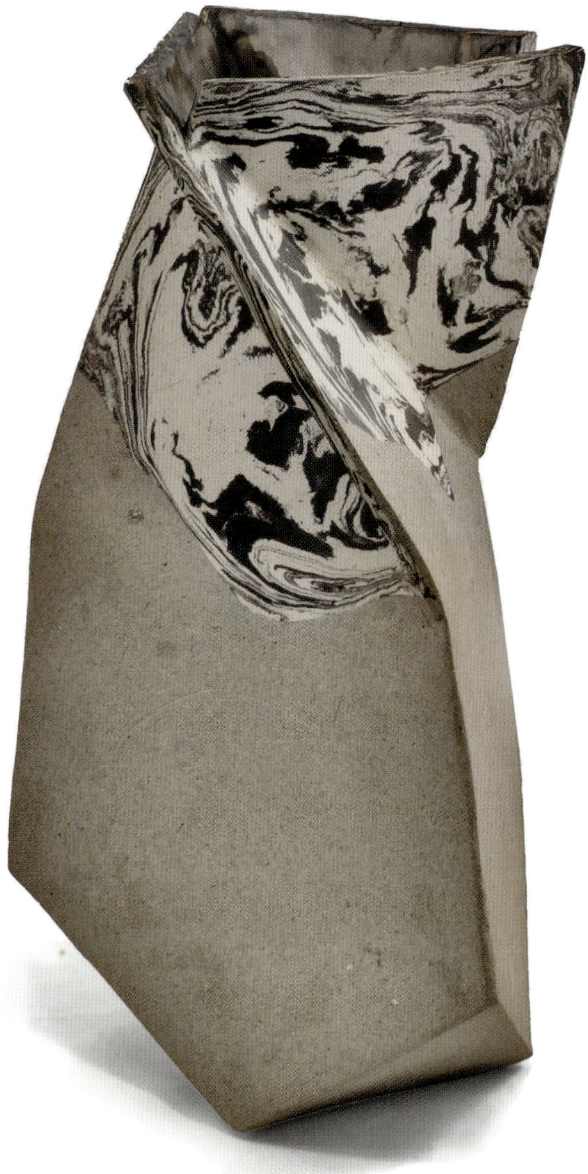

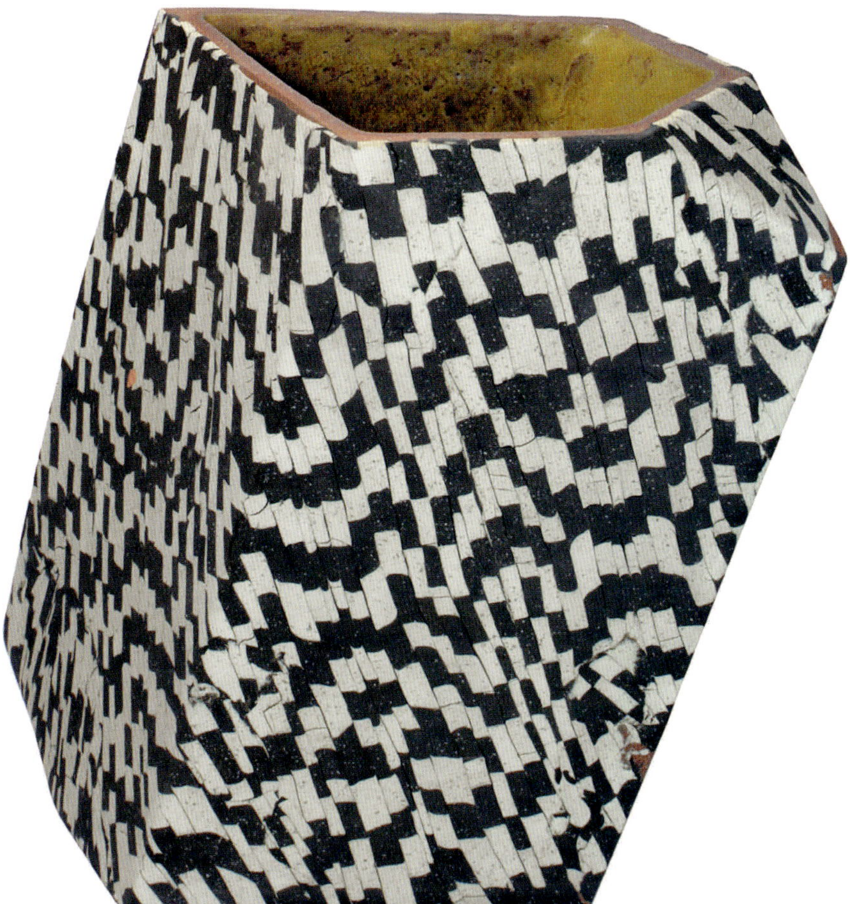

GRIT CERAMICS

LOCATION New South Wales, Australia
YEAR OF BIRTH 1982
MATERIALS Porcelain, stoneware
EXPERIENCE 10 years

Grit Ceramics is a small homewares brand created by Leia Sherblom, a North American transplant. Leia designs and produces a handmade porcelain range of tableware and lighting in her Northcoast NSW beachside studio. Each piece is made from a uniquely formulated blend of porcelain casting clay slip and stained with neutral toned pigments. The interior is glazed clear, and the outside left raw — the smooth finish of the porcelain remains. Each piece is infused with guts, resilience, initiative, and tenacity, and finished with a deep appreciation of everything it takes to be a modern maker.

'Originals of each product are made in a variety of ways — either on the wheel, handbuilt, found objects, or d) all of the above. The clay that I use is my own porcelain recipe that I have formulated over time, mixing from Australian raw materials and pigment stains — the satin glaze recipe is also a product of my own development. My partner makes the plaster moulds (an art in itself!) and after days of mixing, we pour the liquid clay into them — slipcasting each piece. The piece is removed from the mould, fettled, dried slowly and sanded. After the first firing, it is cleaned and sanded again, glazed, screen printed and fired again to stoneware temperatures. It goes through a final process of sanding before it is finished and ready for use.'

Photos courtesy of the artist

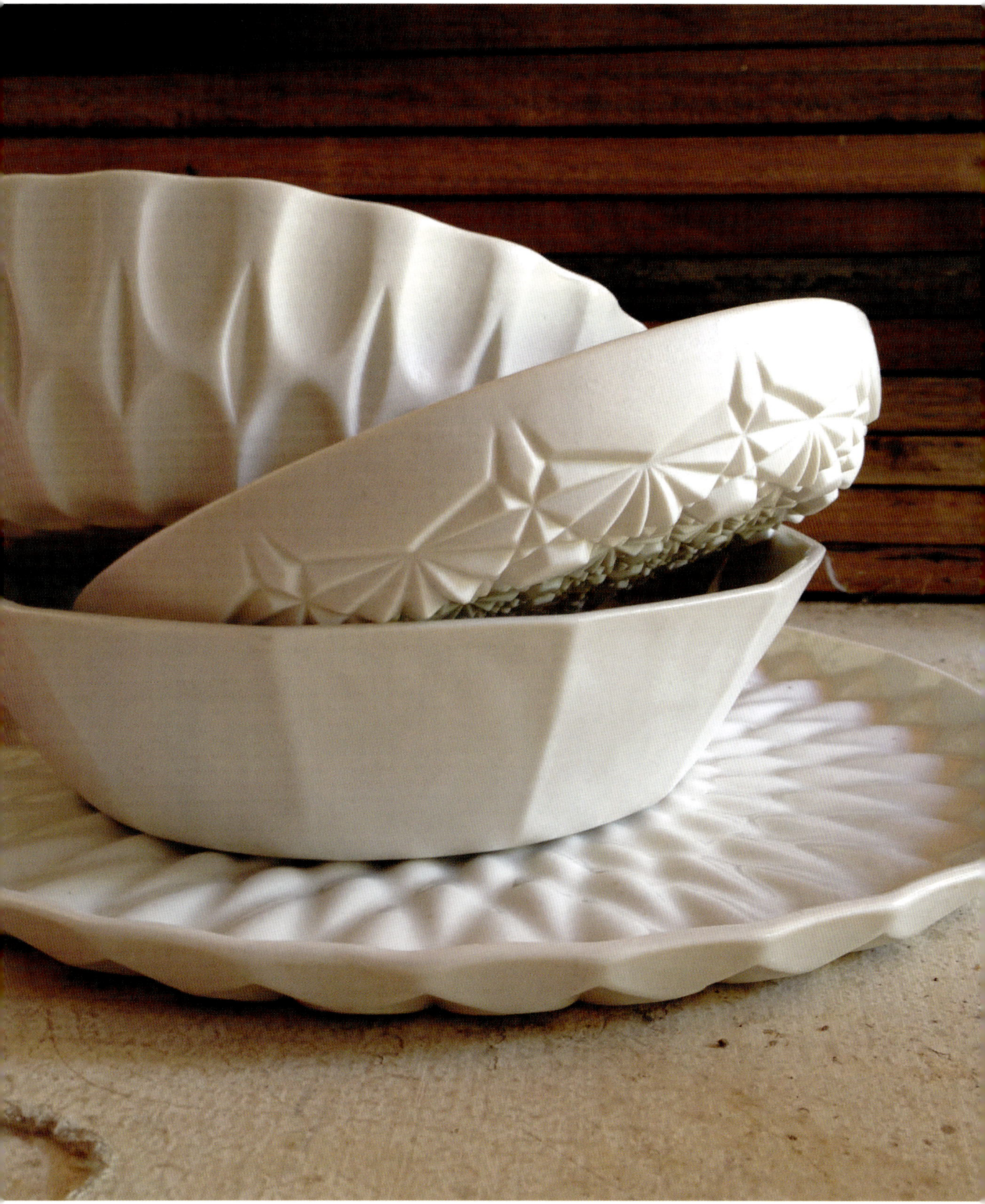

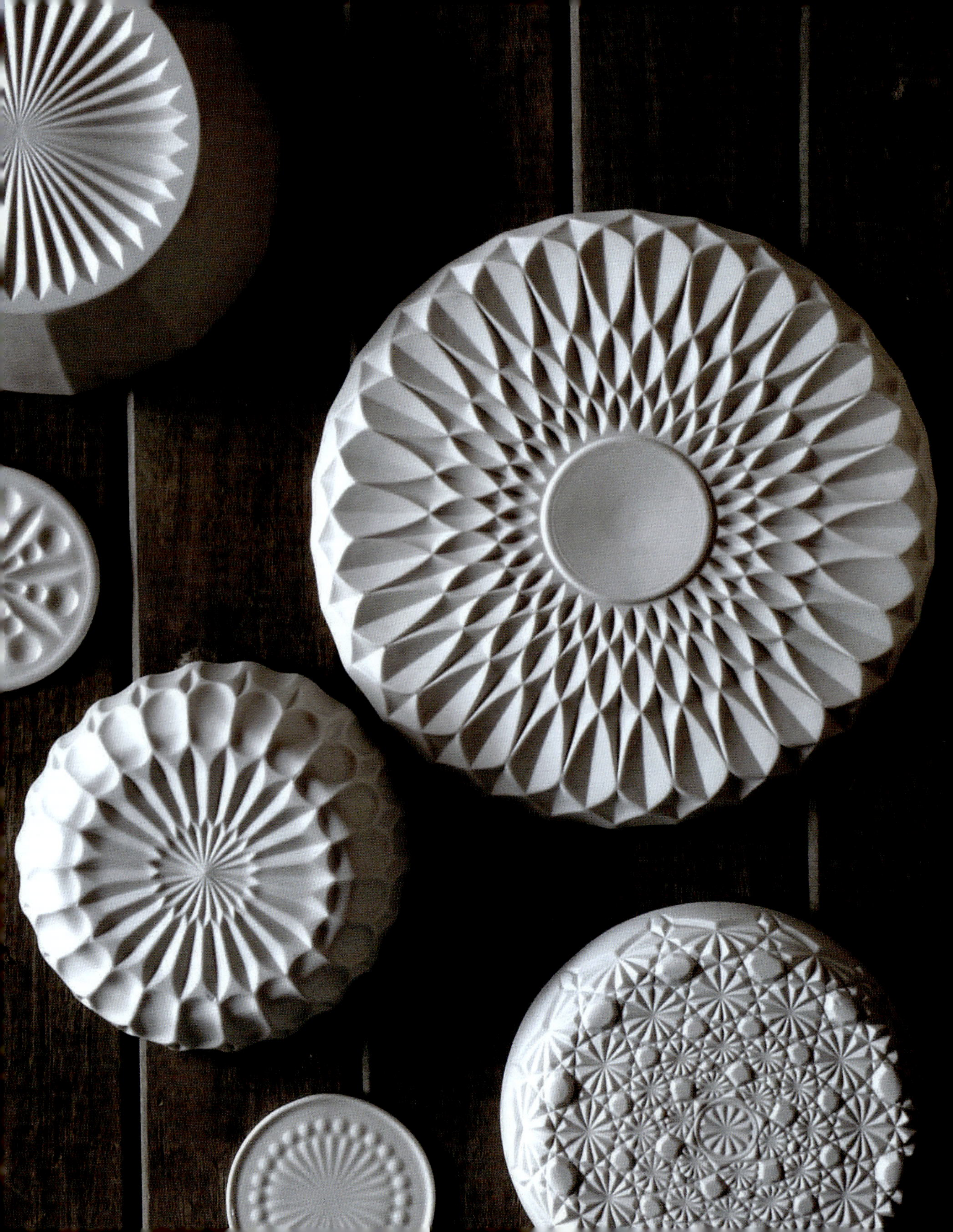

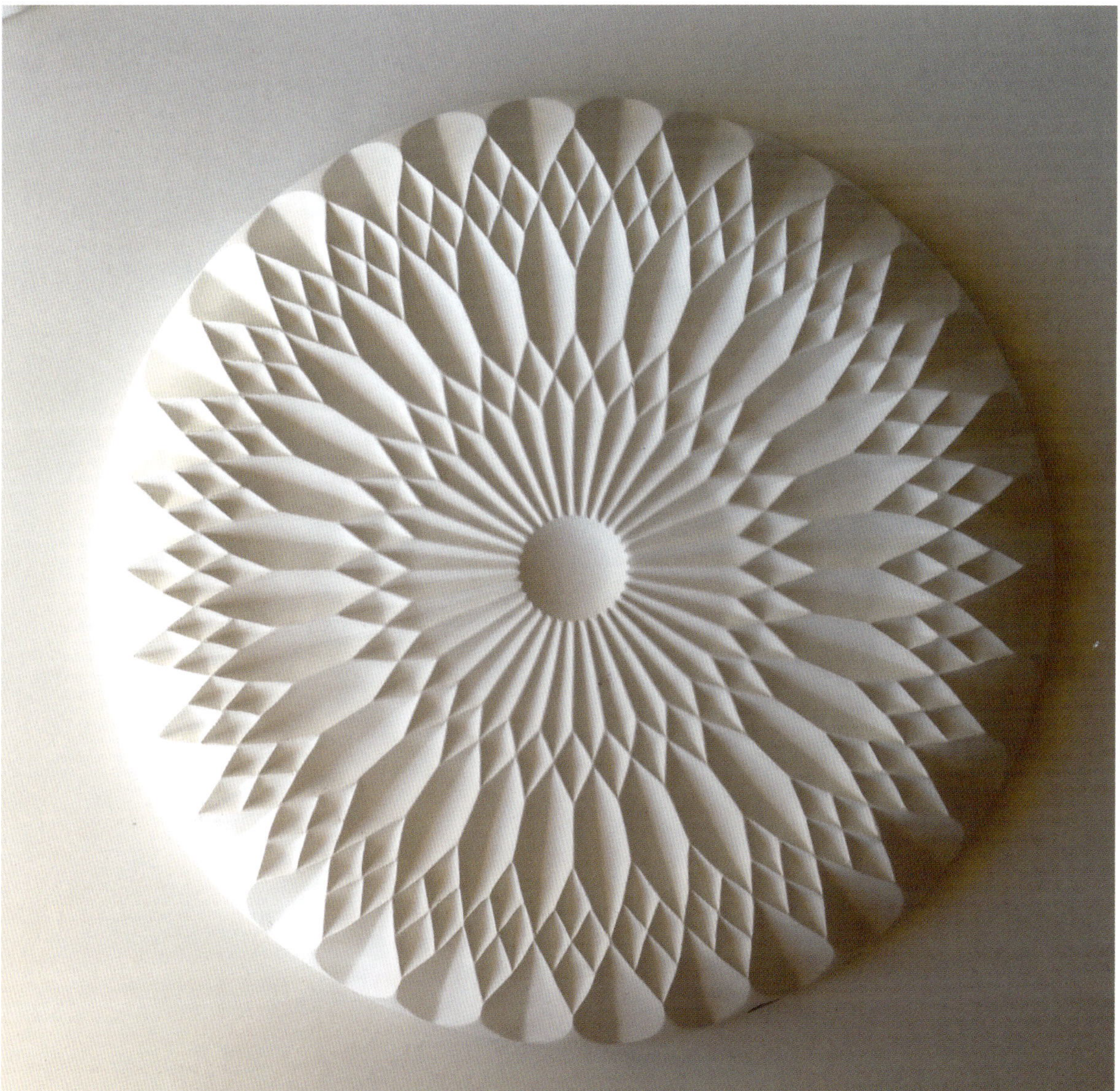

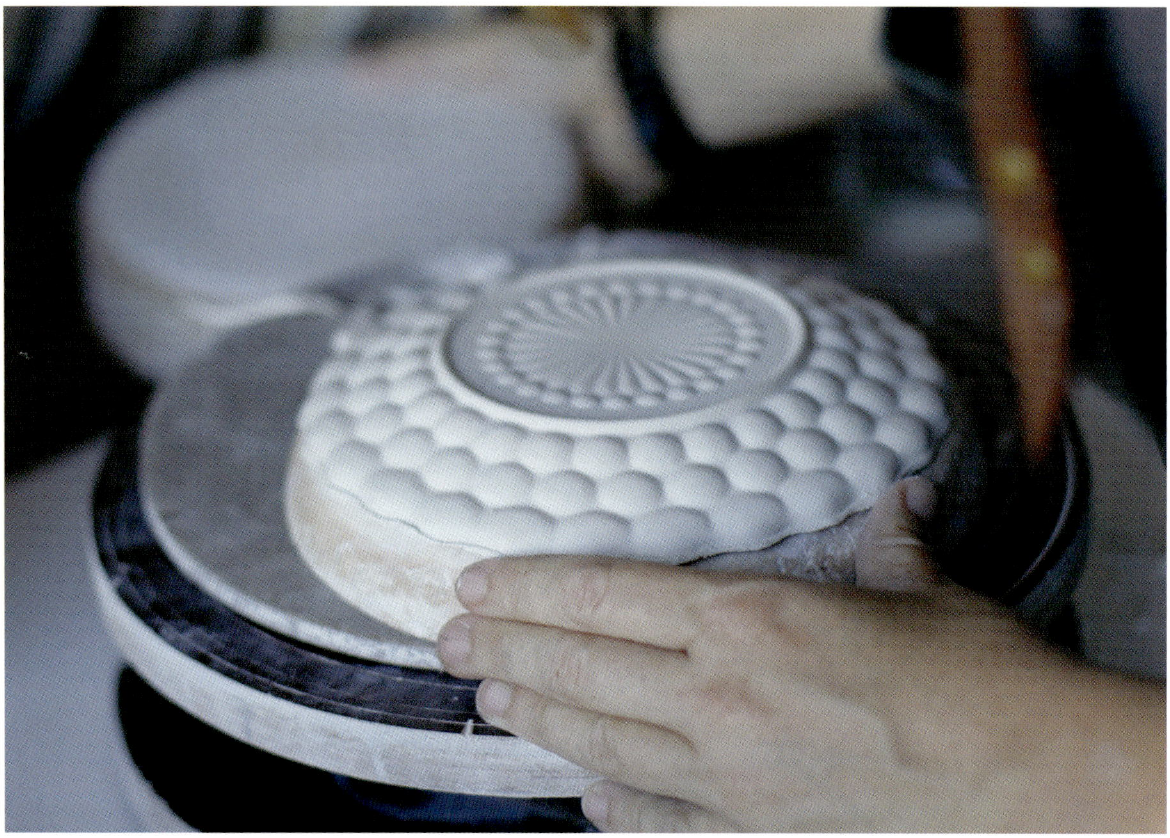

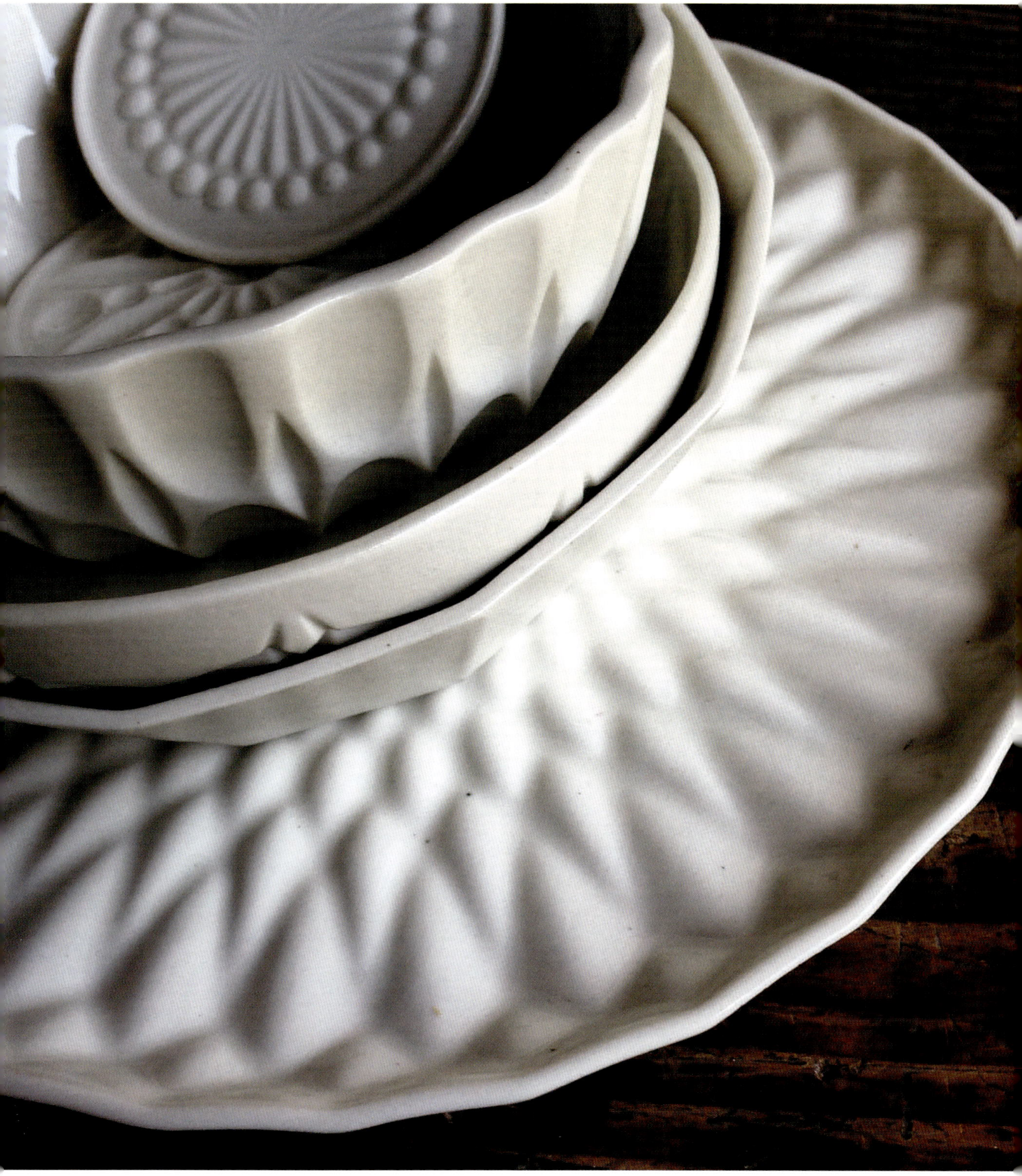

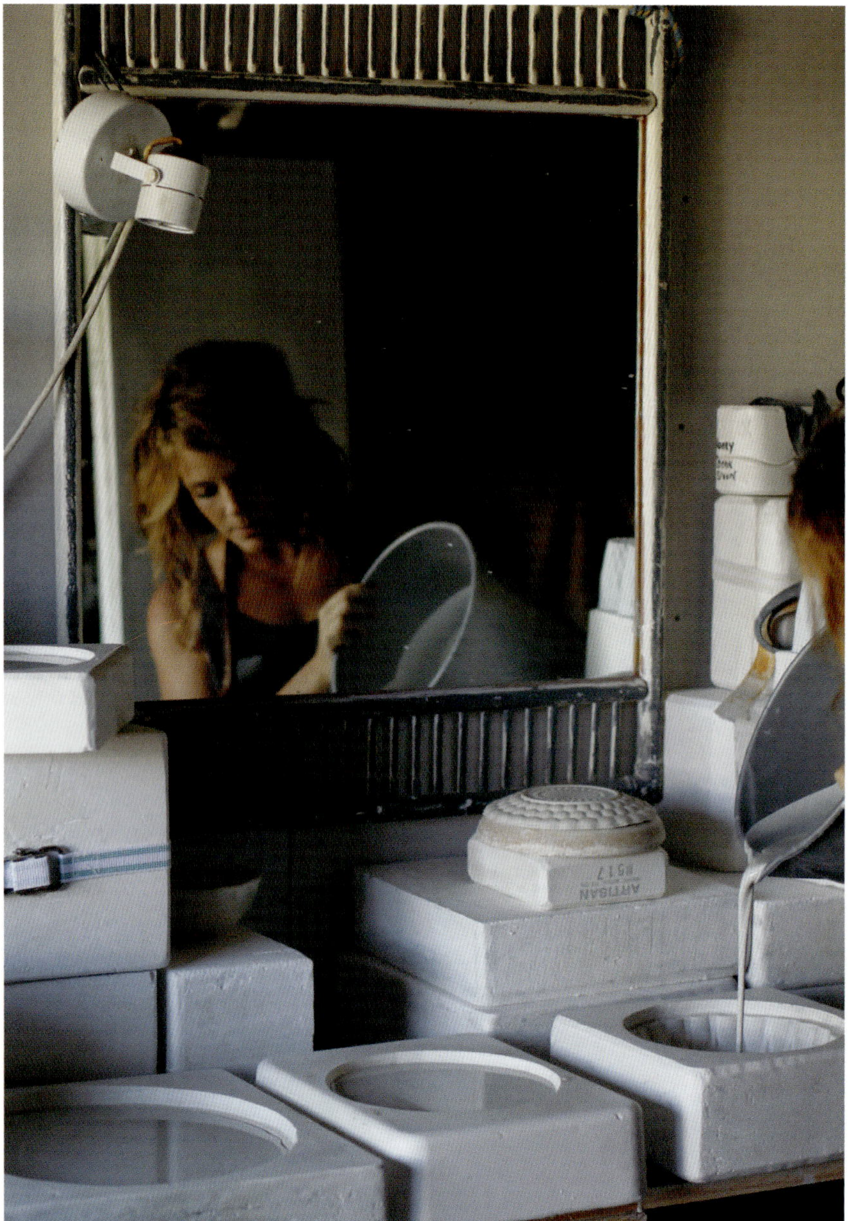

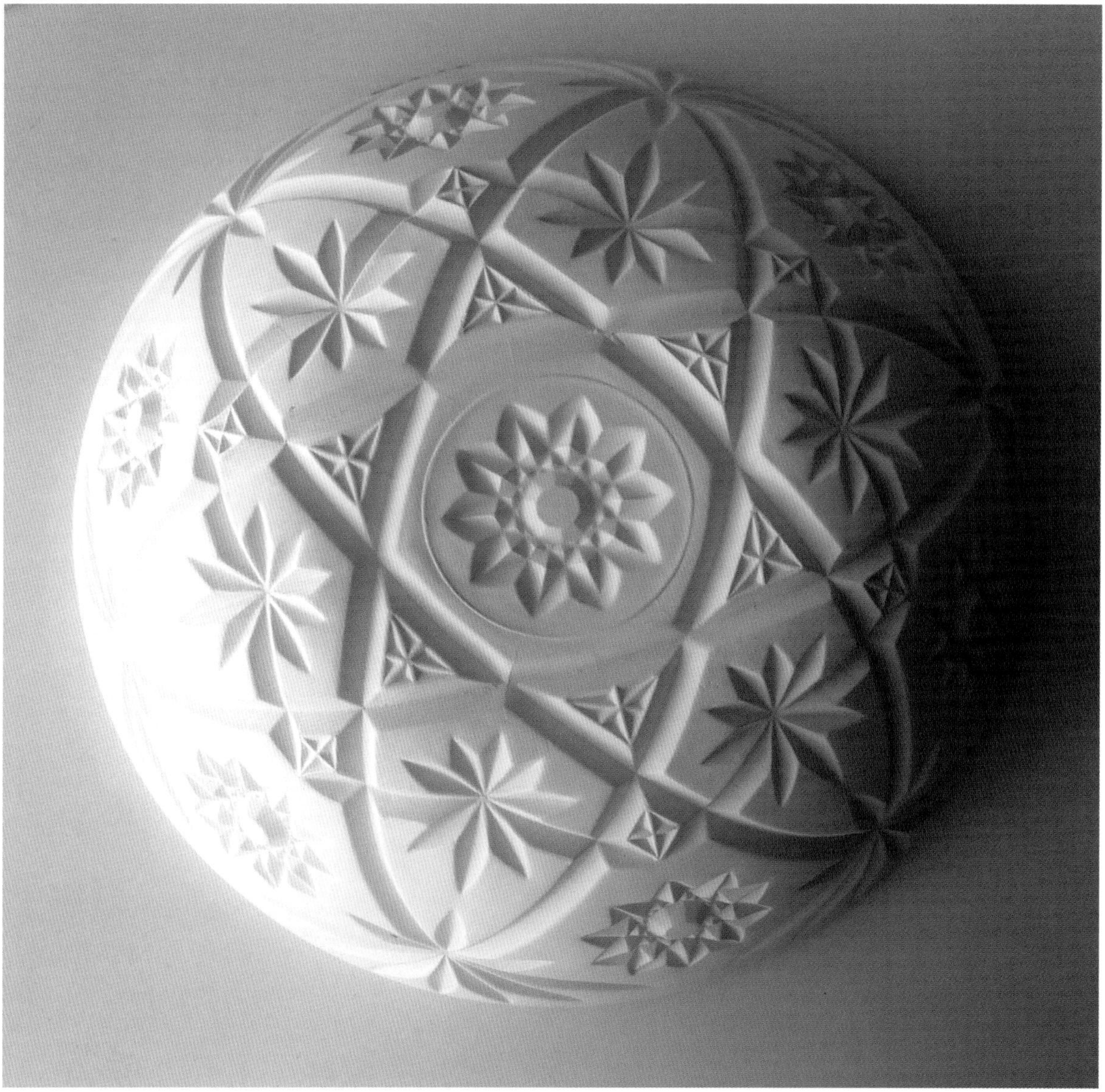

CHRIS HAWKINS

LOCATION Tamar Valley, Devon, Cornwall Borders, UK
YEAR OF BIRTH 1957
MATERIAL Hand thrown, raku fired ceramics
EXPERIENCE 41 years

Chris Hawkins started working with clay at the age of 16, and after leaving Art College in 1976, went on to train as a production thrower at West Pottery, Flete, Devon. This training provided him with the ability to work accurately and precisely with clay. He then spent the next 22 years producing his own work and working as a part-time ceramics and glass technician, as well as lecturing at Plymouth College of Art, UK. He now makes one-off thrown pieces which are fired using the raku process from his studio on the banks of the River Tamar in the Tamar Valley, an area of outstanding natural beauty with a rich mining heritage on the borders of Devon & Cornwall.

His work consists of Copper Matt fumed pots, white-glazed and naked raku pieces. The Copper Matt process is notoriously difficult to control, and he has spent many years of experimentation perfecting a recipe and firing process that relies on post production techniques to enable him to consistently achieve the rich iridescent colours seen on his Copper Matt pieces. He pushes this process to its limits and will sometimes fire a piece several times to achieve the desired results.

Photos courtesy of the artist

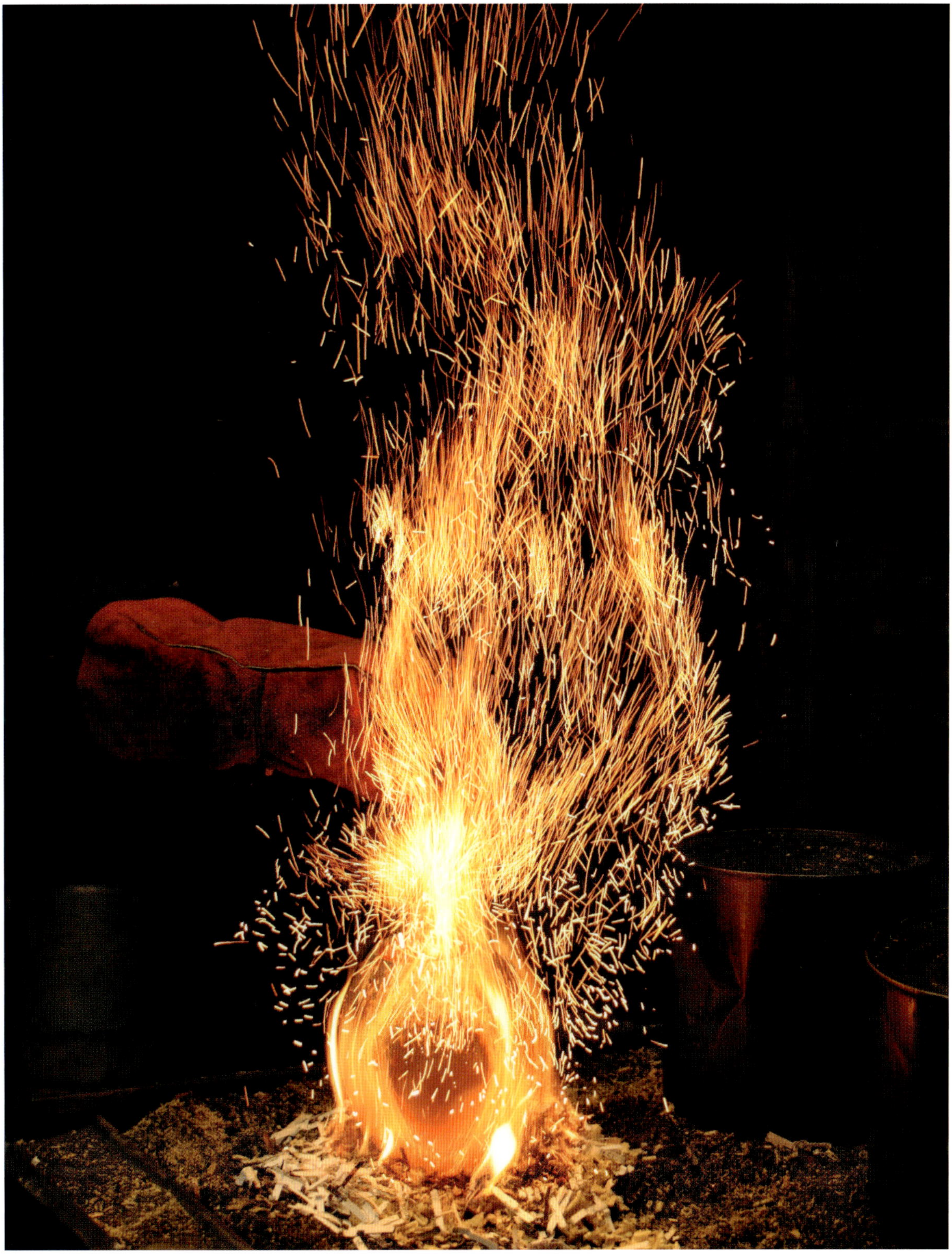

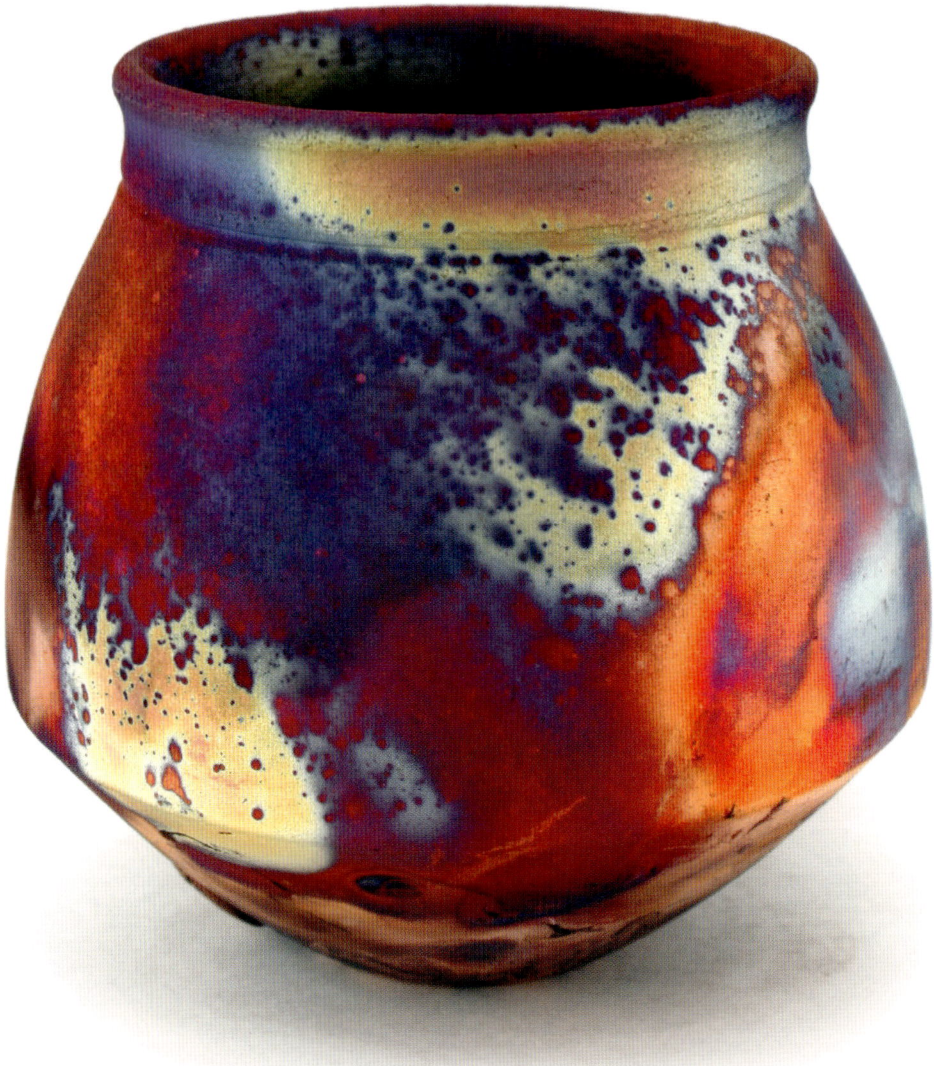

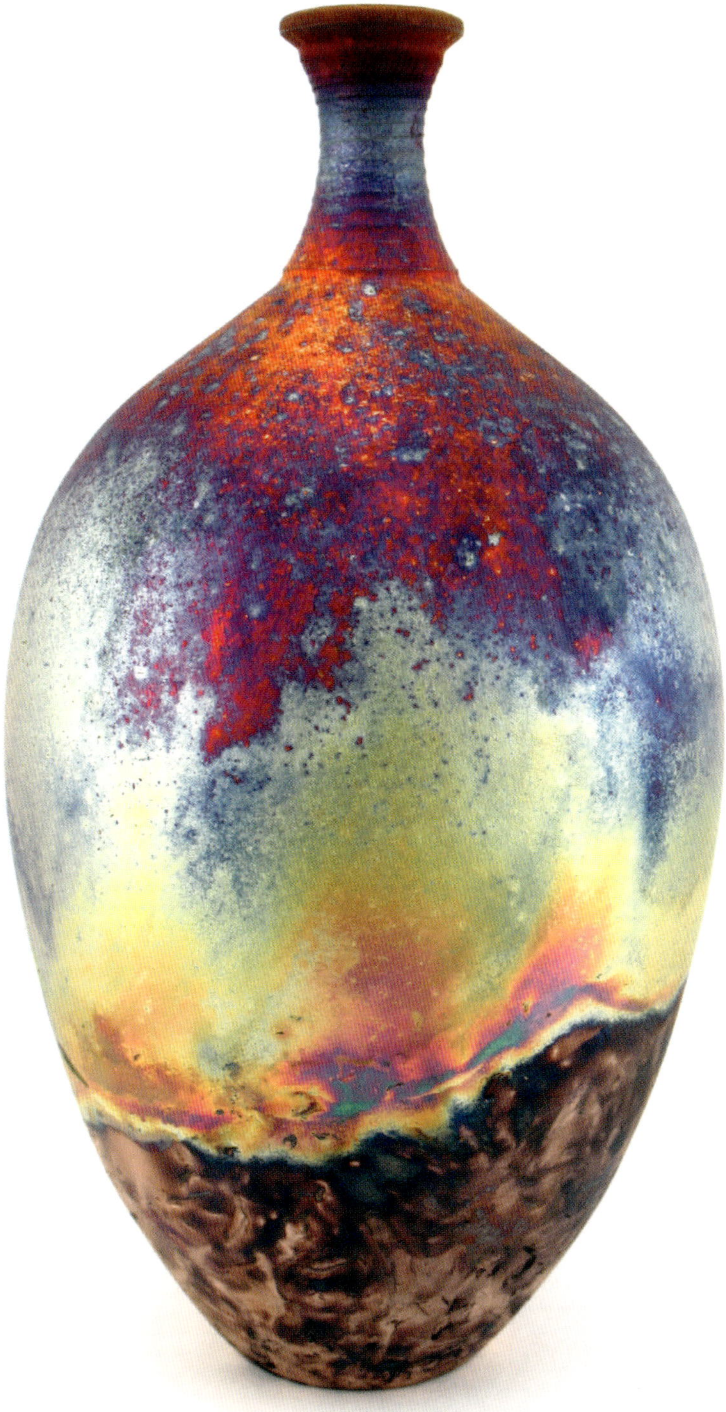

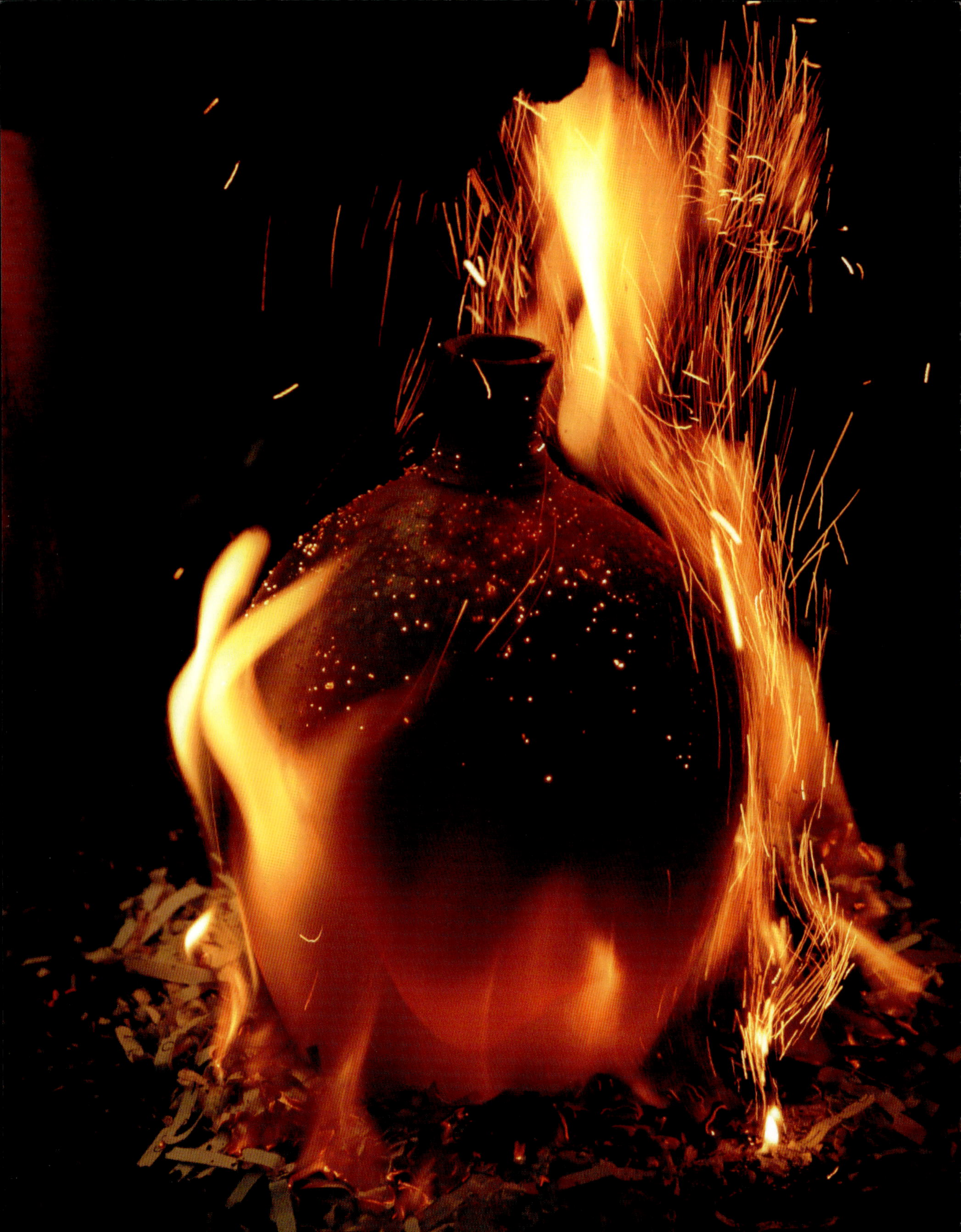

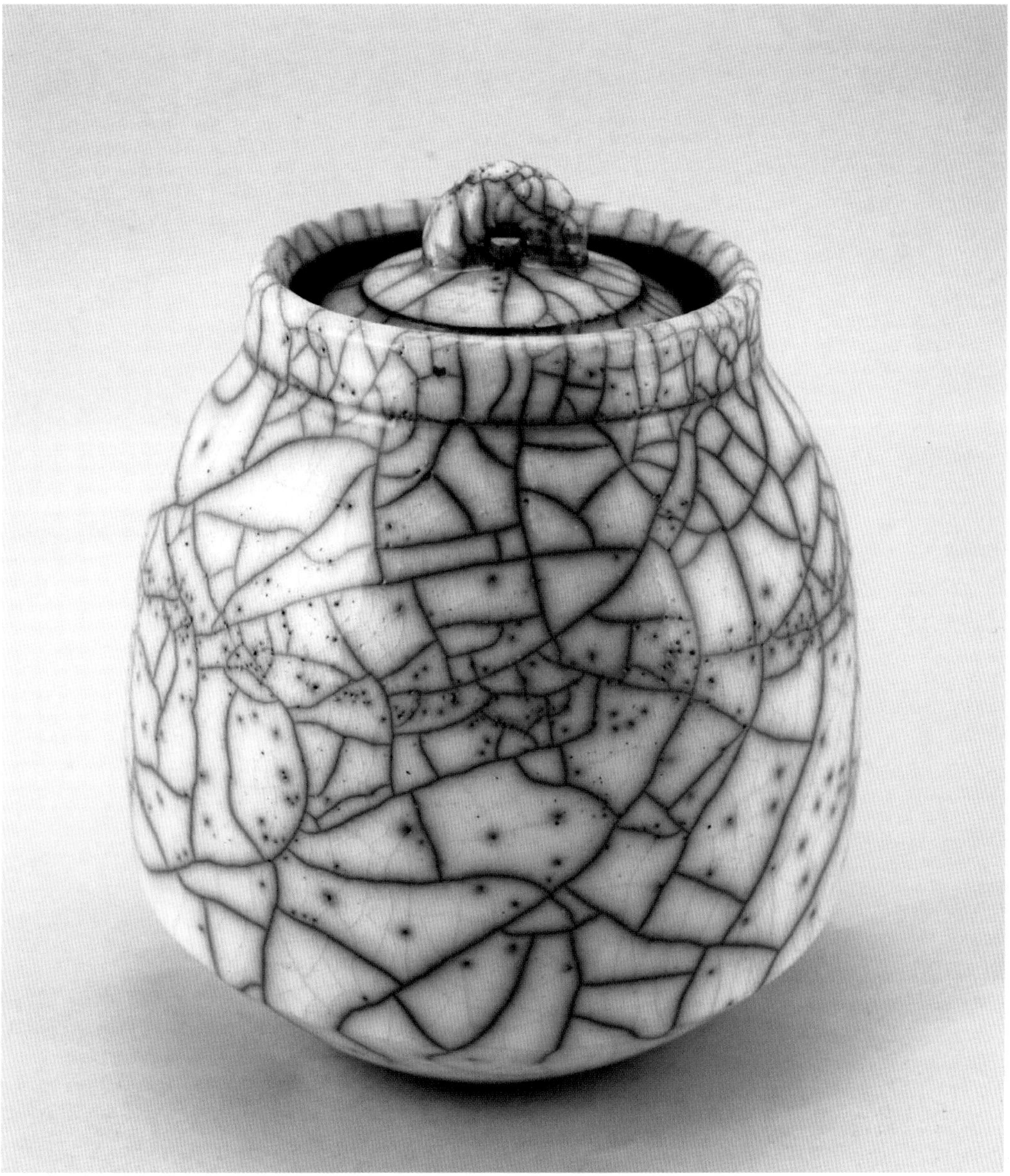

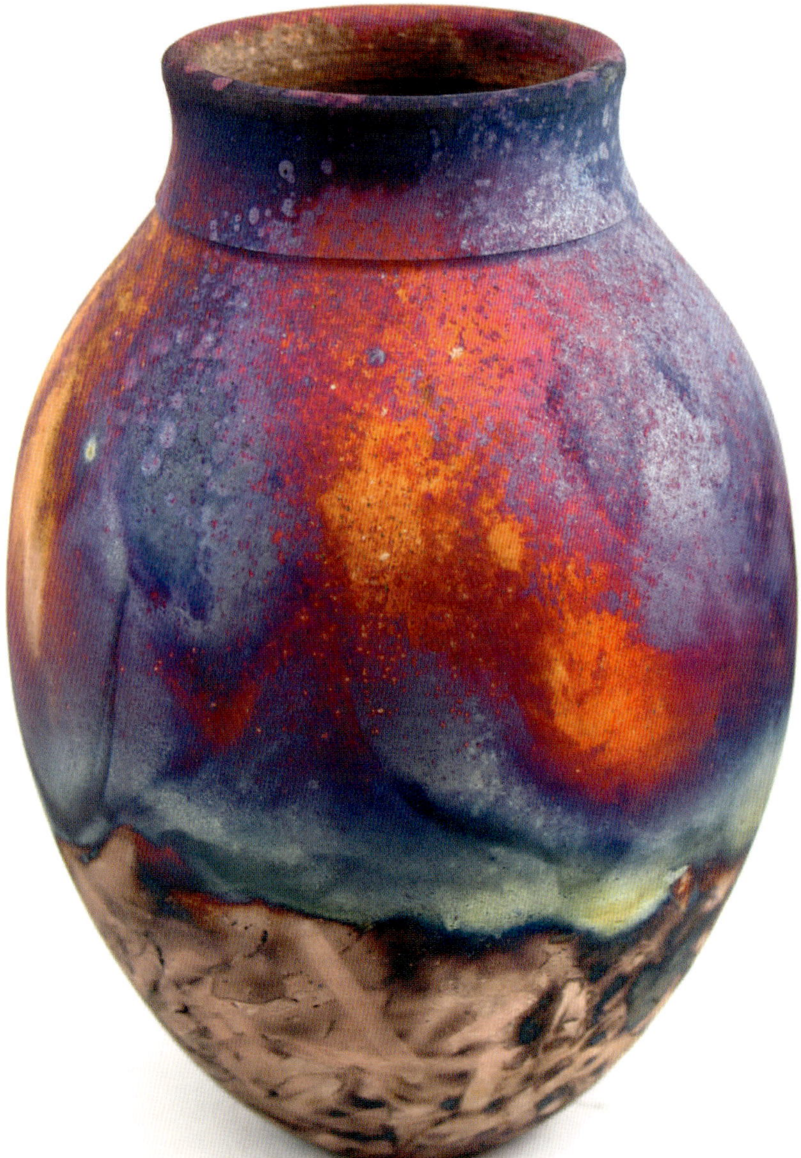

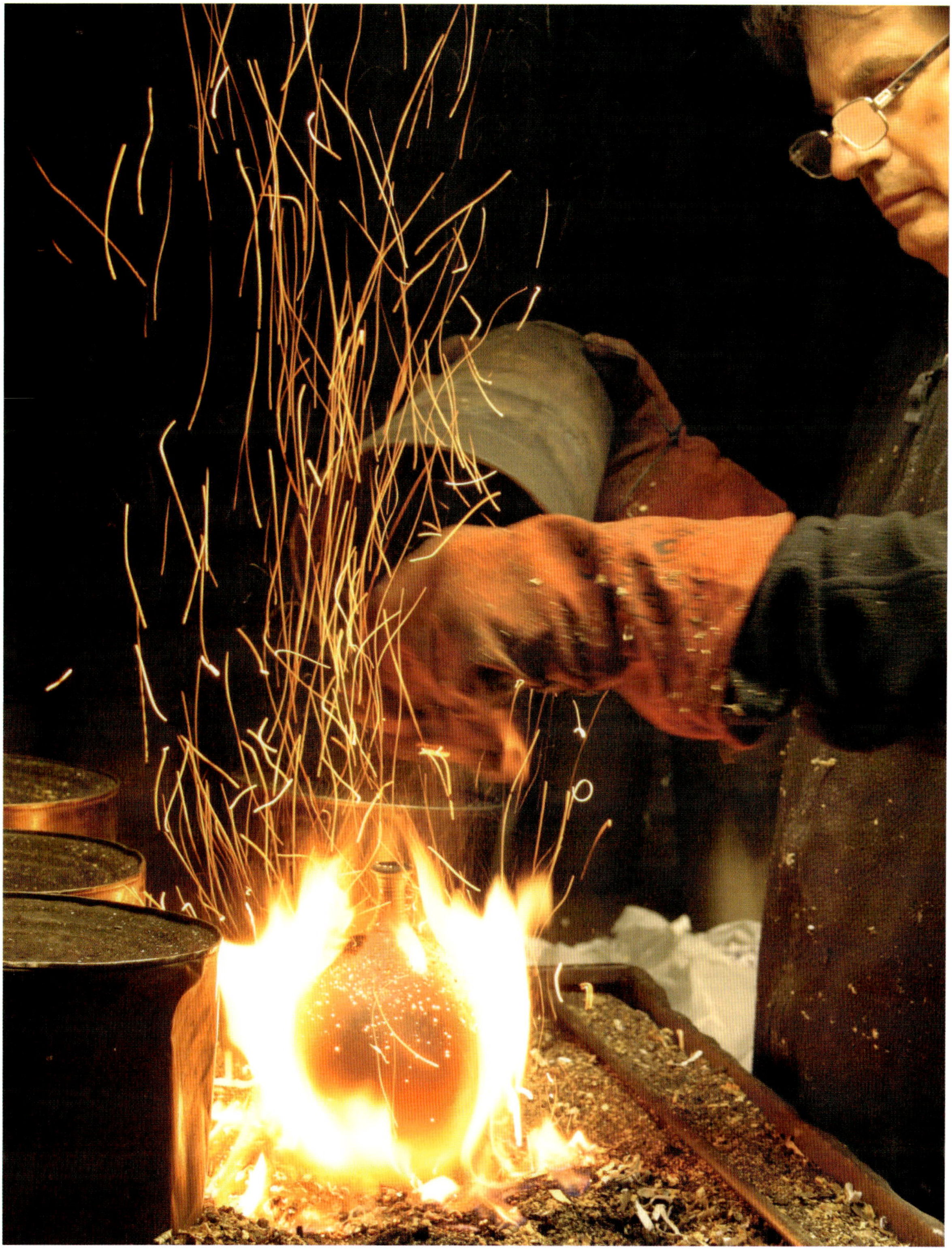

BODIL MANZ

LOCATION Denmark
YEAR OF BIRTH 1943
MATERIALS Ceramic, cast porcelain
EXPERIENCE 40 years

Bodil Manz has masterfully developed her ceramic arts for 40 years, the highlight of which are her paper-thin and transparent cylinders. The cylinders are cast in white, translucent ceramic, which is so thin that the designs on the innter and outer surfaces merge into one composition. Bodil Manz has experimented and sought new challenges in expression and technique throughout her work, and works frequently with gypsum slabs, sand-cast porcelain, and handmade paper.

Photos courtesy of Clark/Del, Santa Fe, New Mexico, Erik Brahl and the artist

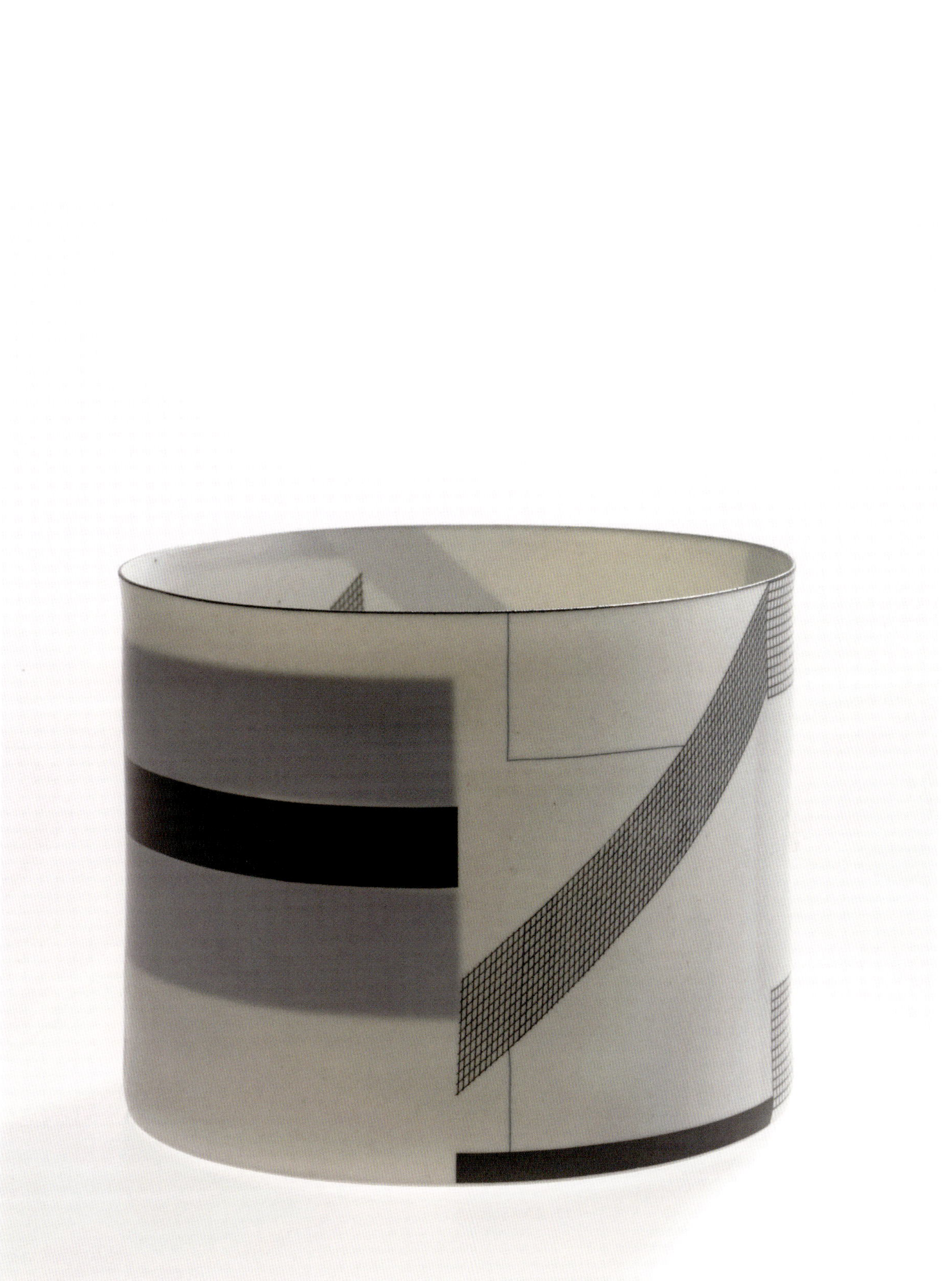

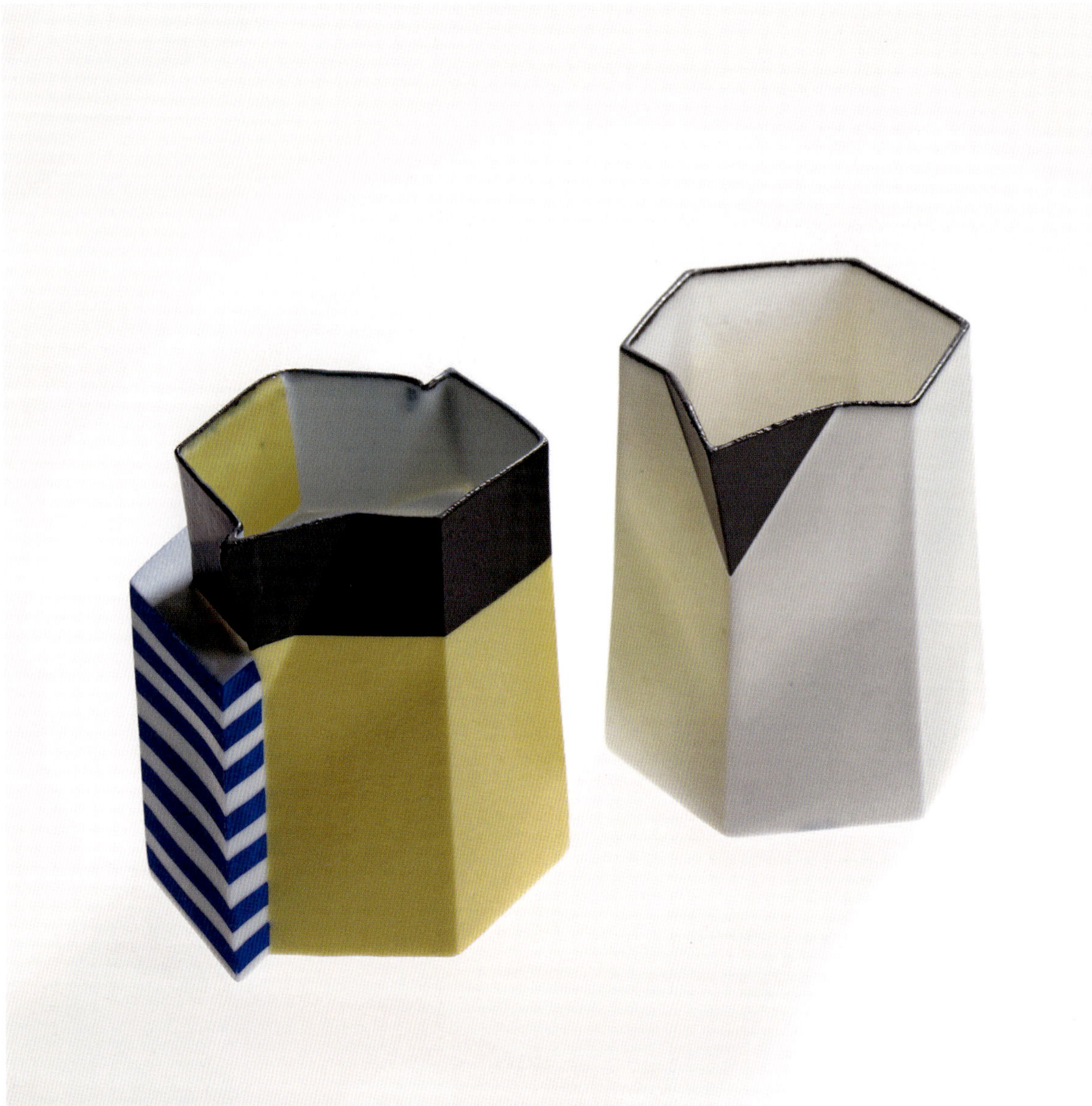

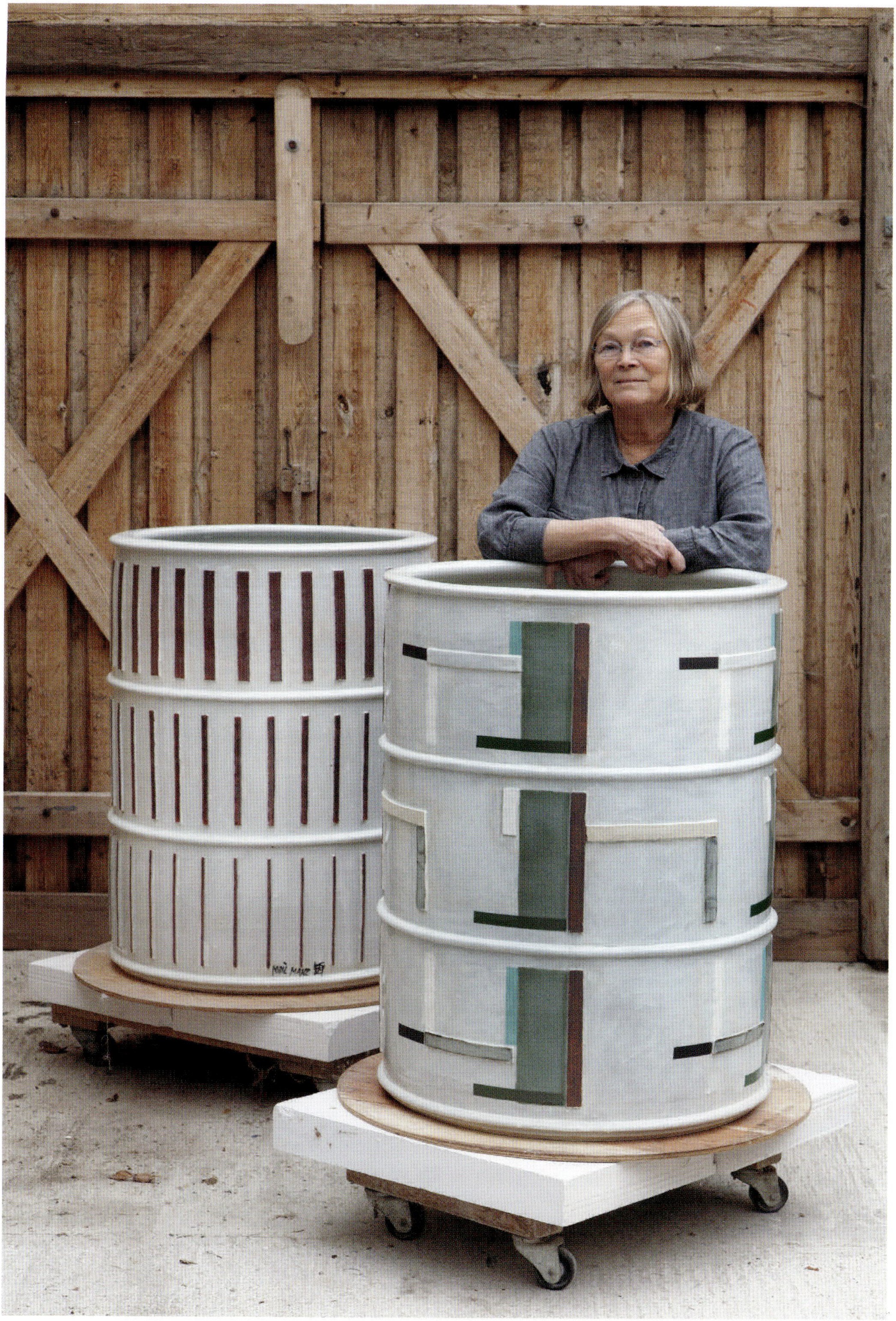

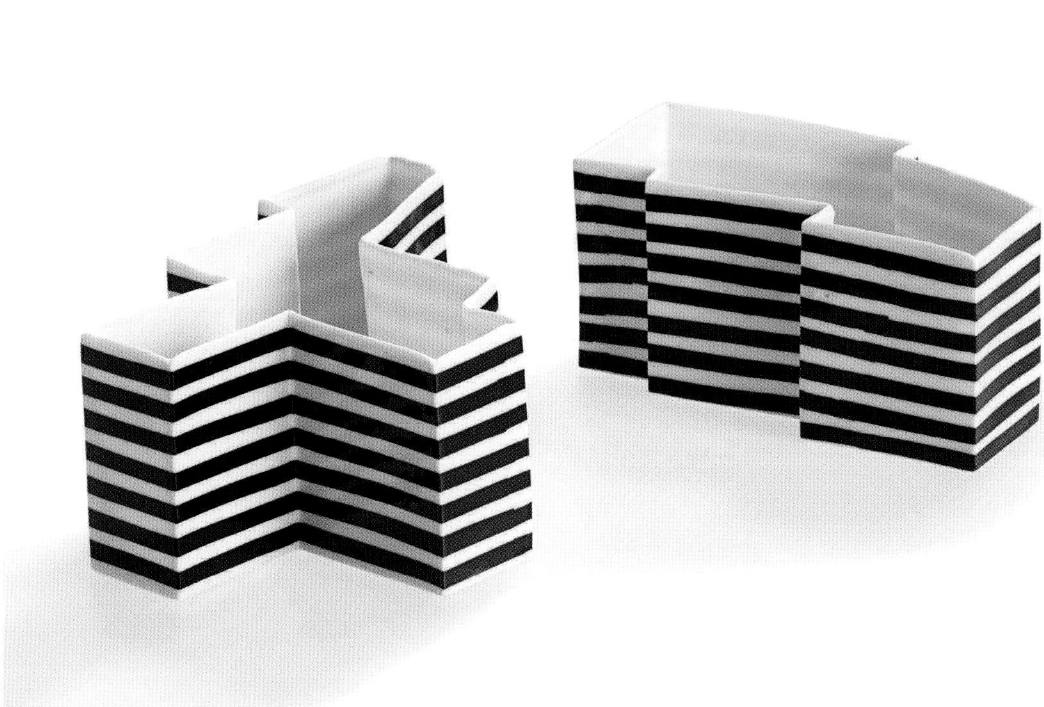

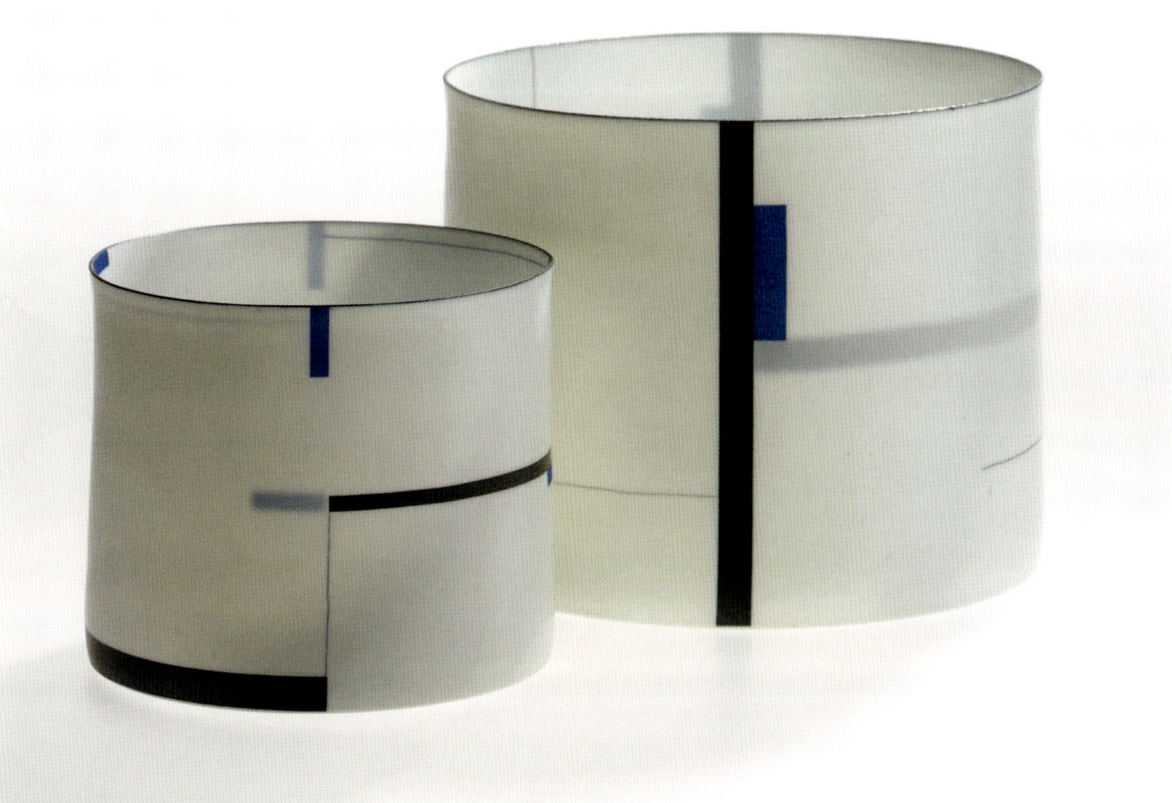

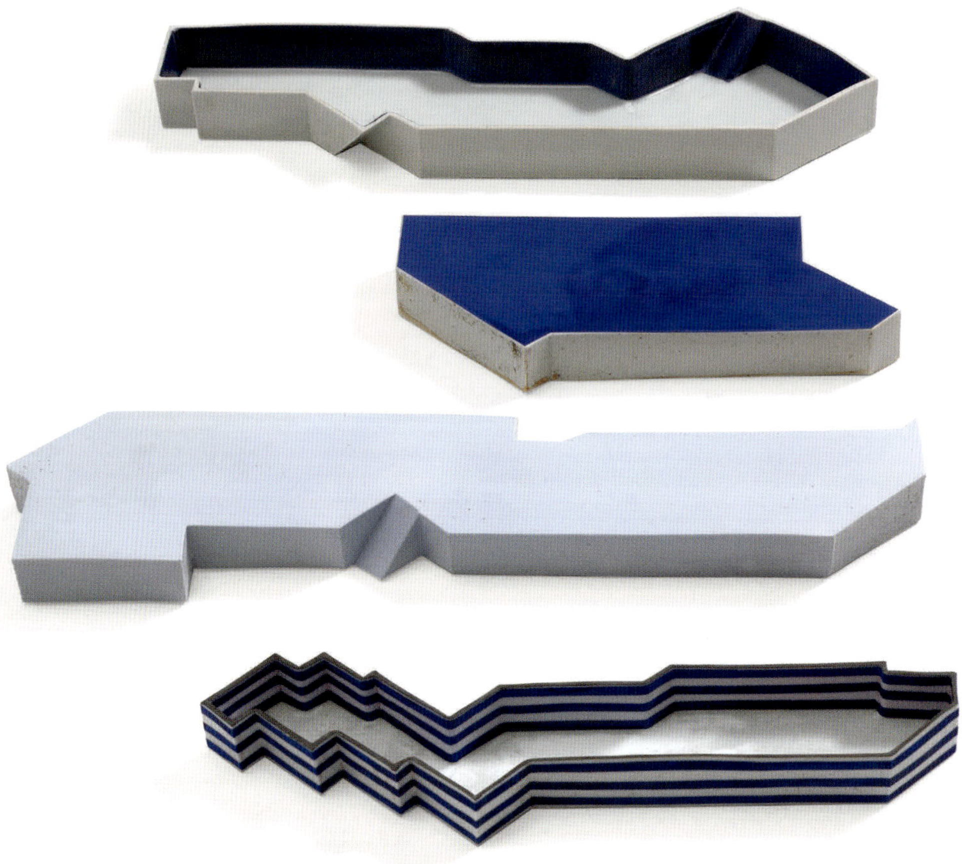

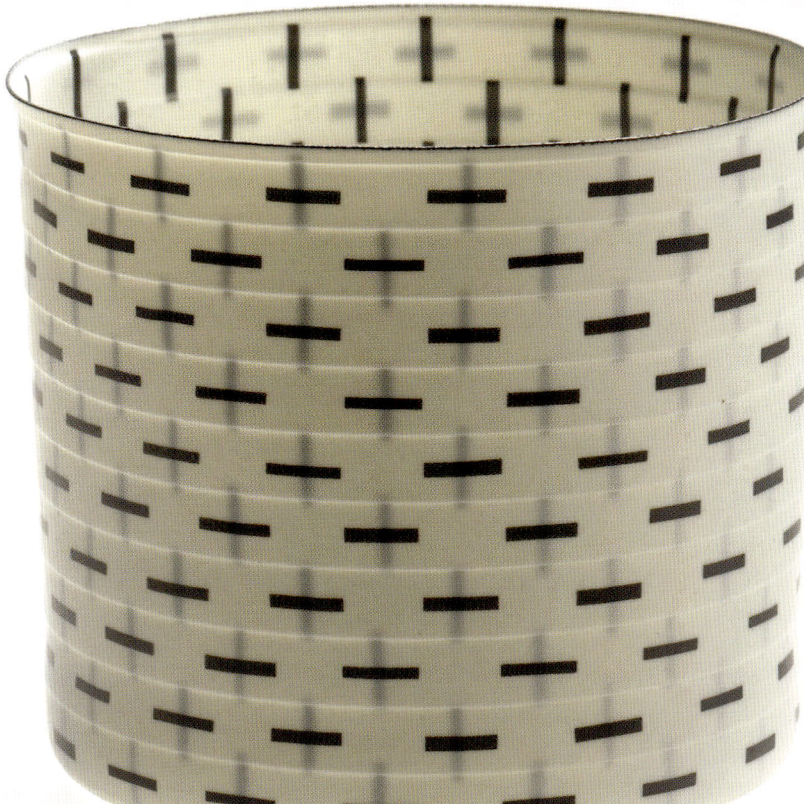

HEATH CERAMICS

LOCATION San Francisco, CA
YEAR OF BIRTH 1948
MATERIALS Ceramics
EXPERIENCE 66 years

Once upon a time, two designers stumbled upon a curious factory — one that made ceramics in an unlikely place: Sausalito, California. It was Heath, of course, a company founded by Edith and Brian Heath in 1948. The couple, Robin Petravic and Catherine Bailey, saw in Heath the potential for a holistic business model, one that integrated designing, making, and selling.

The pair bought Heath Ceramics in 2003 and quickly set about revitalizing the company. They boosted productivity, streamlined offerings, and created new products, collaborated with artists, mavericks, and companies whose vision complemented theirs, and began carrying products from independent makers.

They expanded too, opening showrooms in Los Angeles and in San Francisco's Ferry Building, and a new tile factory and showroom in San Francisco's Mission District in 2012. Along the way, they gained legions of fans as their products found their way into hotels, restaurants, museums, and homes worldwide. They've even won some prestigious design awards, including the Cooper Hewitt's National Design Award in 2015. That same year, Robin and Catherine authored "Tile Makes The Room," which explores tile's integral role in good design in projects across the world.

Photos courtesy of Heath Ceramics

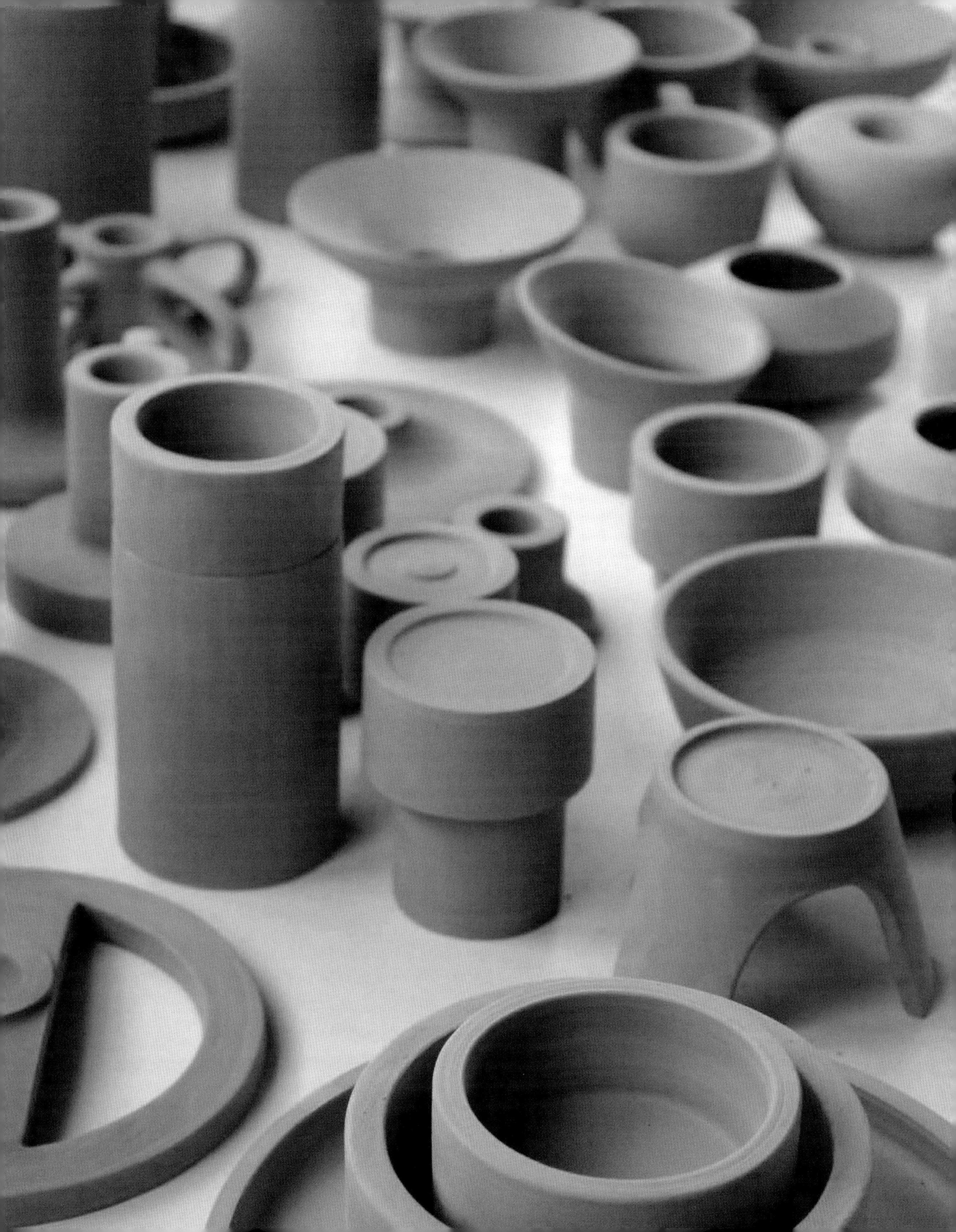

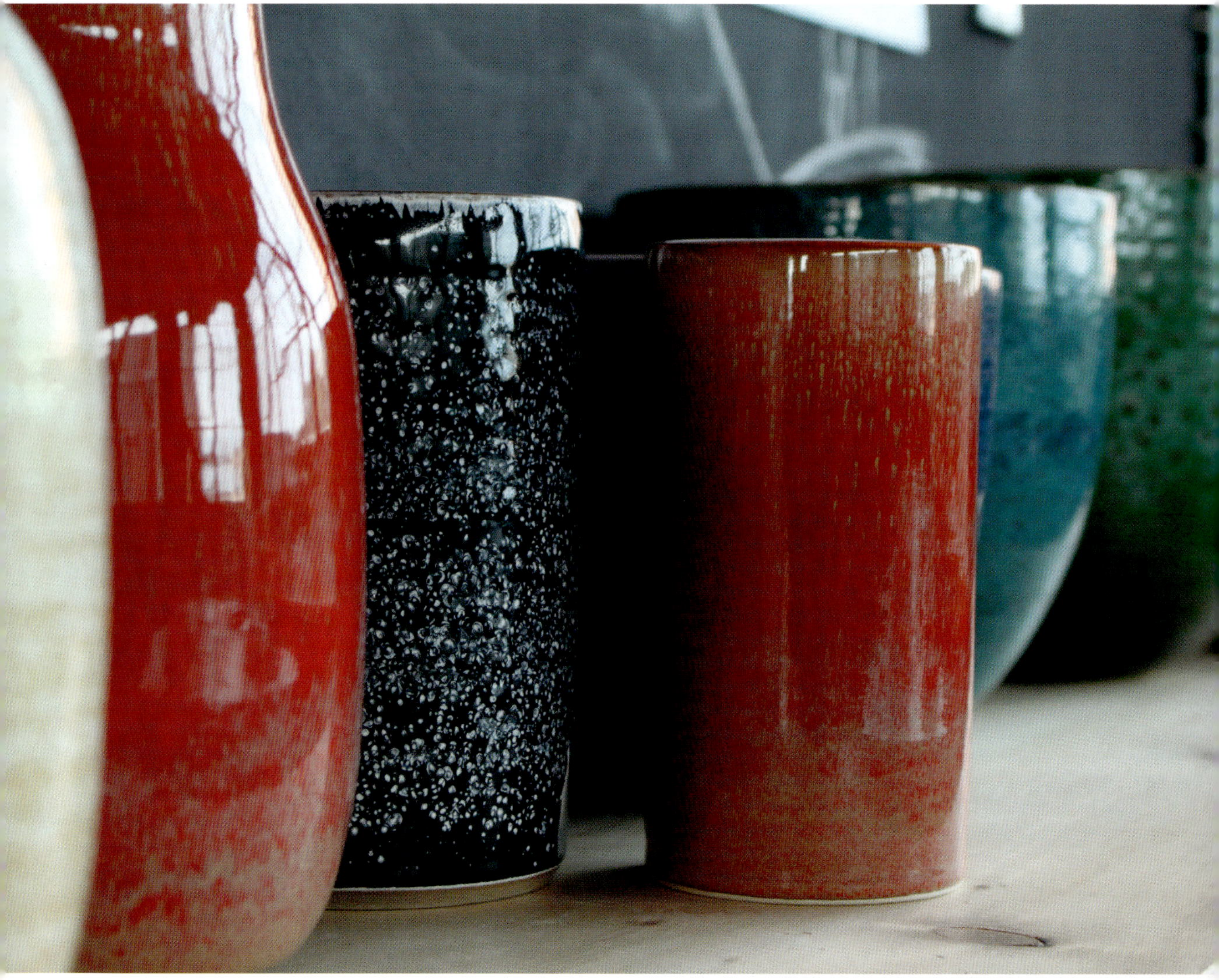

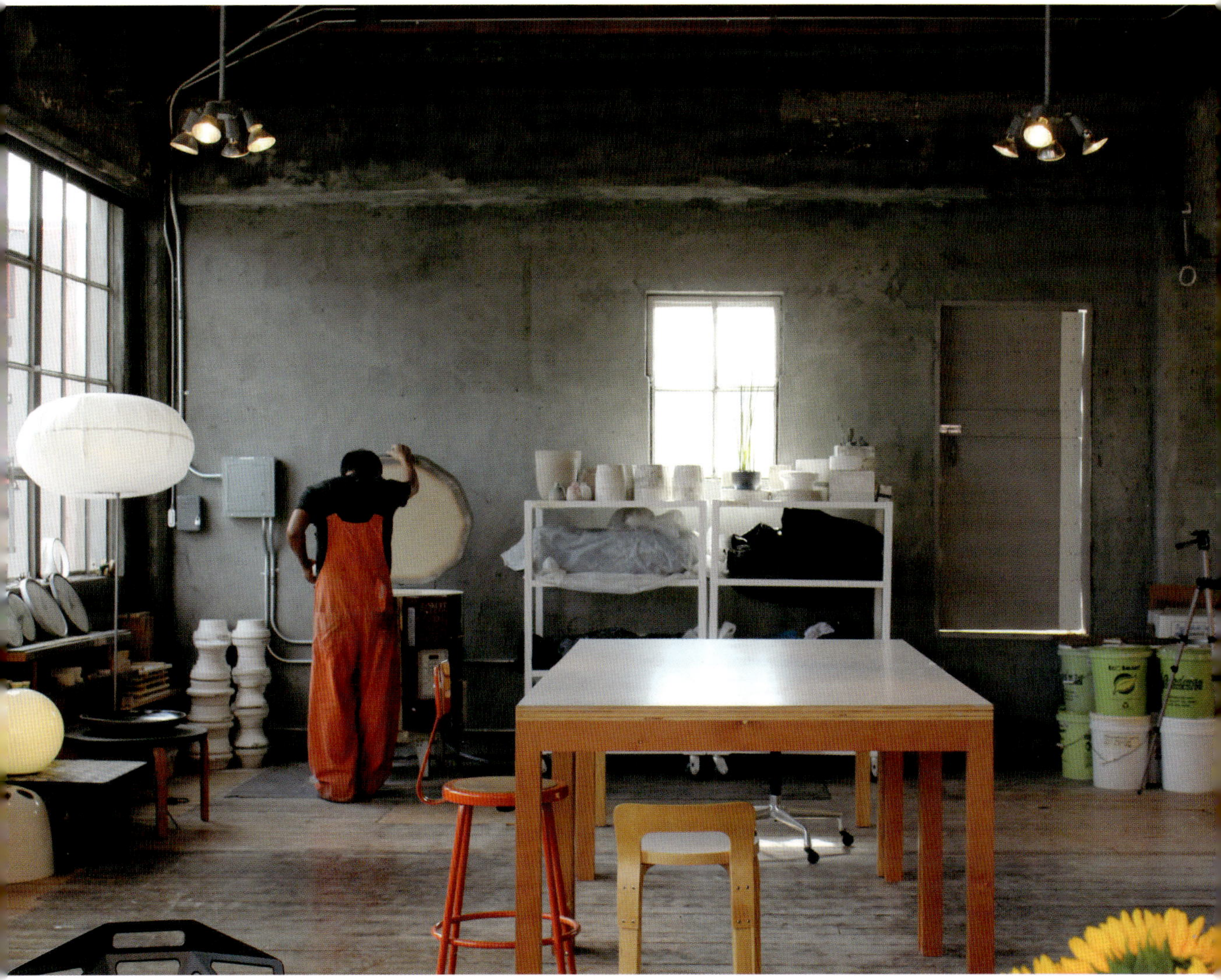

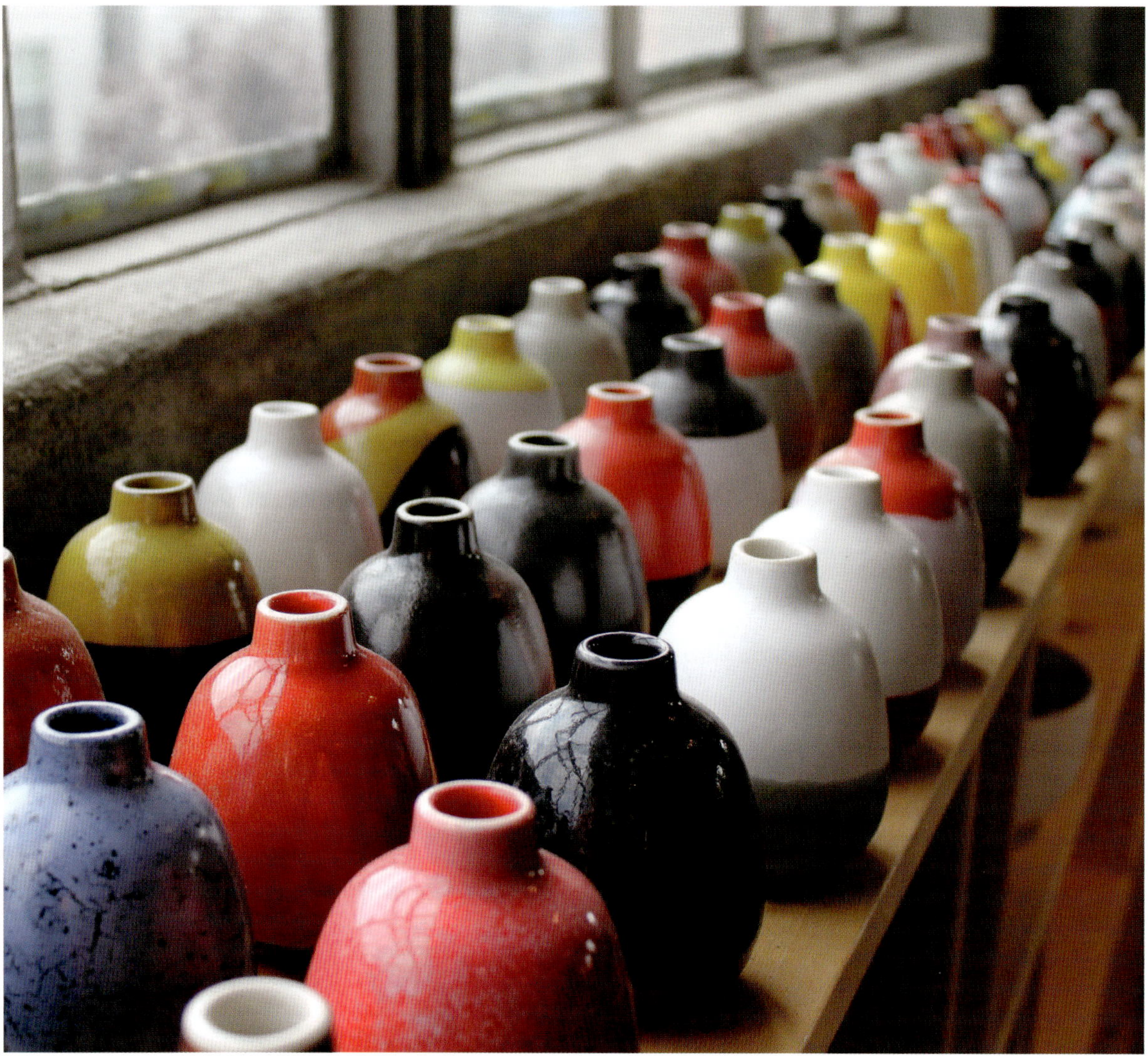

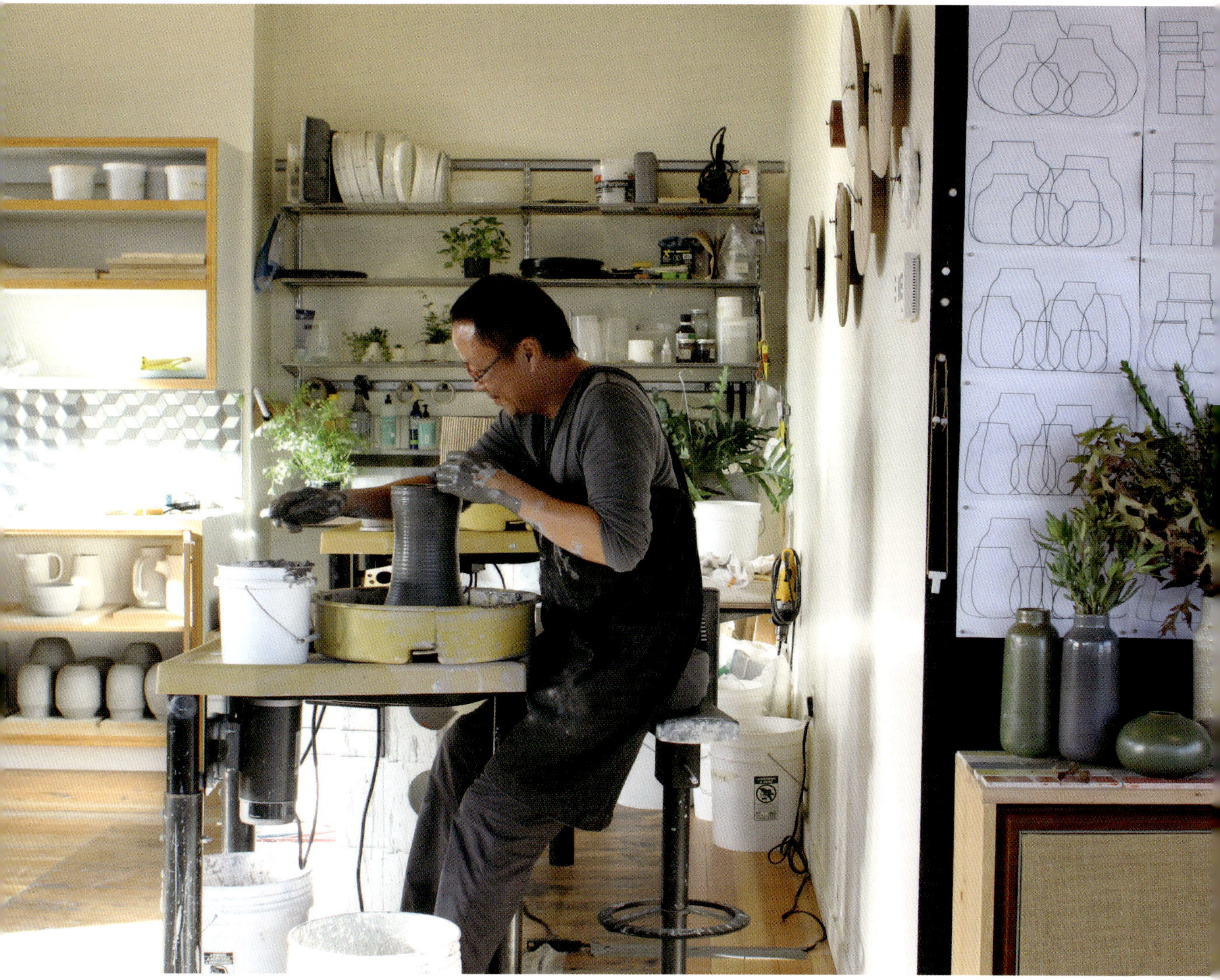

DECORATIVE VESSELS

HITOMI HONOSO

THEONE CERAMICS

JON ALMEDA

JENNIFER MCCURDY

CHE JEN

HITOMI HOSONO

LOCATION London, UK
YEAR OF BIRTH 1978
MATERIALS Porcelain
EXPERIENCE 20 years

'My ceramics experience is rooted in both Japanese and European traditions. Before arriving in the UK, I studied Kutani Pottery in Kanazawa city, Japan, as well as ceramic product design in Copenhagen, Denmark.

The subjects of my current porcelain work are shapes inspired by leaves and flowers. I study botanical forms in the garden. I find myself drawn to the intricacy of plants, examining the veins of a leaf, how its edges are shaped, the layering of a flower's petals.'

Photos courtesy of Adrian Sassoon and the artist

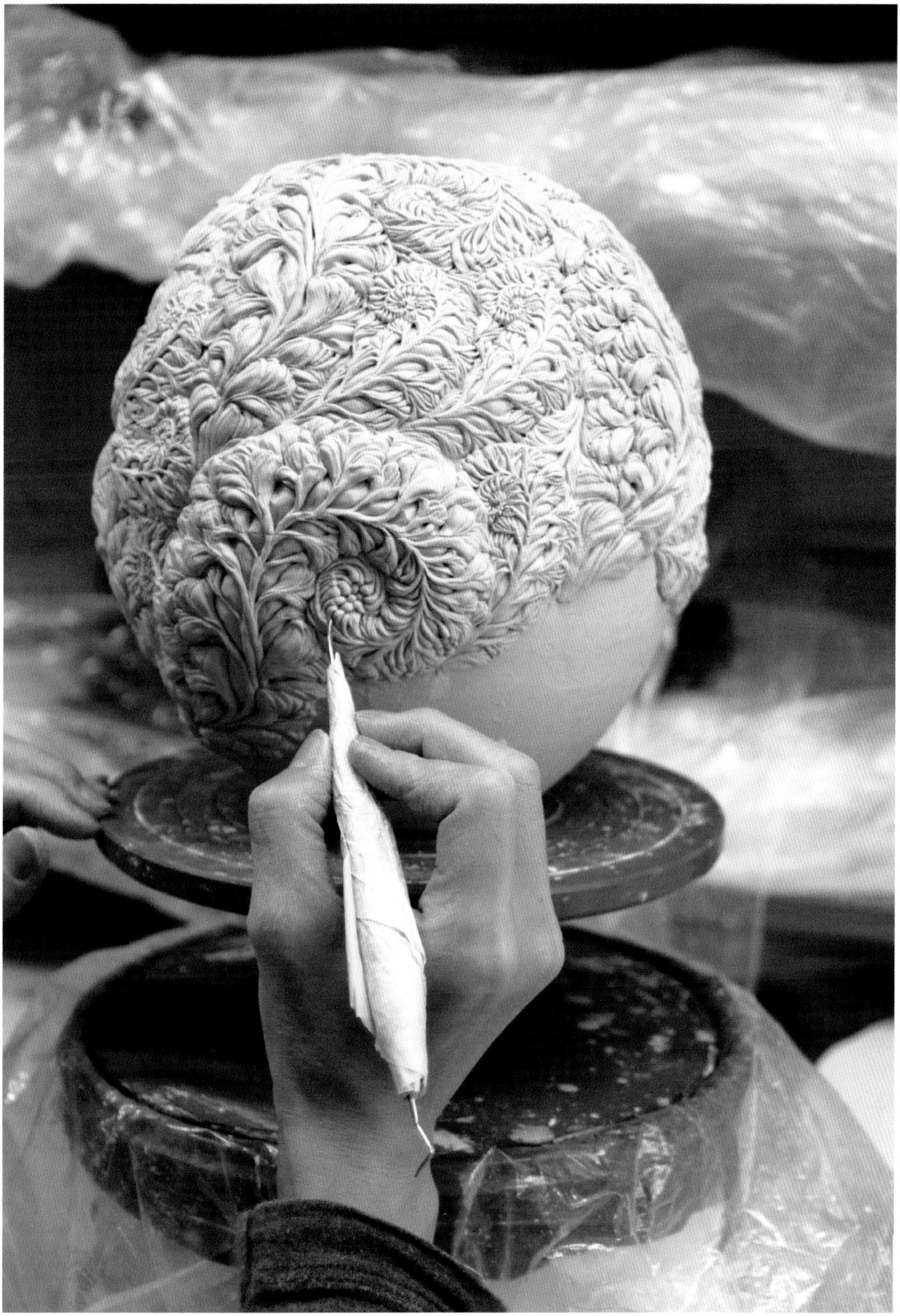

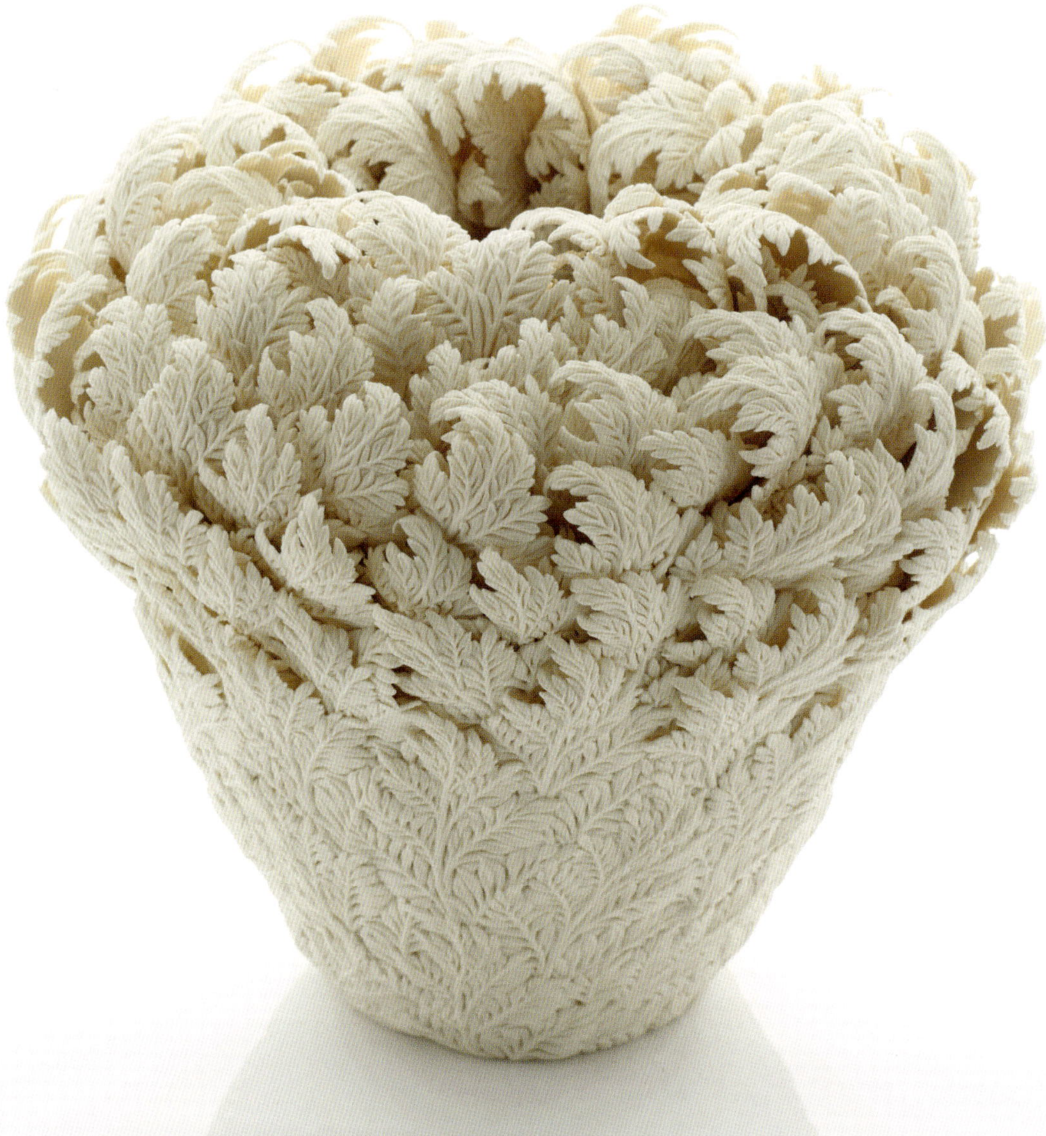

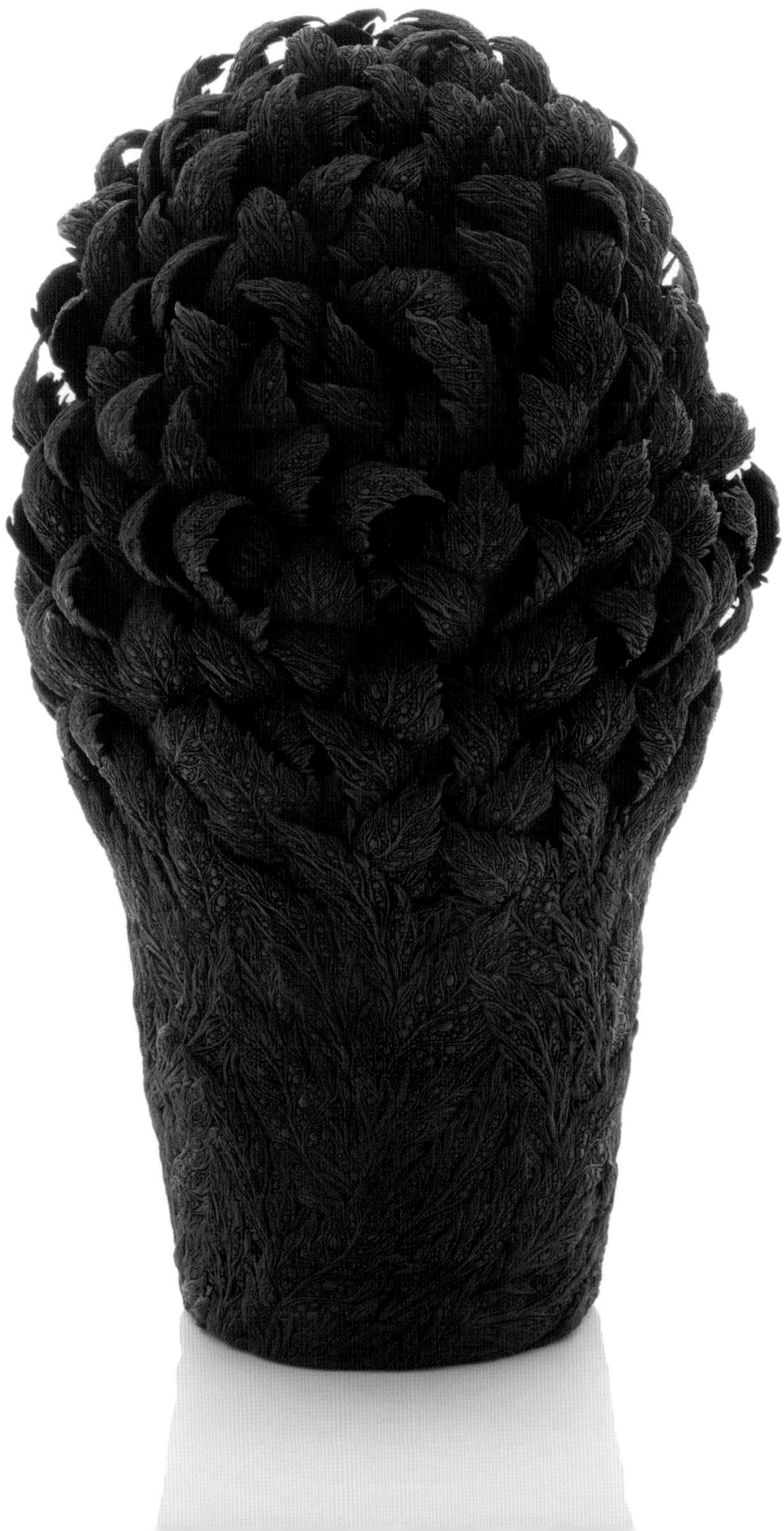

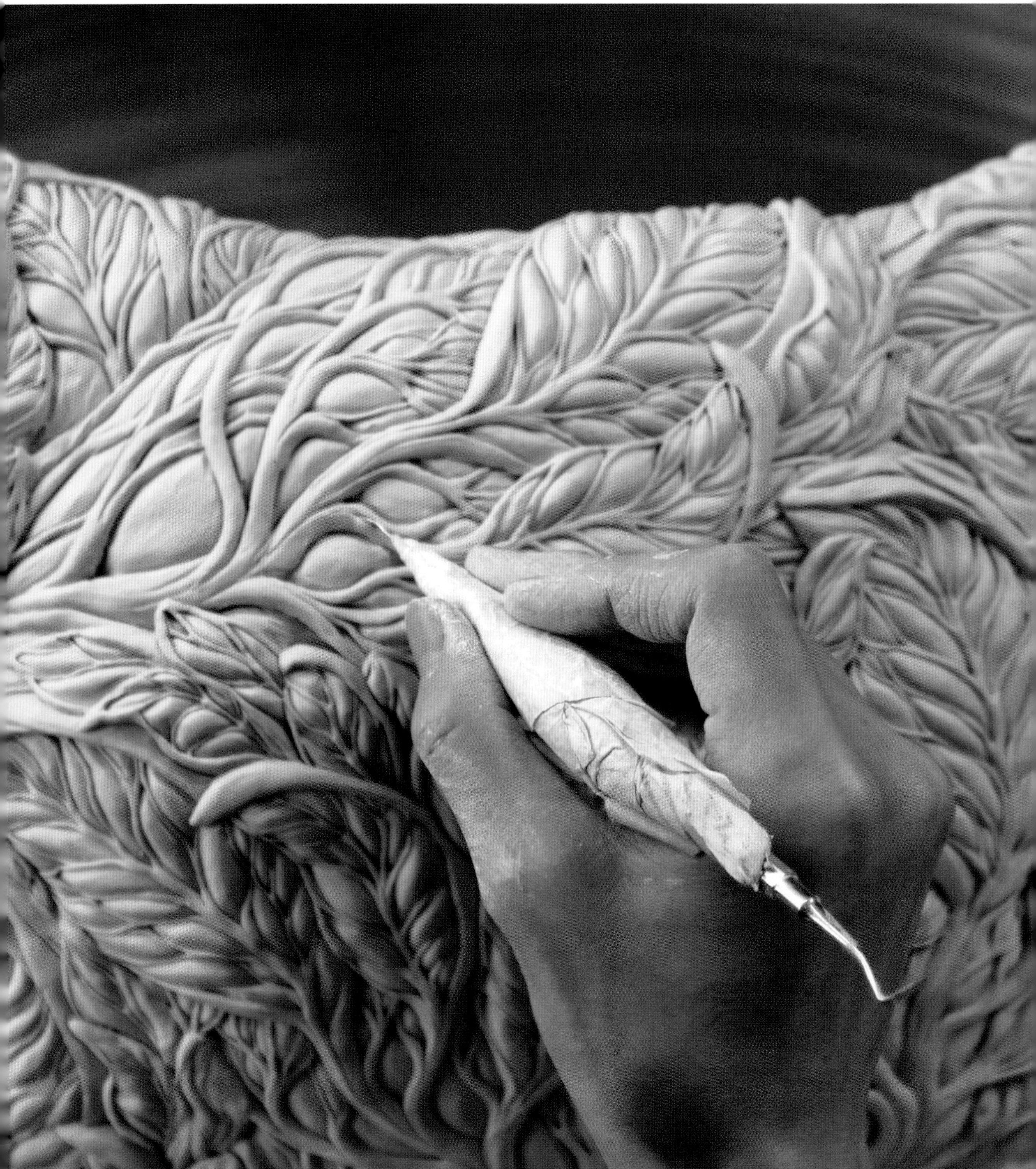

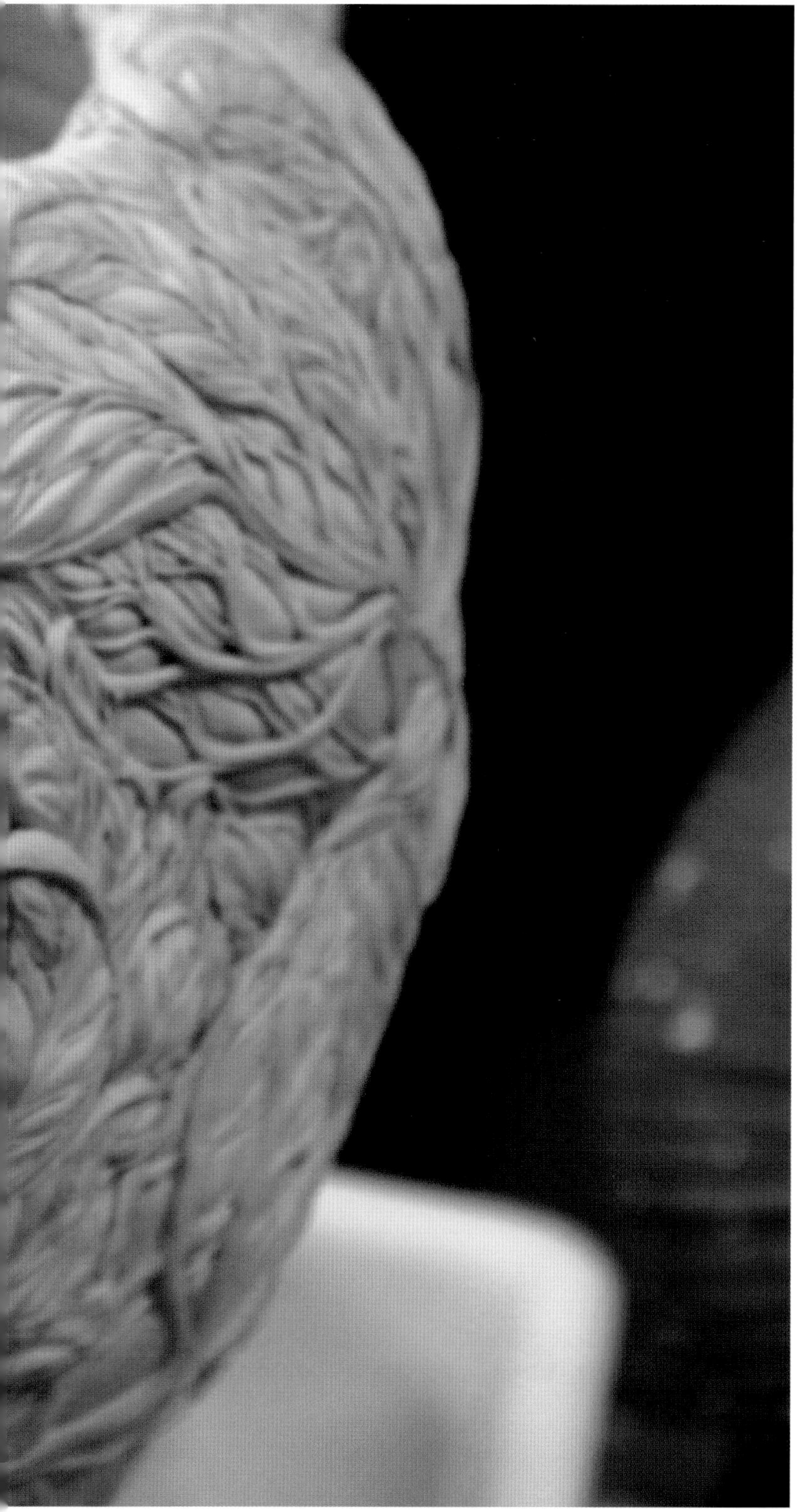

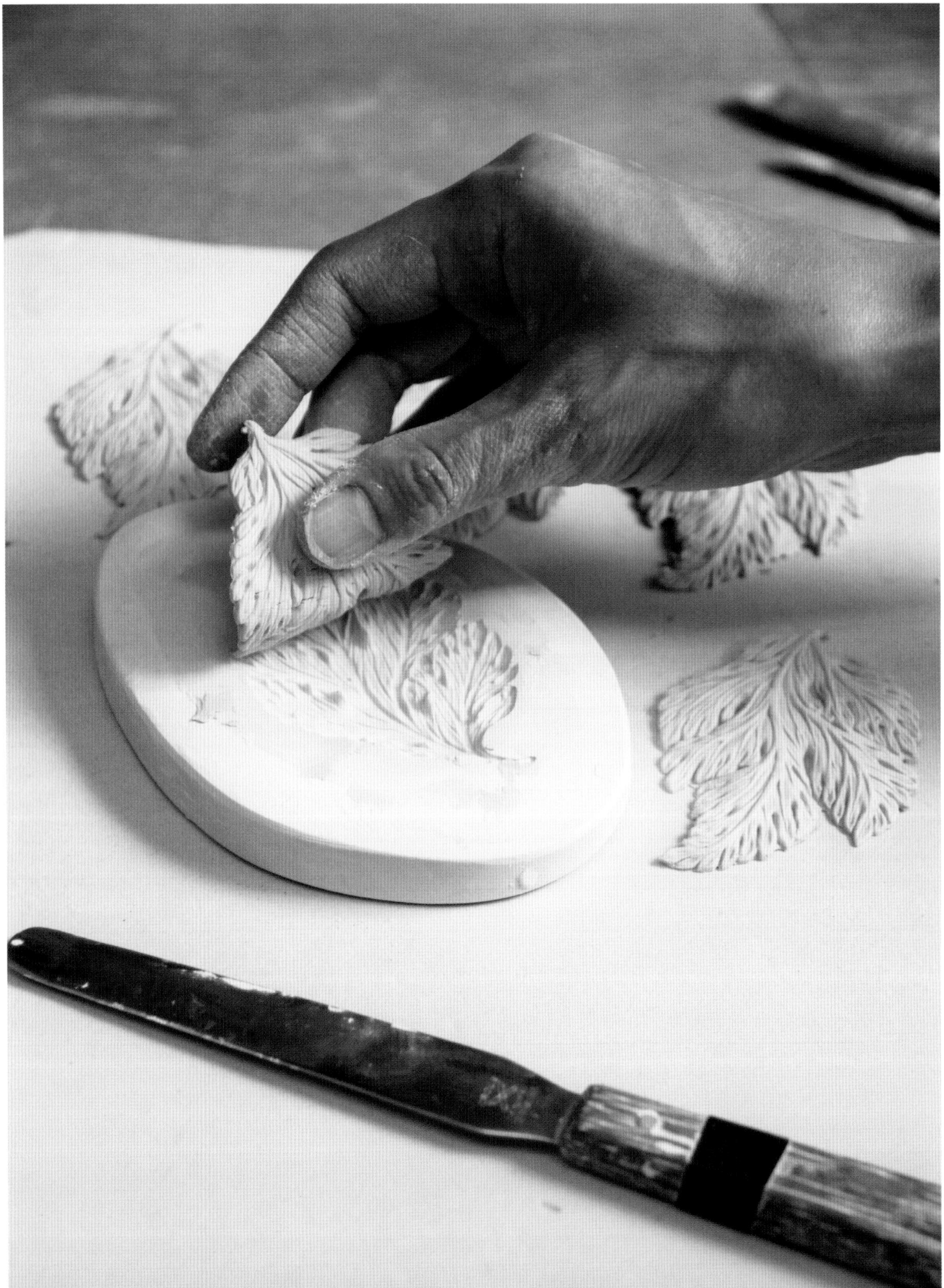

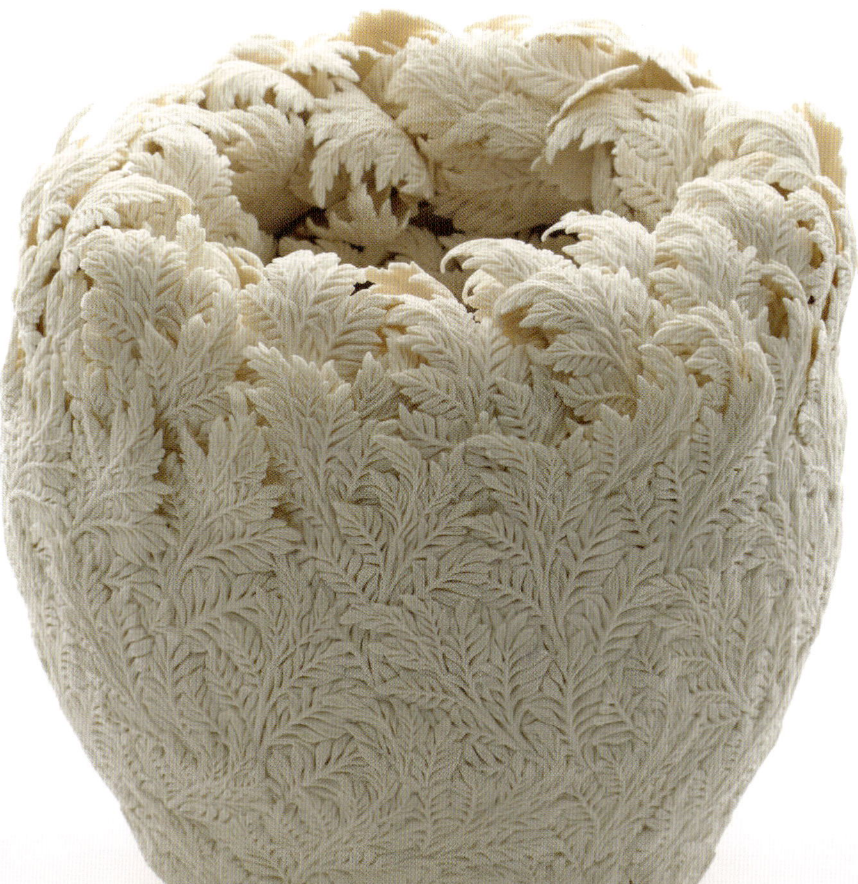

THEONE CERAMICS

LOCATION Los Angeles, CA
YEAR OF BIRTH 1989, 1981
MATERIALS Frost porcelain and black cassius
EXPERIENCE 10 years

THEONE Ceramics is a collective effort between two artists that allows ceramicist Jesse Figueroa and painter Hannah Stouffer to combine their creative strengths with a common goal of creating decadent porcelain pieces. Hailing from Santa Monica, CA, Figueroa has been working with the ceramic medium for nearly 10 years. She is currently teaching group and private classes in Los Angeles, in addition to creating work for THEONE. With 15 years experience in the contemporary arts fields, Stouffer brings her hand to THEONE, painting intricate glazed embellishments on the vessels, urns, and daggers that the duo design collaborativley.

With Figueroa handbuilding and sculpting their pieces on the wheel and Stouffer adding fine lined decadence, each piece is as delicate as it is unique. Working primarily with frost porcelain and black velvet underglaze, THEONE is a modern take on an exquisite tradition.

Photos courtesy of the artist

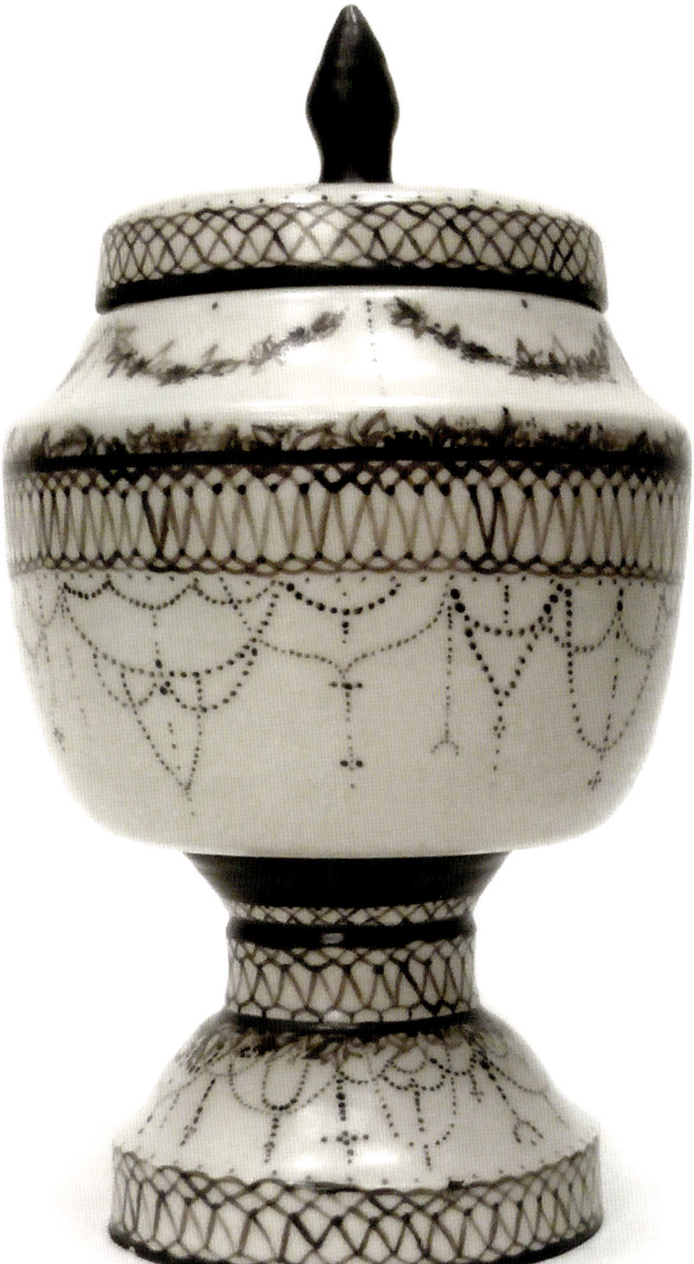

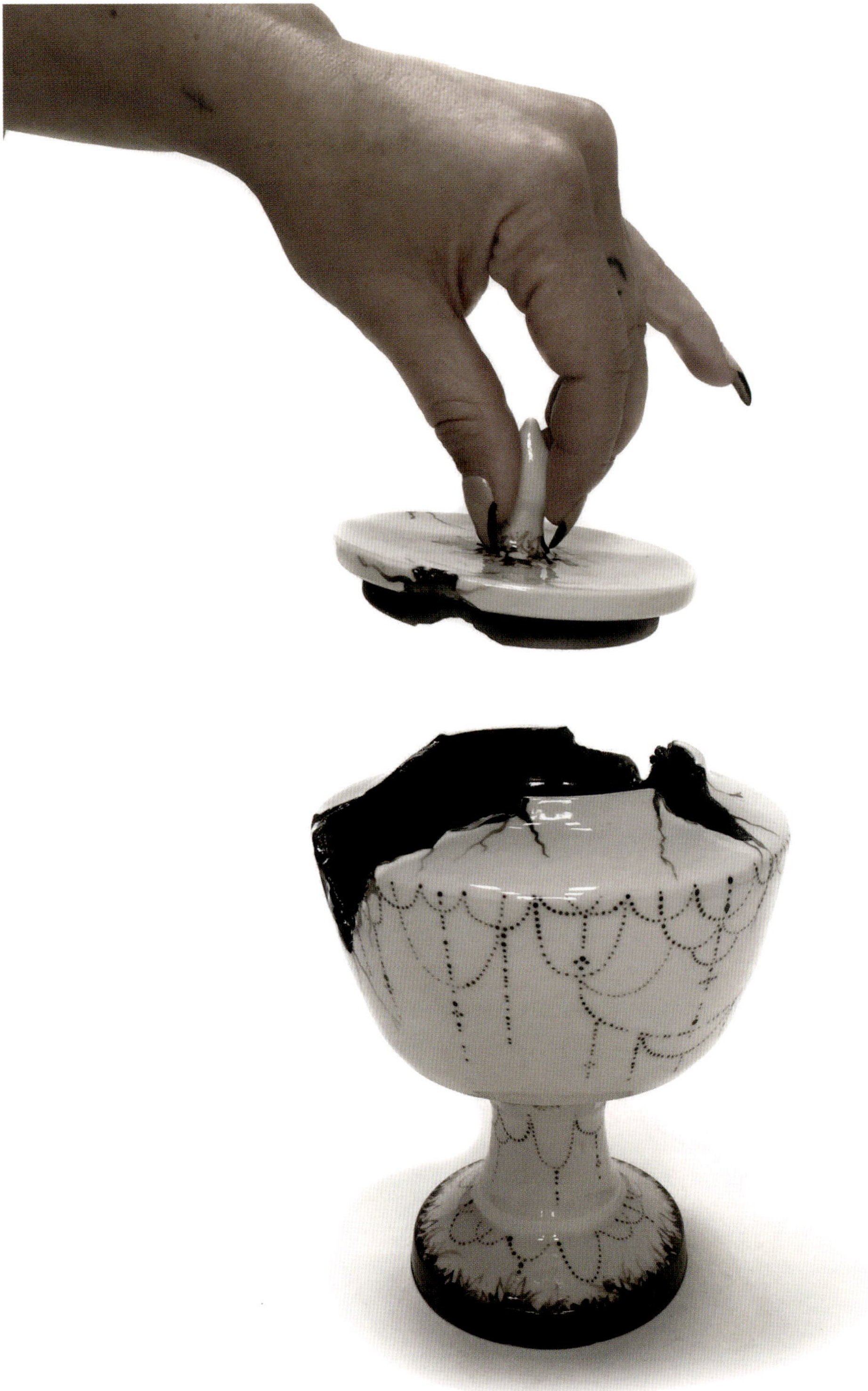

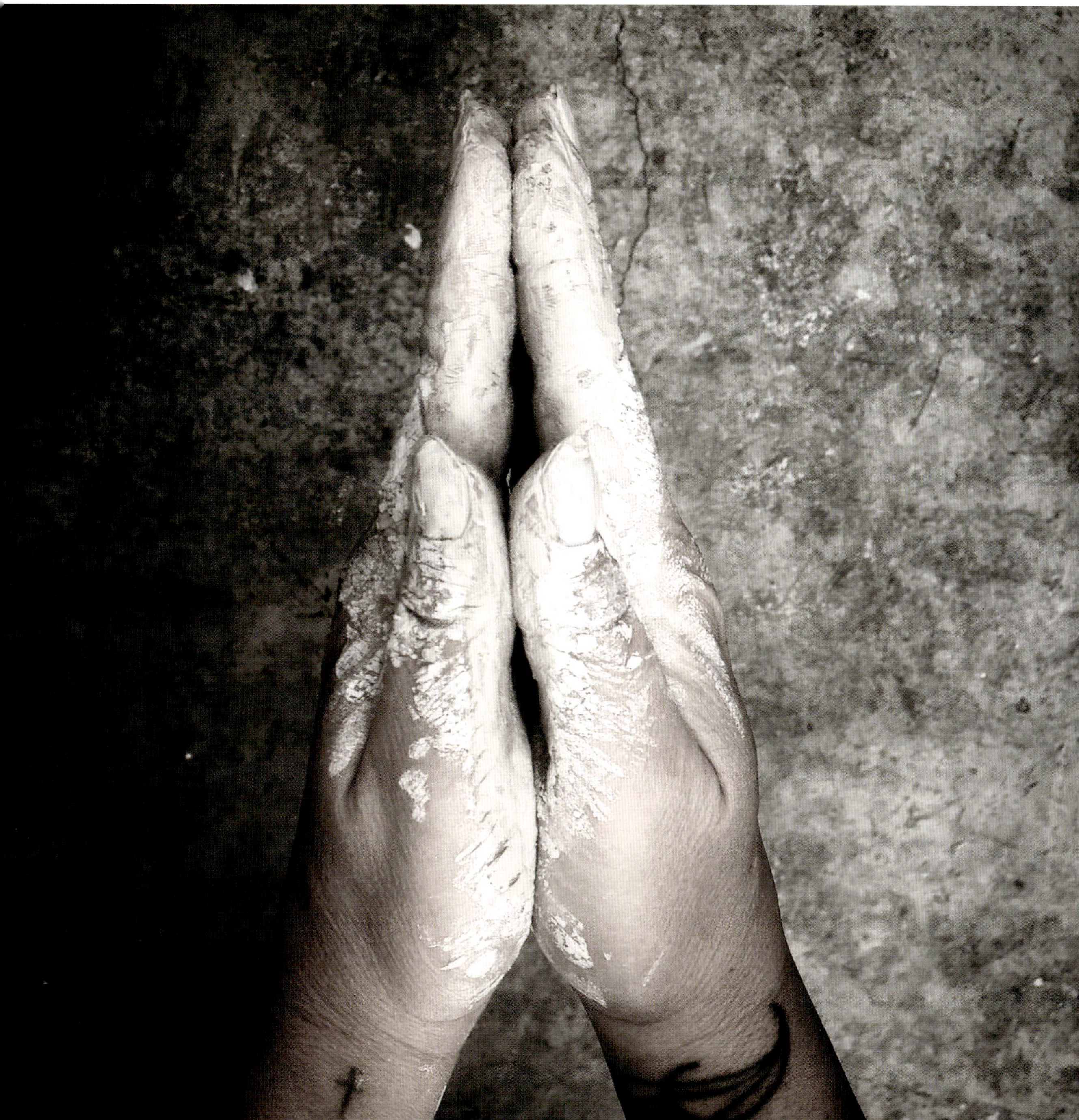

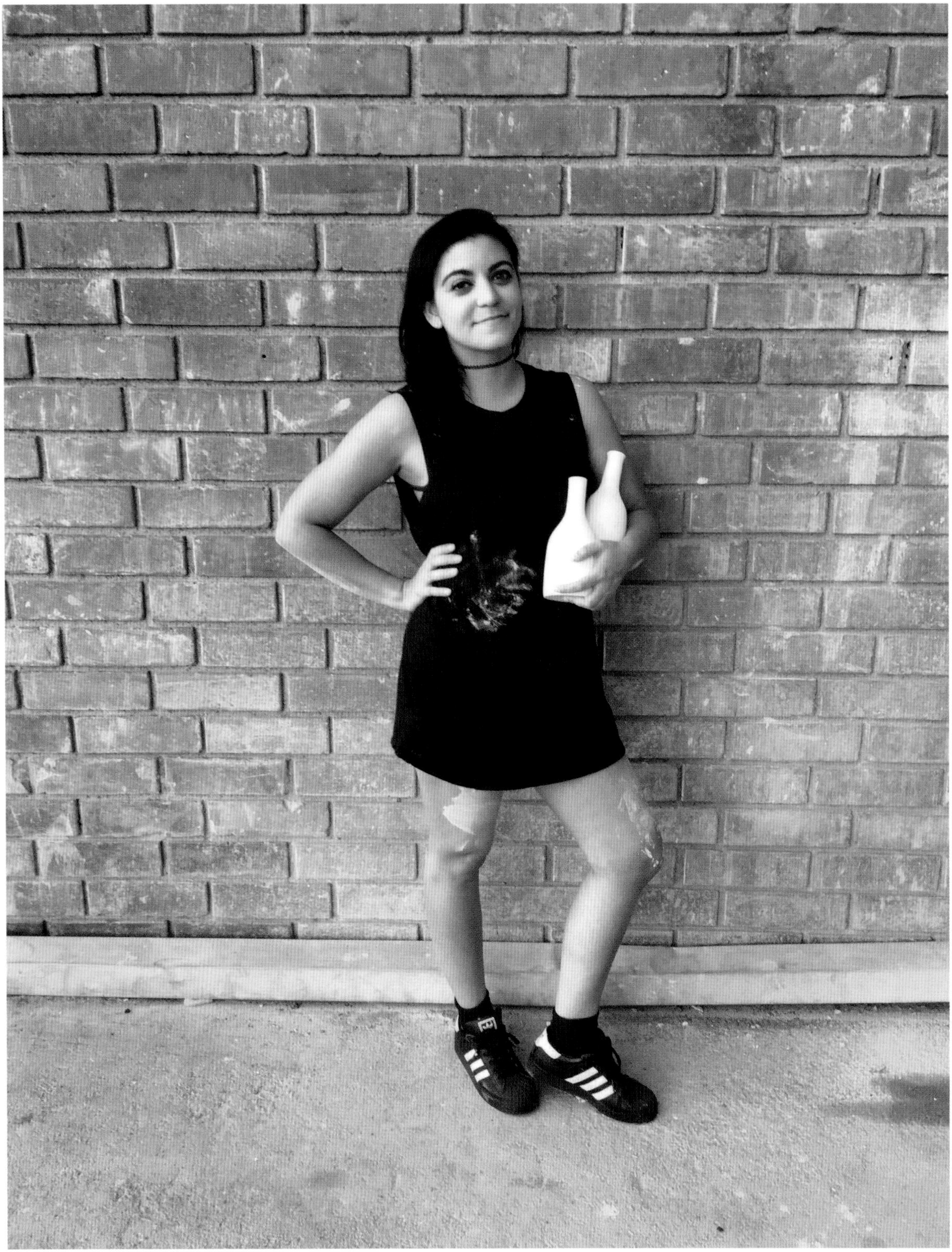

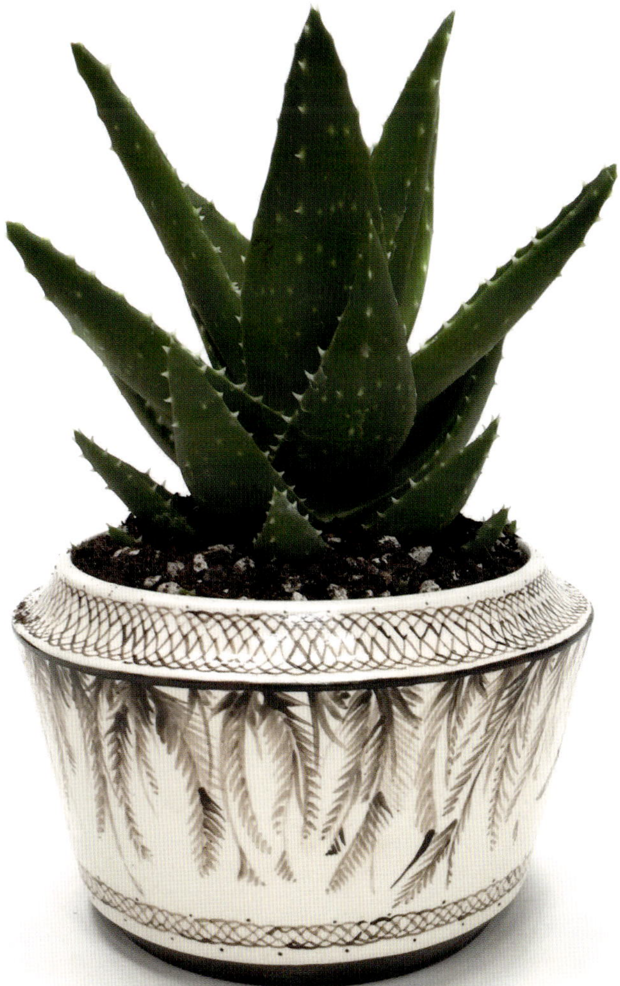

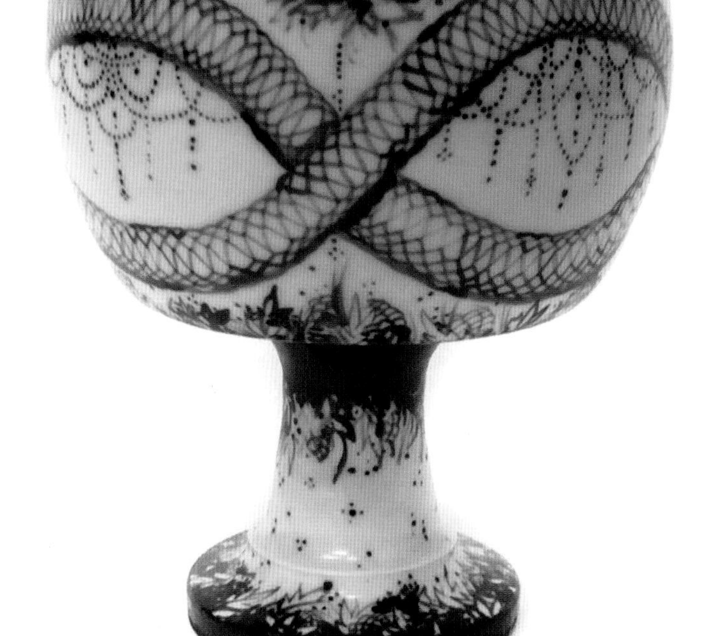

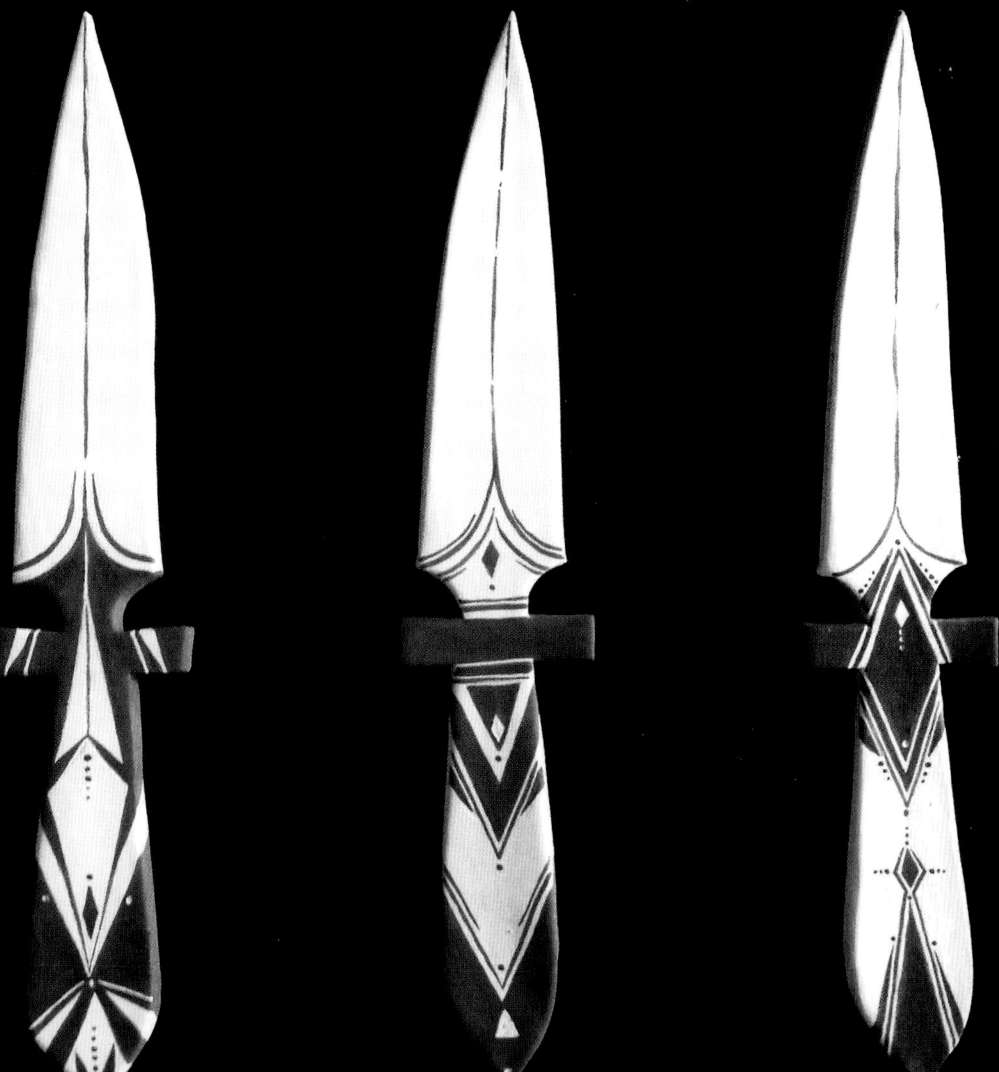

JON ALMEDA

LOCATION Tacoma, WA
YEAR OF BIRTH 1976
MATERIALS Ceramics
EXPERIENCE 22 years

'I have always been drawn to art, sculpture, and objects that play with scale, and in the past I tended to focus more on objects that were over-sized. However everything shifted in 2001 when I first discovered the world of miniatures. The challenge and process of working in 1" scale really captivated and drew me in. I tend to be detail-oriented, so miniatures were an easy transition for me. Working small forces me to look at things from all angles.

I try to produce most of my work outdoors. My studio is located in the basement of my home and I sometimes find it hard to be down there. In June of 2015 I installed a battery on my Curio Wheel (the miniature pottery wheel I designed) to make it portable and decided to head outdoors for the summer for my throwing sessions. That is where #pleinairpottery was born. I borrowed the term and process from plein air painters who work and create outdoors.'

Photos courtesy of the artist

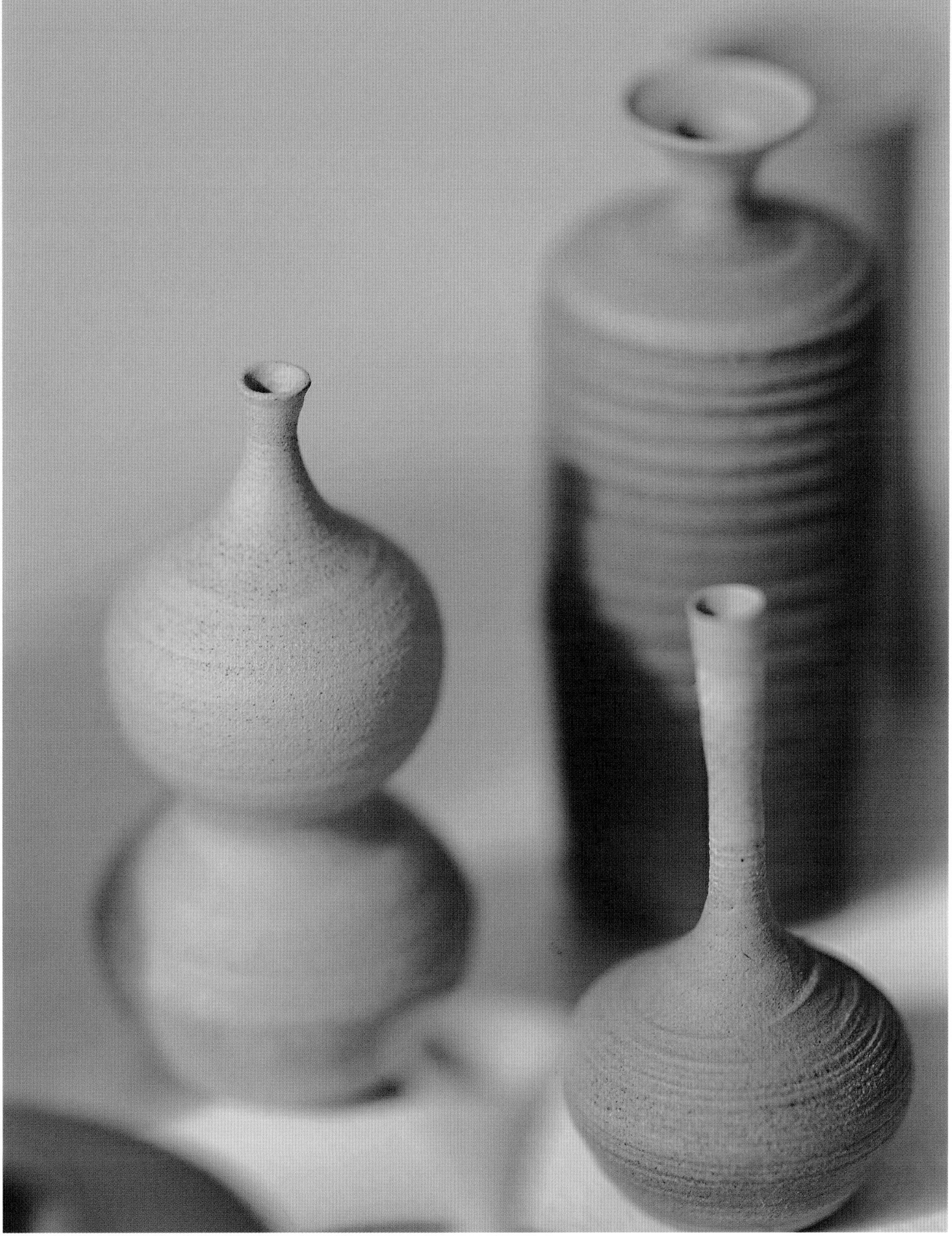

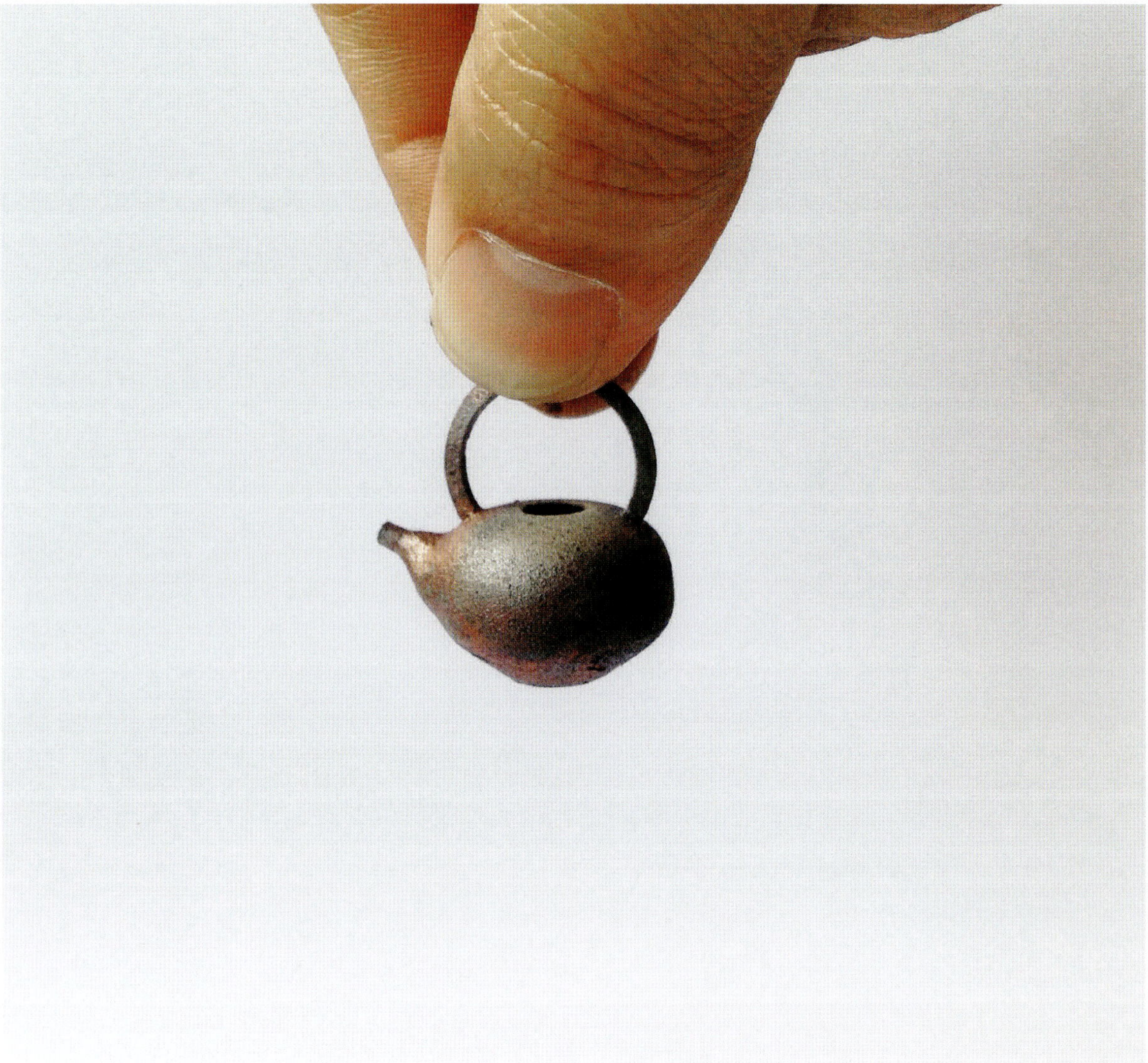

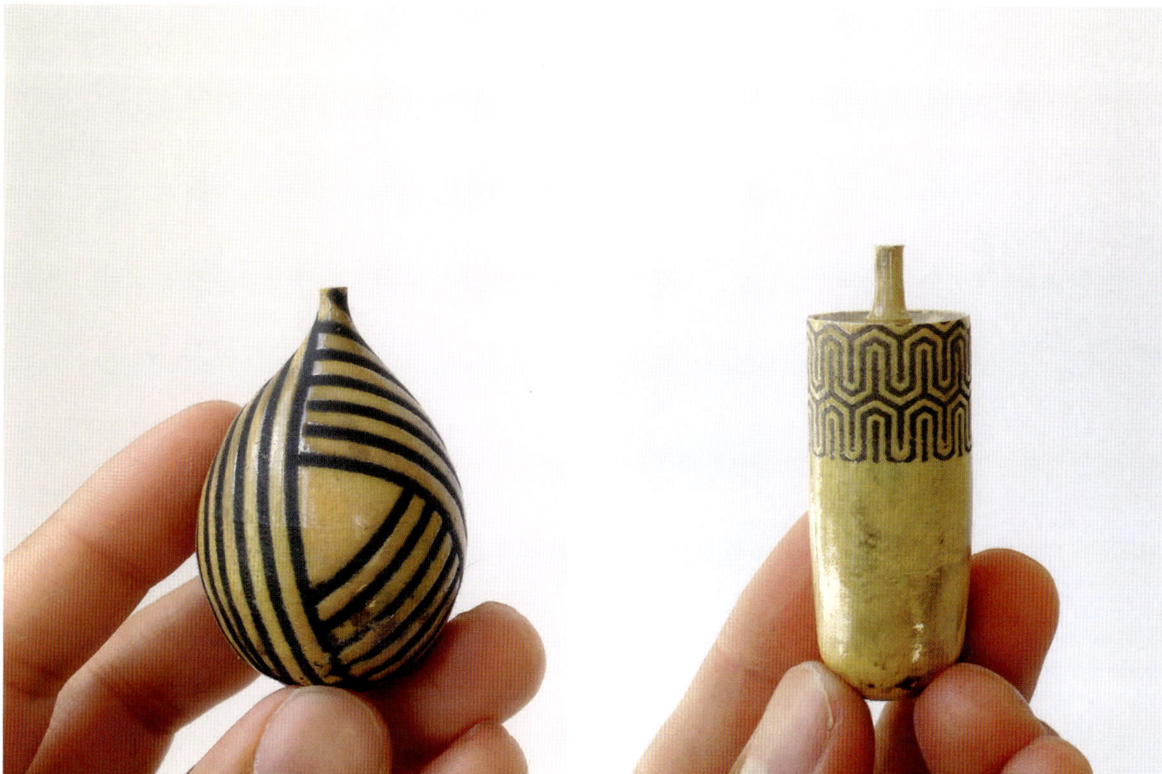

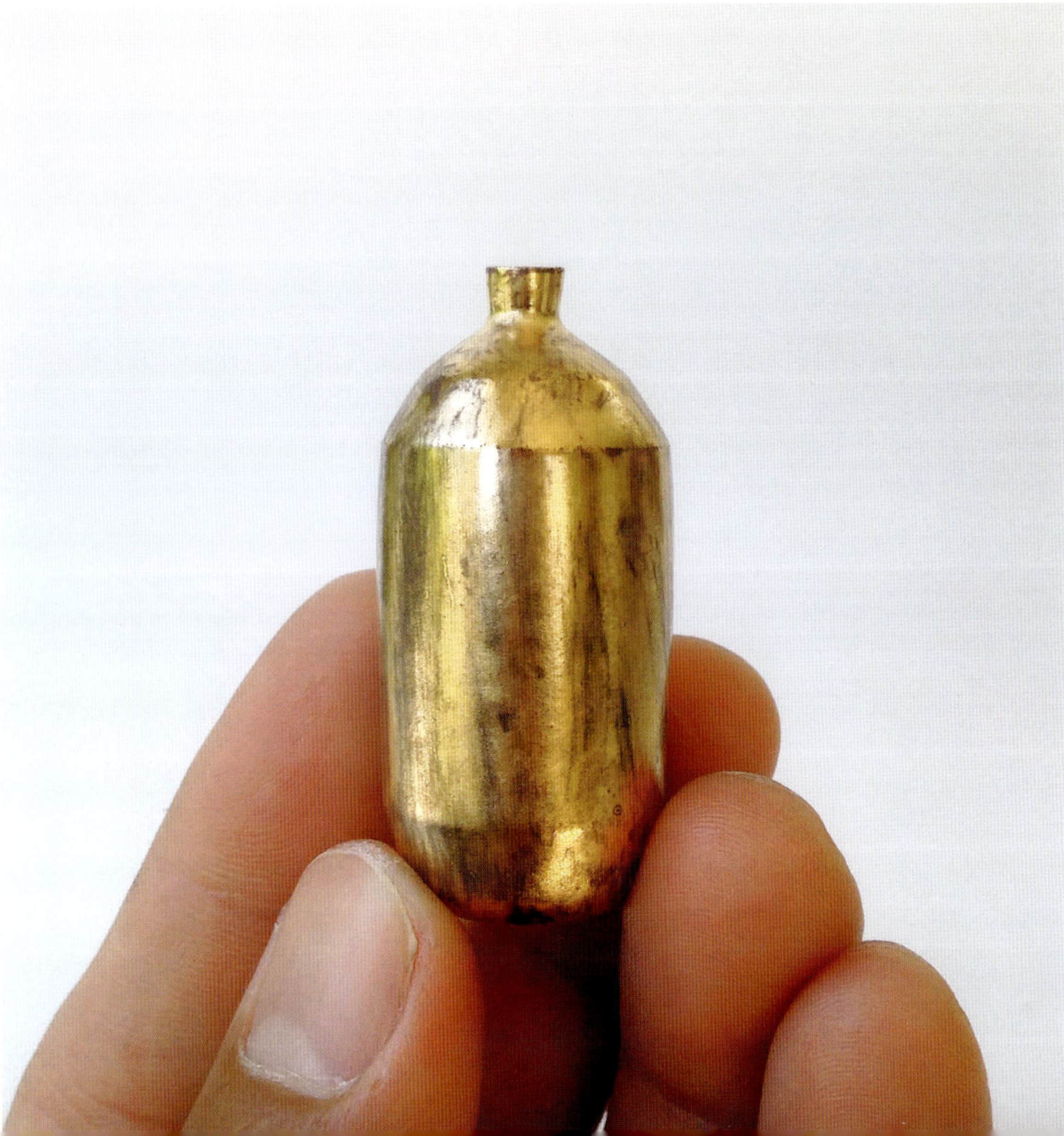

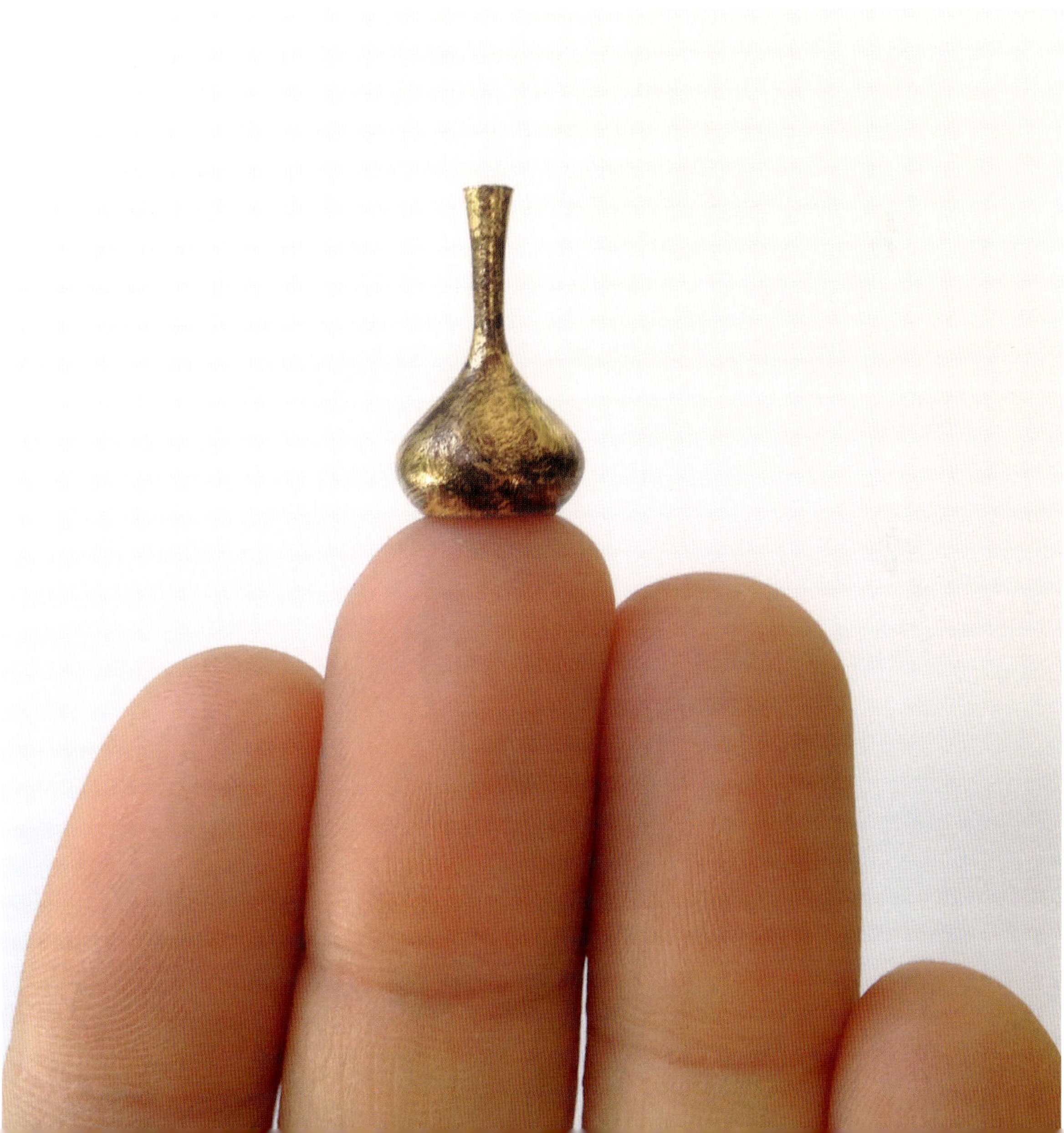

JENNIFER MCCURDY

LOCATION Vineyard Haven, MA
YEAR OF BIRTH 1955
MATERIALS Porcelain, wheel thrown, altered, and carved
EXPERIENCE 40 years

Jennifer McCurdy received a BFA from Michigan State University in 1979. She maintains a studio in Vineyard Haven, Massachusetts. Her focus on the reflective and bone-like quality of bare porcelain has led her to explore the light and shadow of the clay forms, resulting in a unique melding of line and structure. She has been selling her porcelain in art shows and galleries for the last thirty-five years, and her work is included in the collections of several museums, including the Smithsonian's Renwick Gallery in Washington, DC.

Photos courtesy of Tom McCurdy, Gary Mirando and the artist

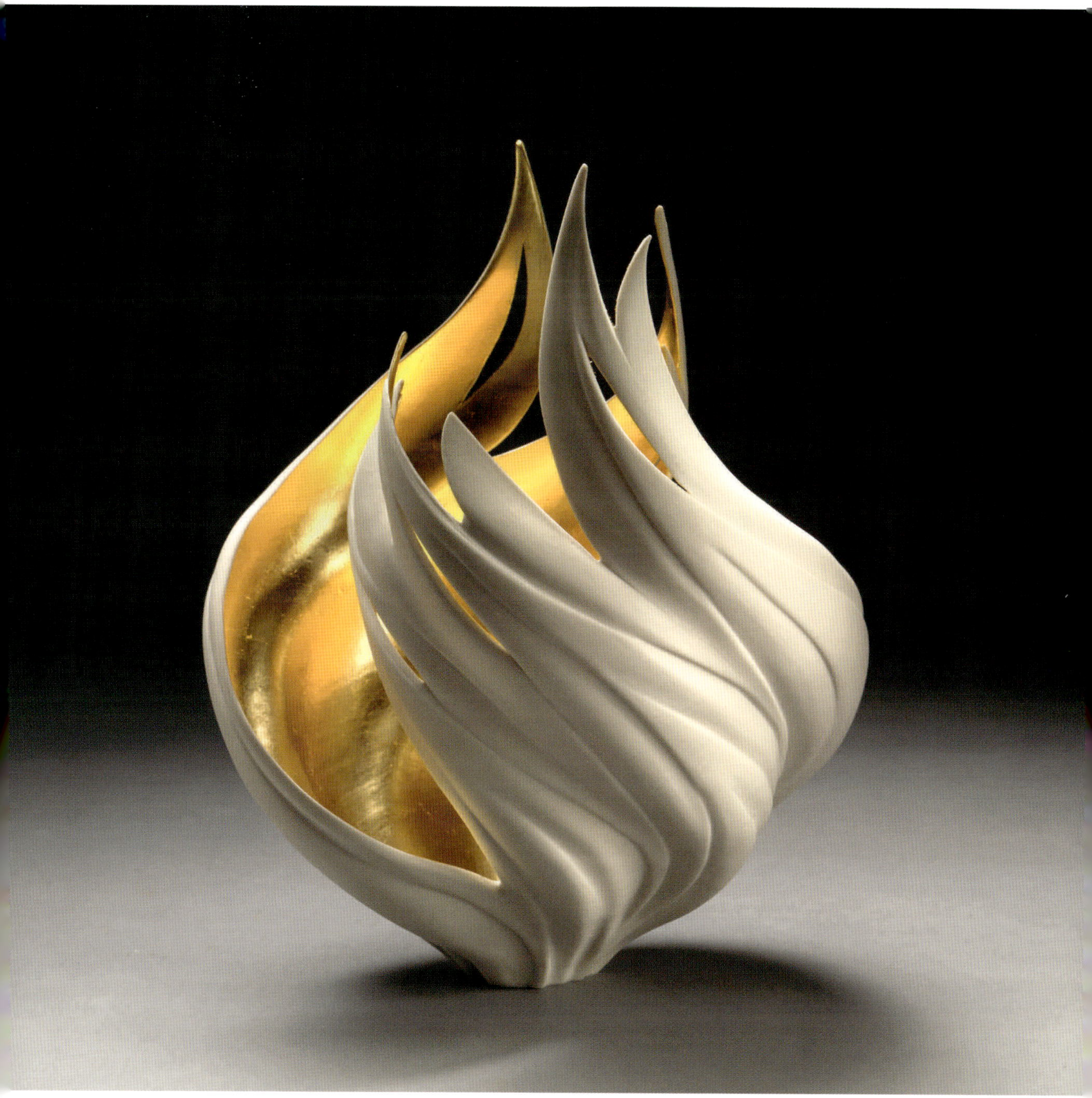

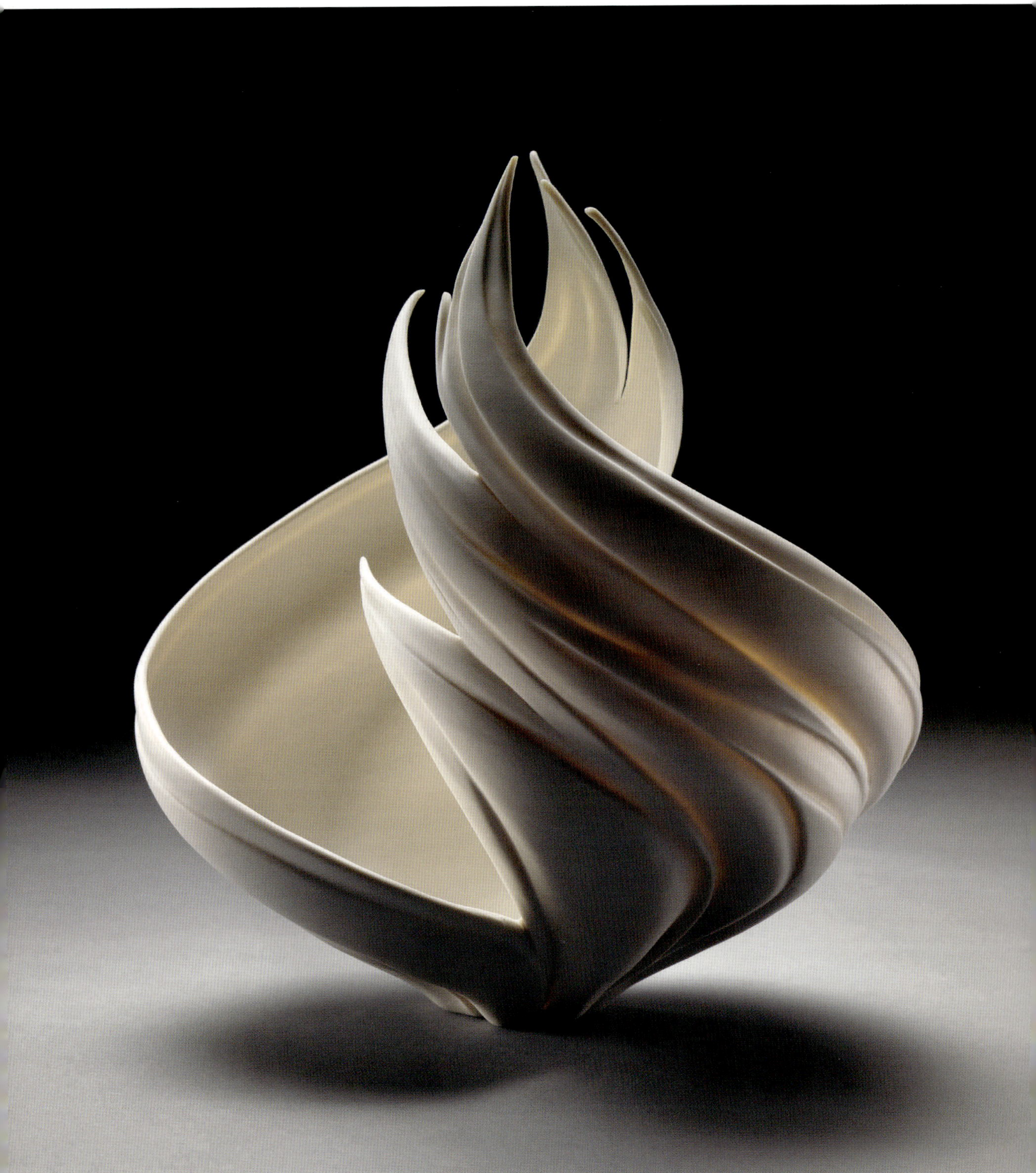

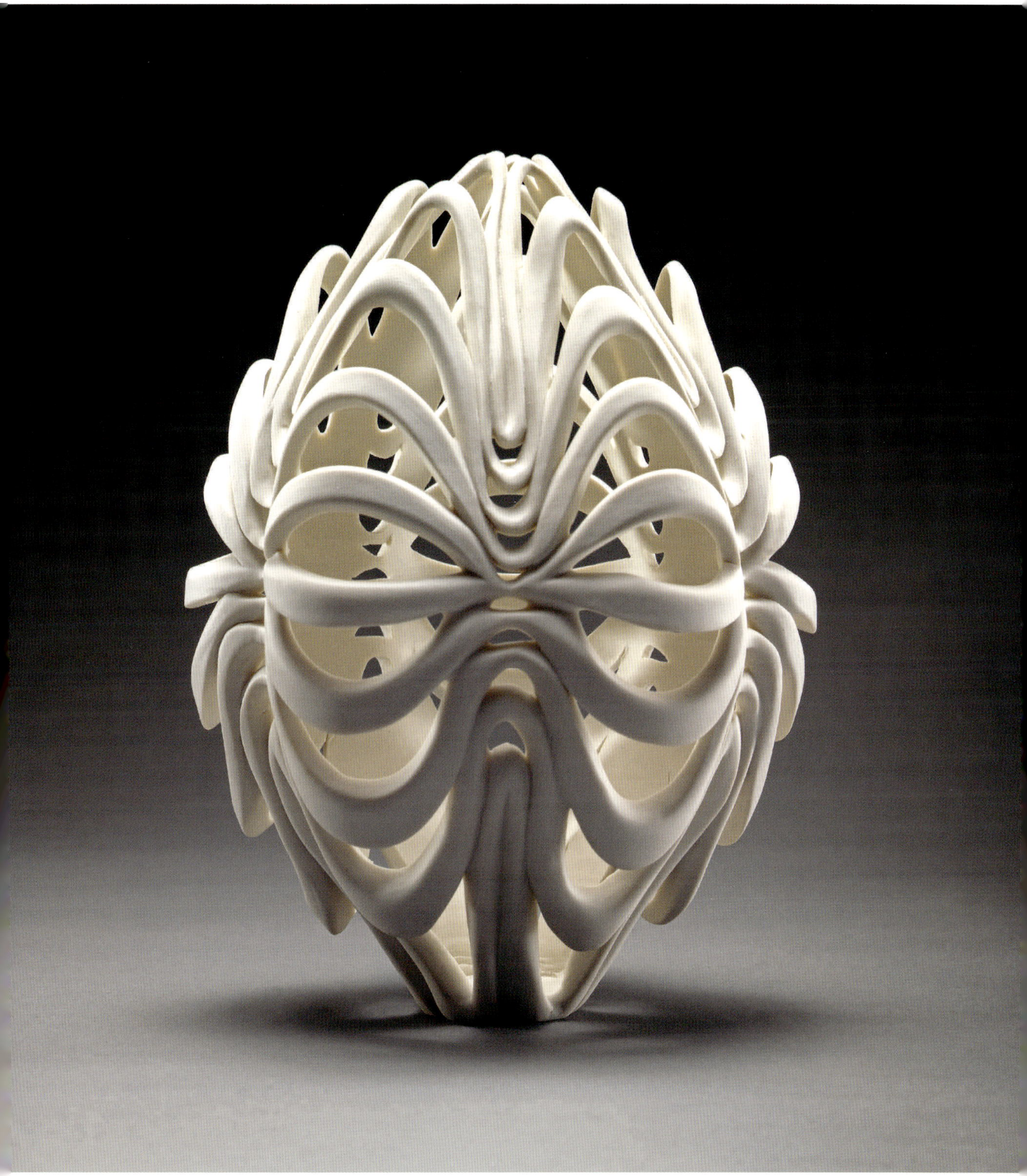

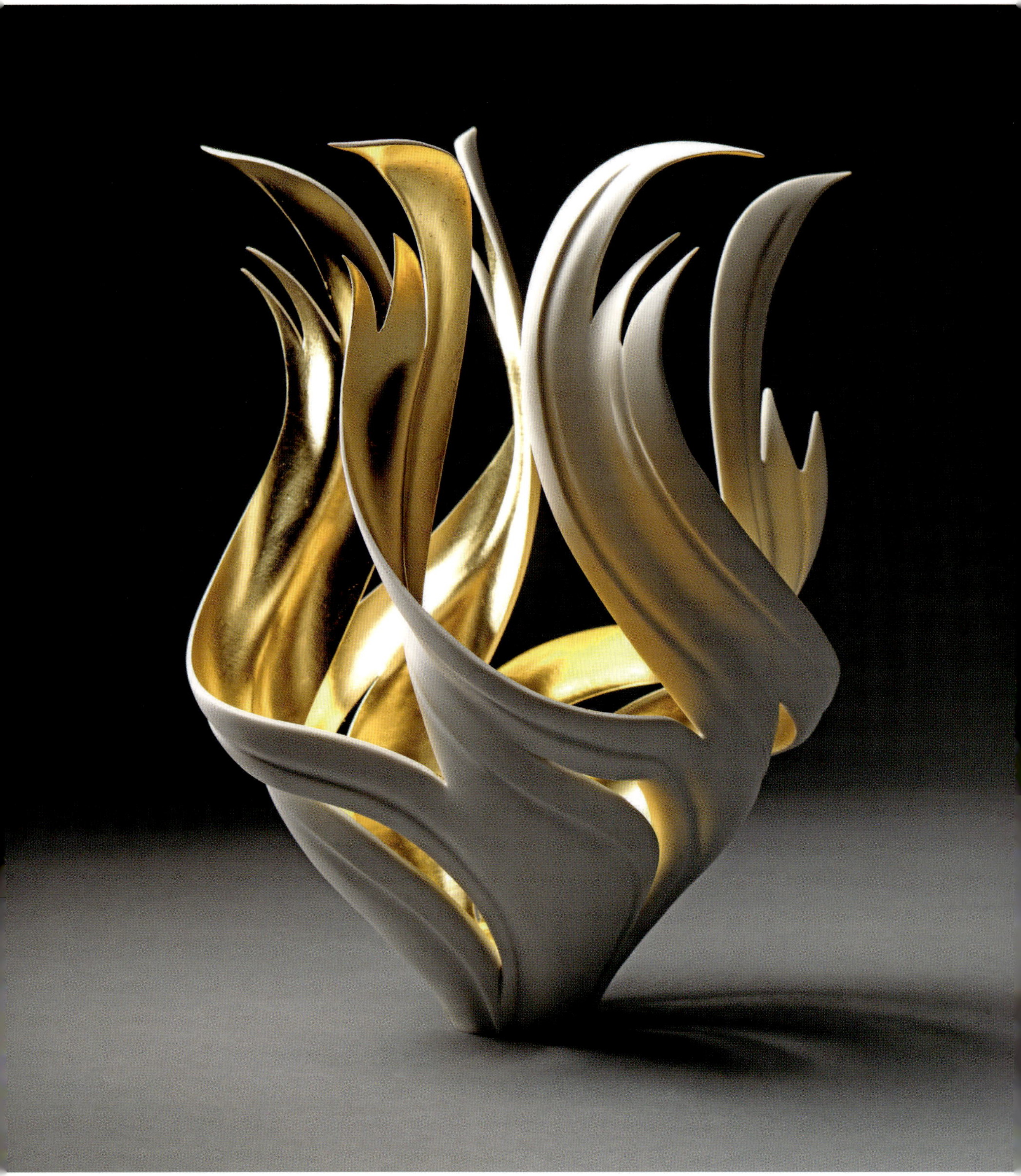

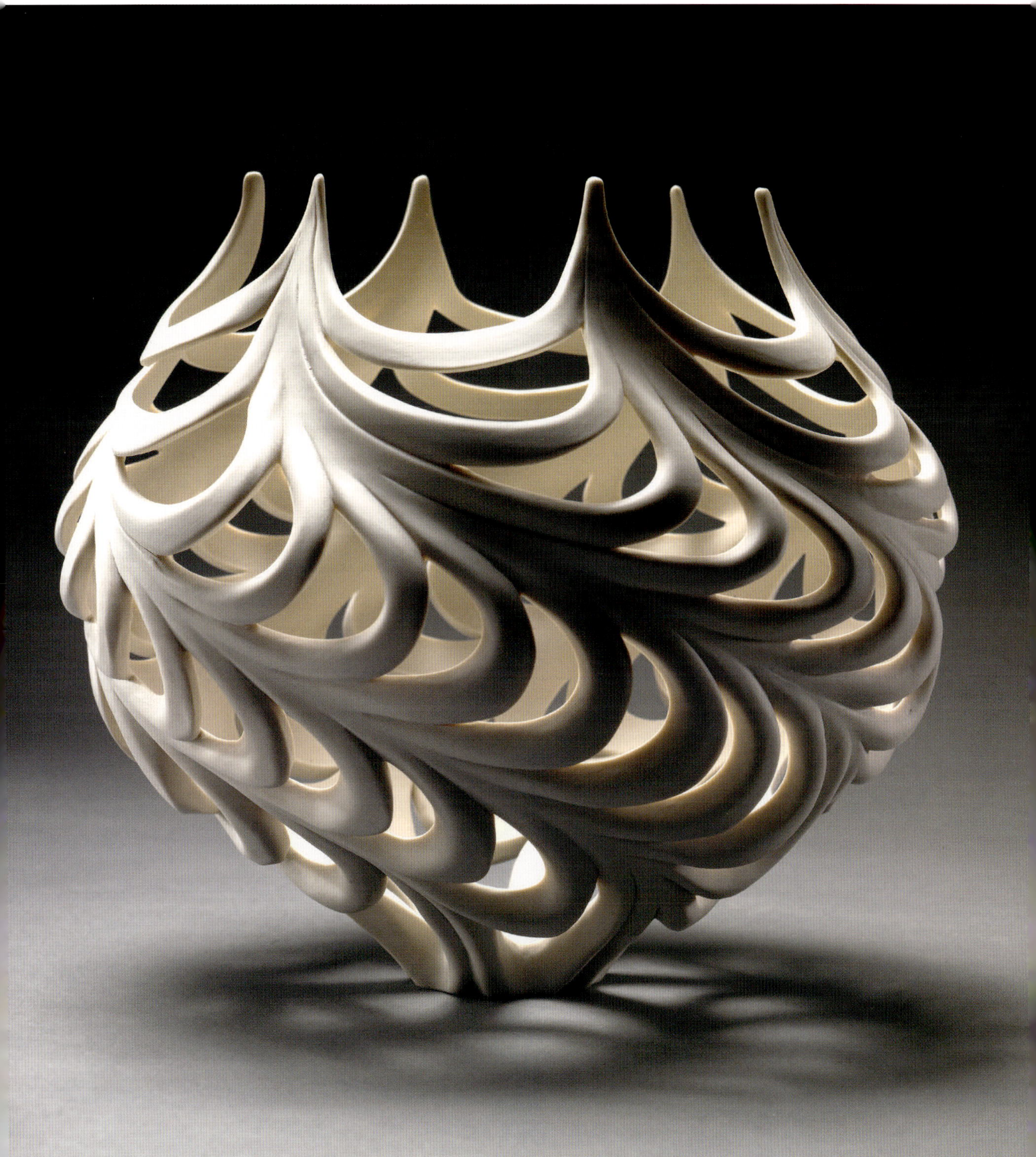

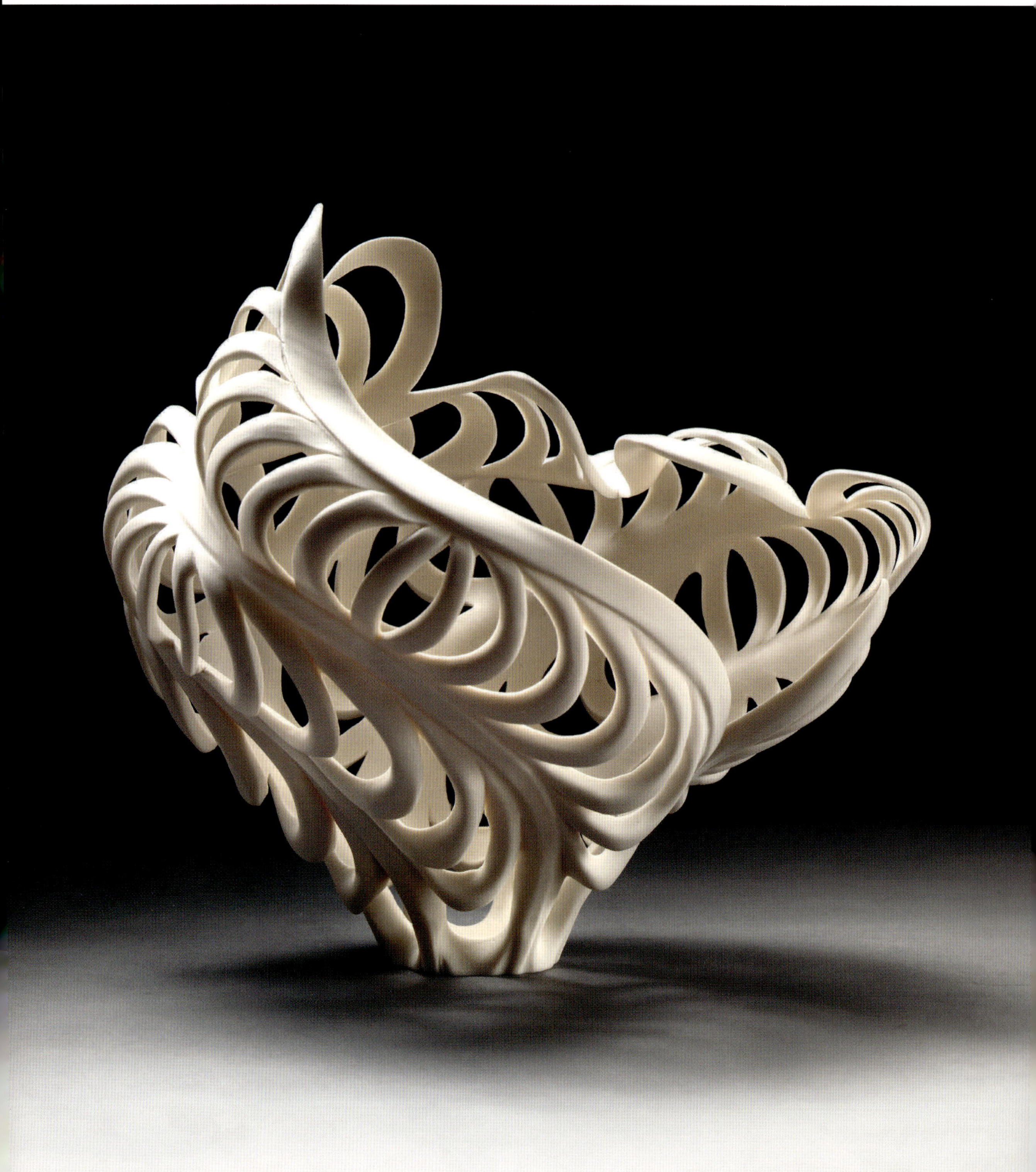

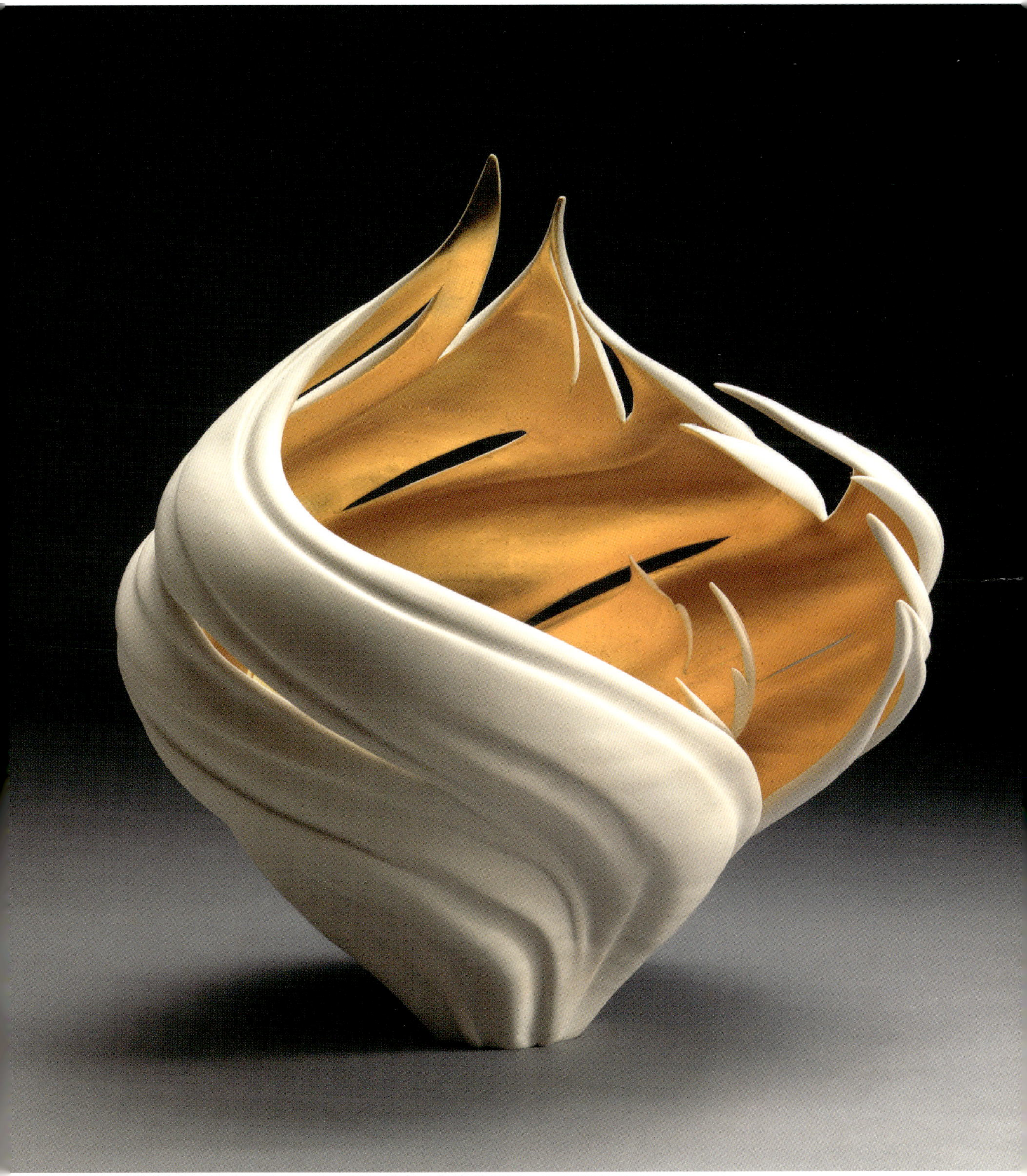

CHE JEN

LOCATION Brooklyn, NY
YEAR OF BIRTH Shhh...
MATERIALS Ceramics, mixed media
EXPERIENCE 3 Years

Che Jen's paintings on ceramic are enigmatic abstractions, from ornate filigree to Eastern calligraphy and tribal markings. Her line work draws upon archetypal mythos and resonances of the deep subconscious.

Drawing upon a social anthropological perspective, she creates a body of work that appears as physical embodiments of rituals and offerings. Her work forms a narrative of objects and the sacred spaces in between.

Photos courtesy of the artist

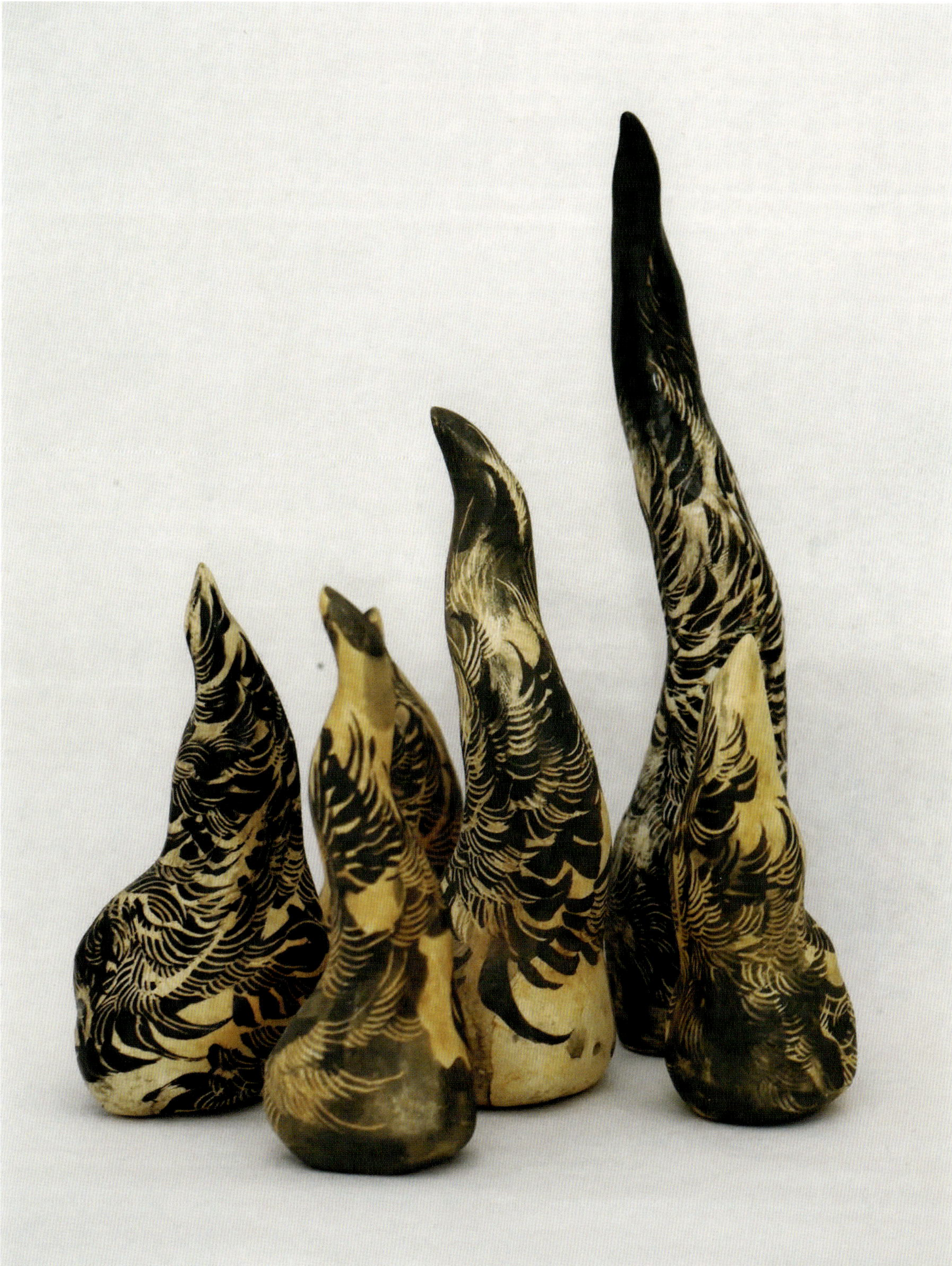

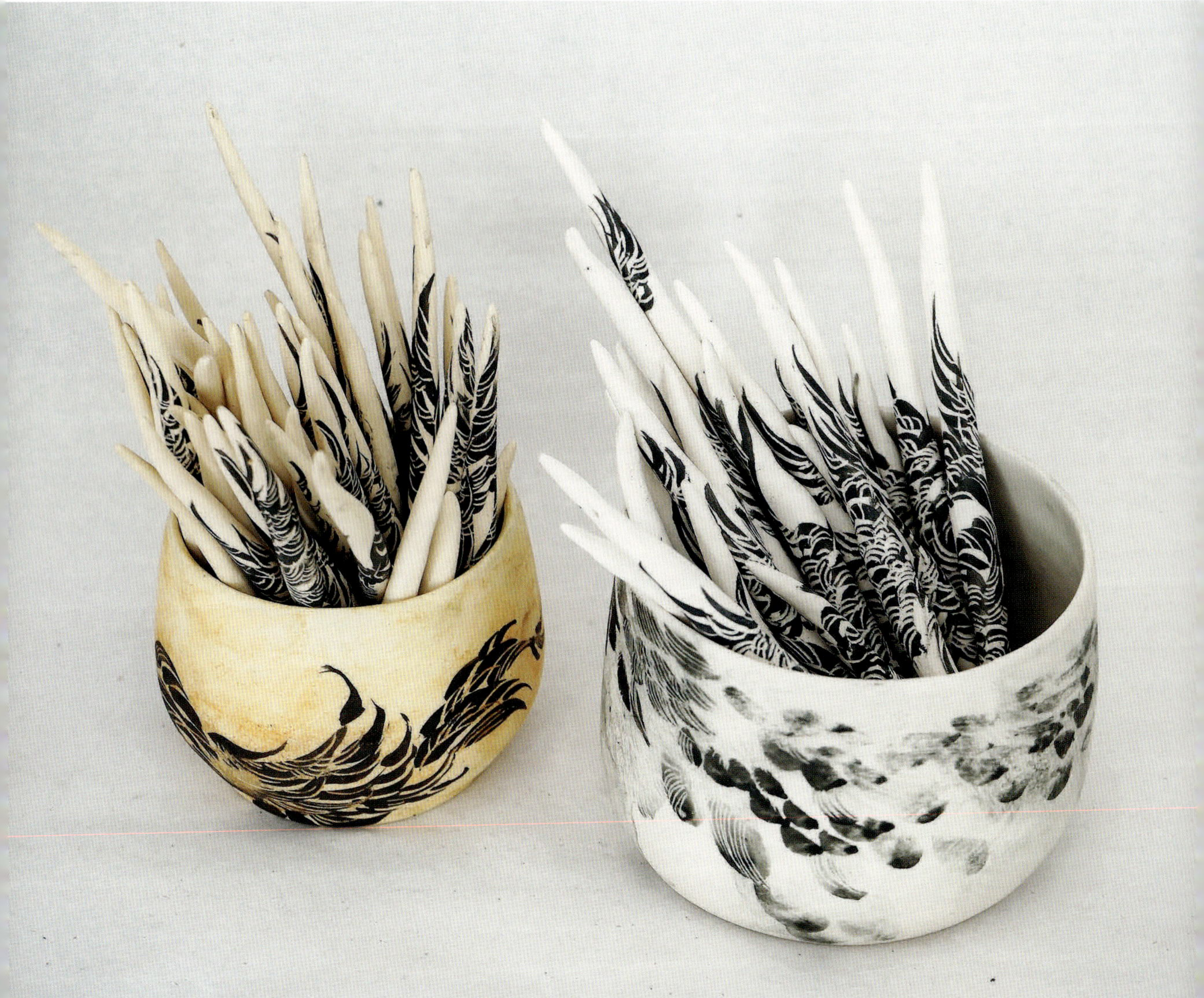

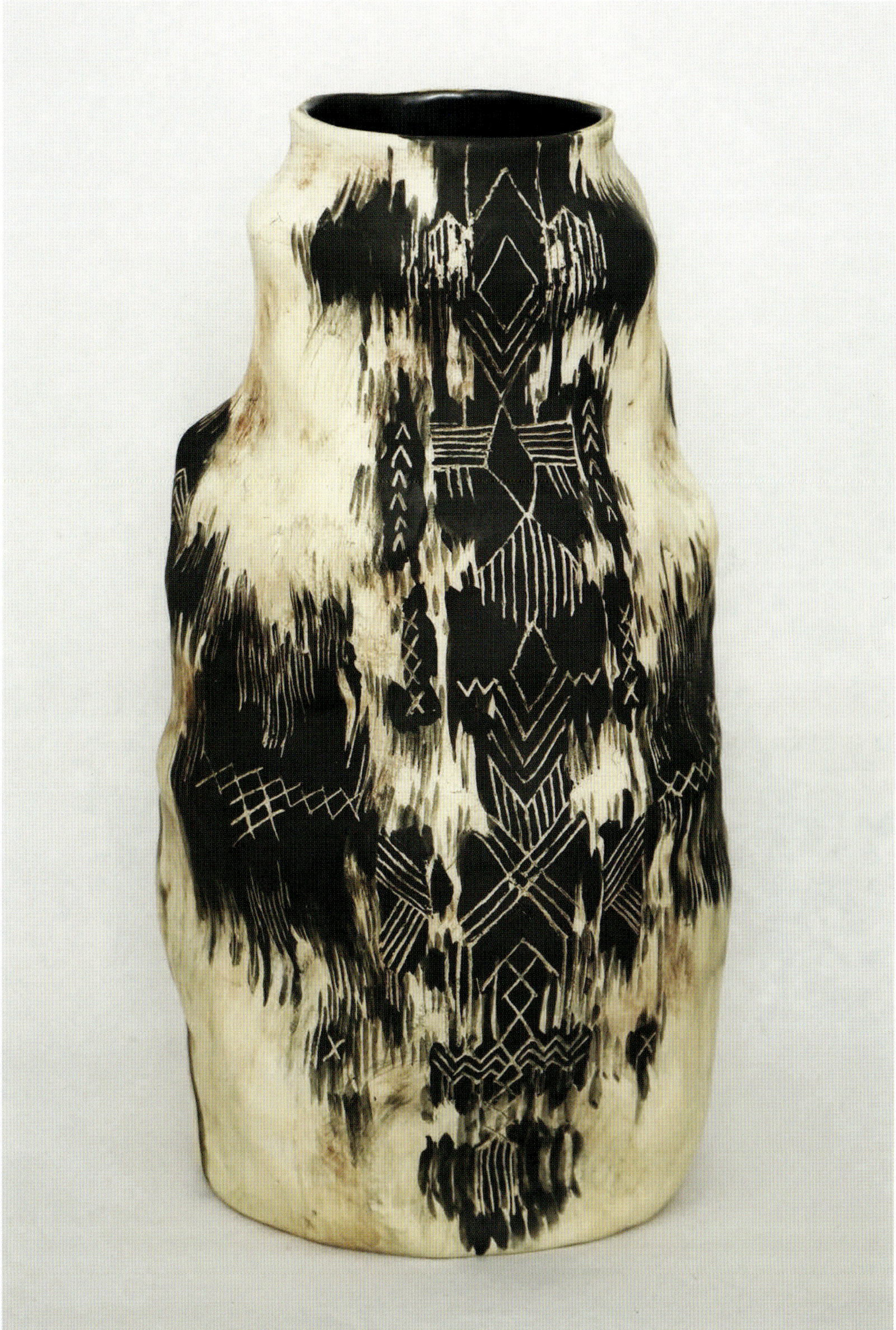

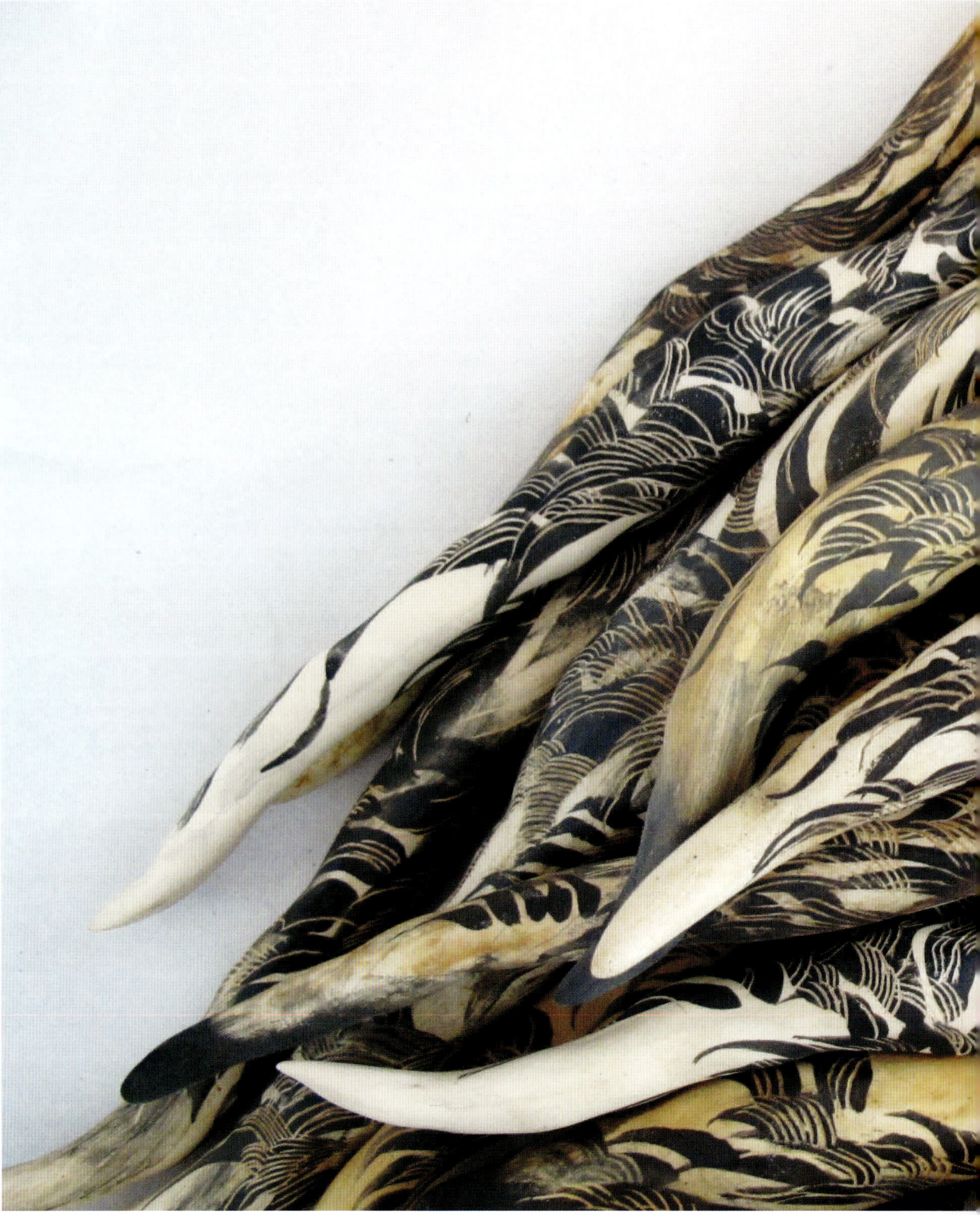

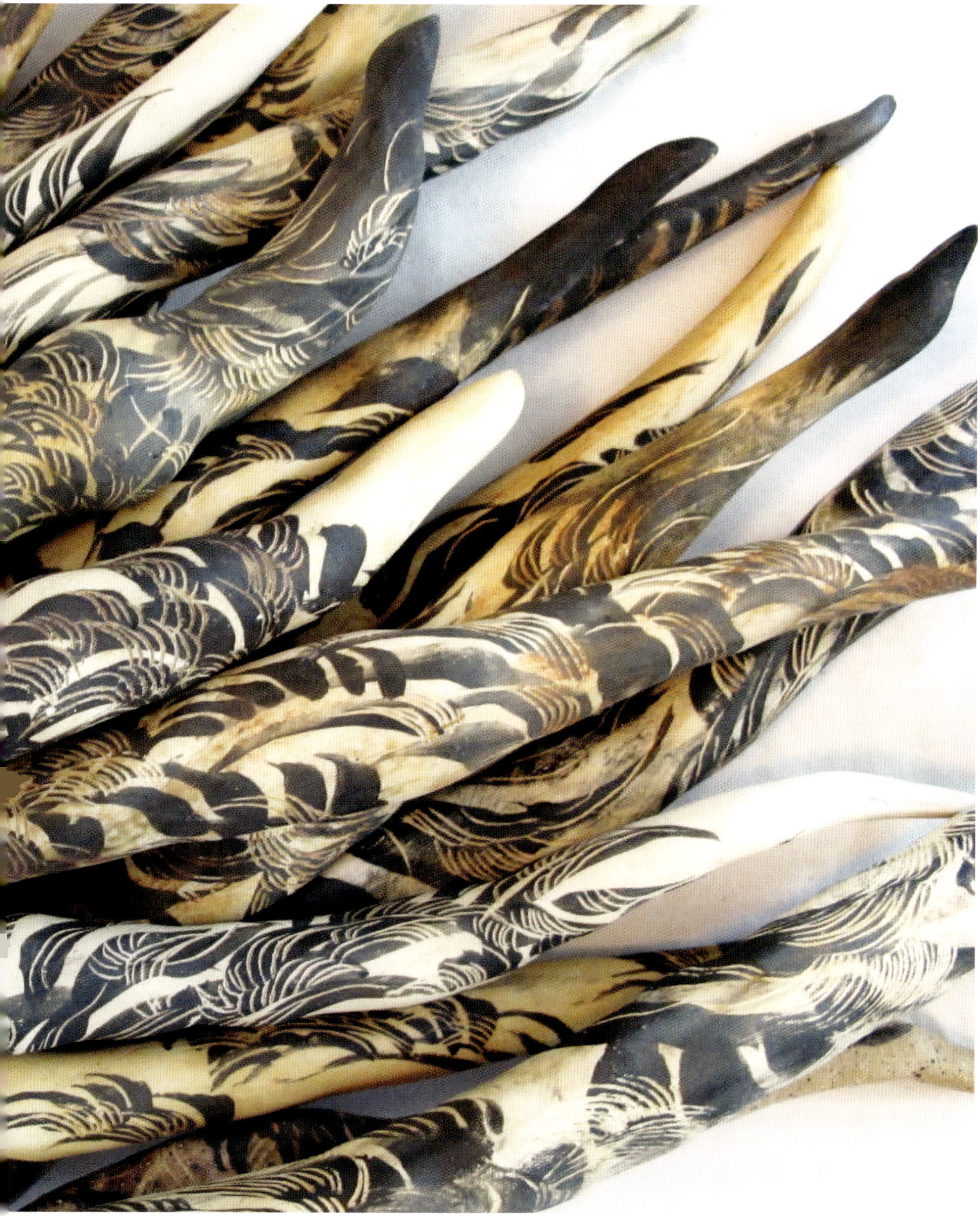

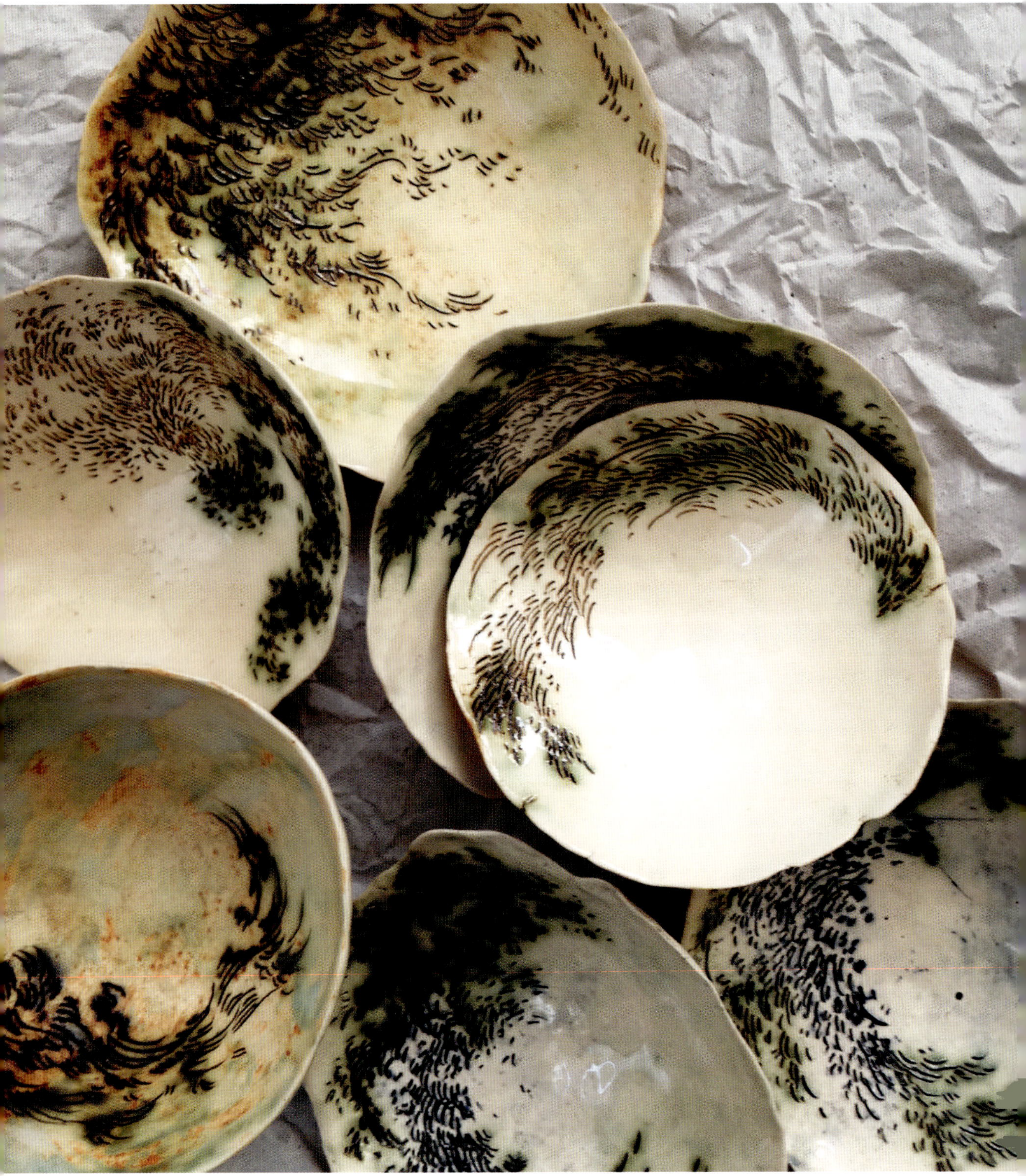

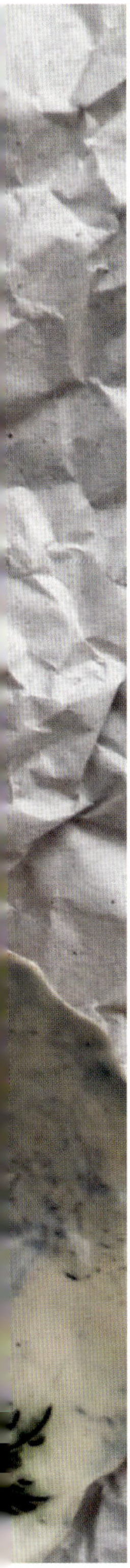

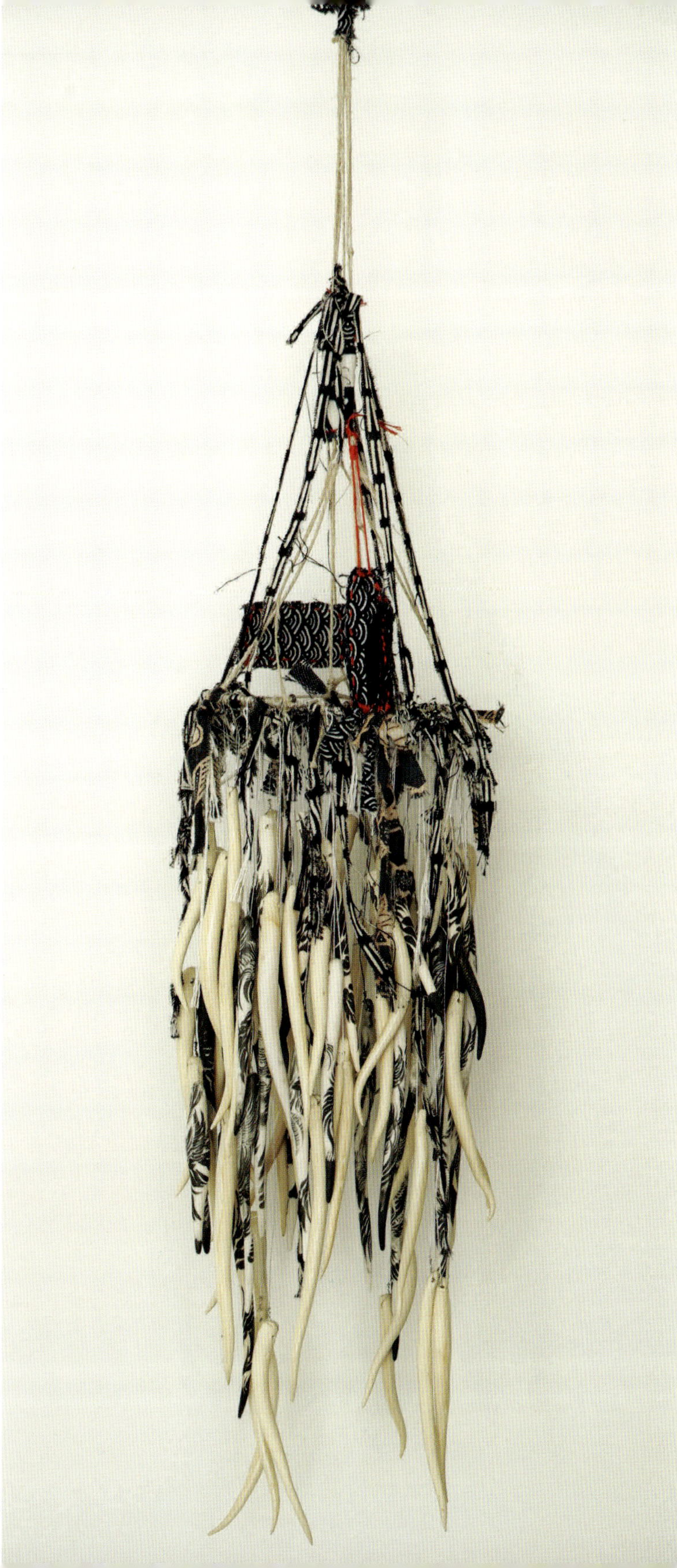

ODE TO TRADITION

STEVEN YOUNG LEE

LIVIA MARIN

LINDSAY SCYPTA

PAUL WOOD

LAURENT CRASTE

STEVEN YOUNG LEE

LOCATION Helena, MT
YEAR OF BIRTH 1975
MATERIALS Ceramics, mixed media
EXPERIENCE 25 Years

'Most of my work is thrown on the wheel. Larger vessels are made in multiple parts and assembled. I primarily work in a high temperature white porcelain but use other clays and materials depending on the idea.'

Steven Young Lee is the Resident Artist Director of the Archie Bray Foundation for the Ceramic Arts in Helena, Montana. Steve has lectured extensively in North America and in China and Korea. His work has appeared in the Duane Reed Gallery (St. Louis), the Ferrin Gallery (Massachusetts), Foster White Gallery (Seattle), and the Northern Clay Center (Minneapolis). His work has been collected by the Renwick Gallery at the Smithsonian Museum, the Daum Museum of Contemporary Art, the Four Seasons Hotel in Seoul, Korea, as well as by many private collectors. Steven received his MFA in Ceramics from the New York State College of Ceramics at Alfred University. He has also taught at Interlochen Center for the Arts in Michigan, the Clay Art Center in New York, Lill Street Studio in Chicago and numerous workshops. Originally from Chicago, he lives in Helena with his wife, Lisa, and their son, Gavin.

Photos courtesy of the artist

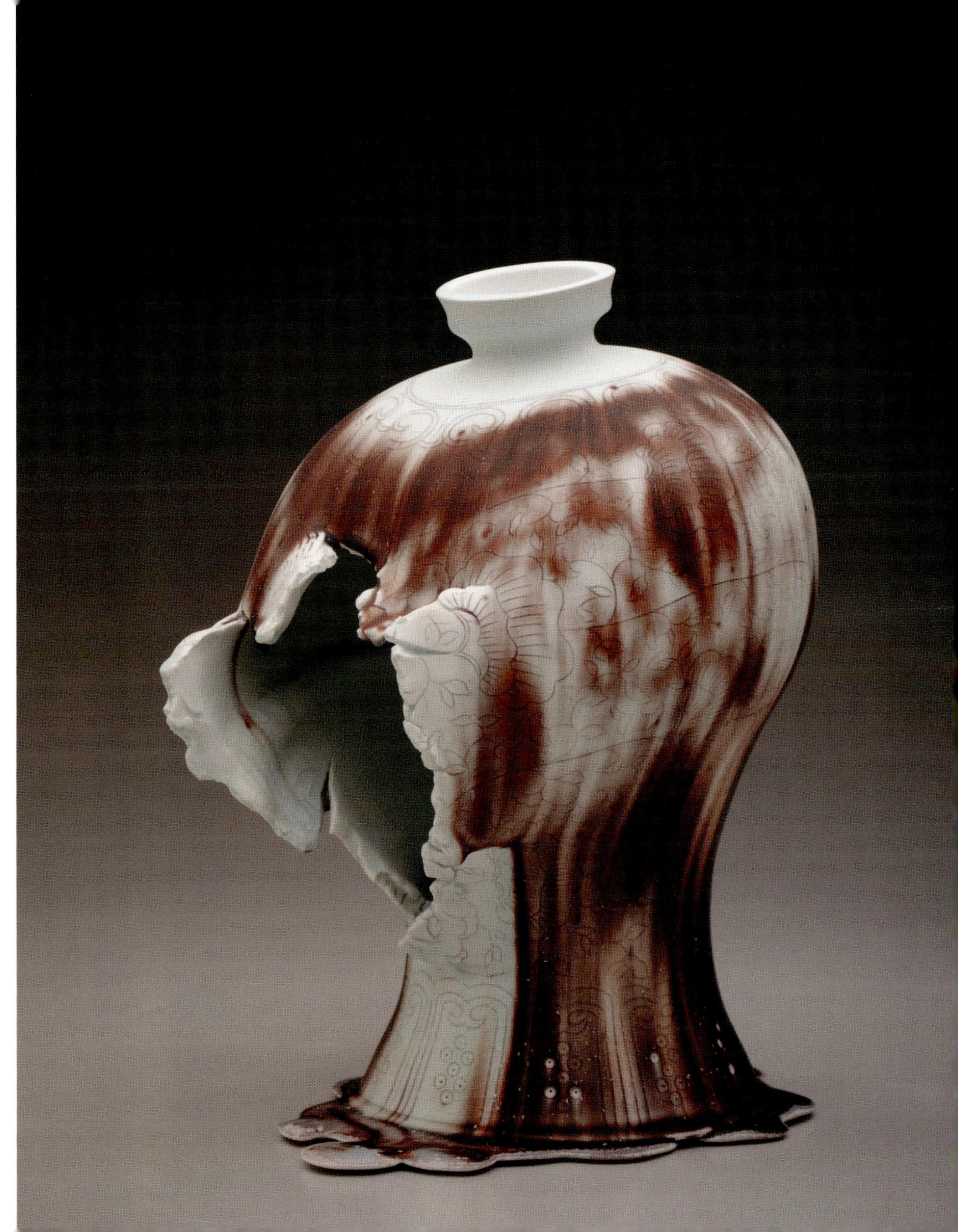

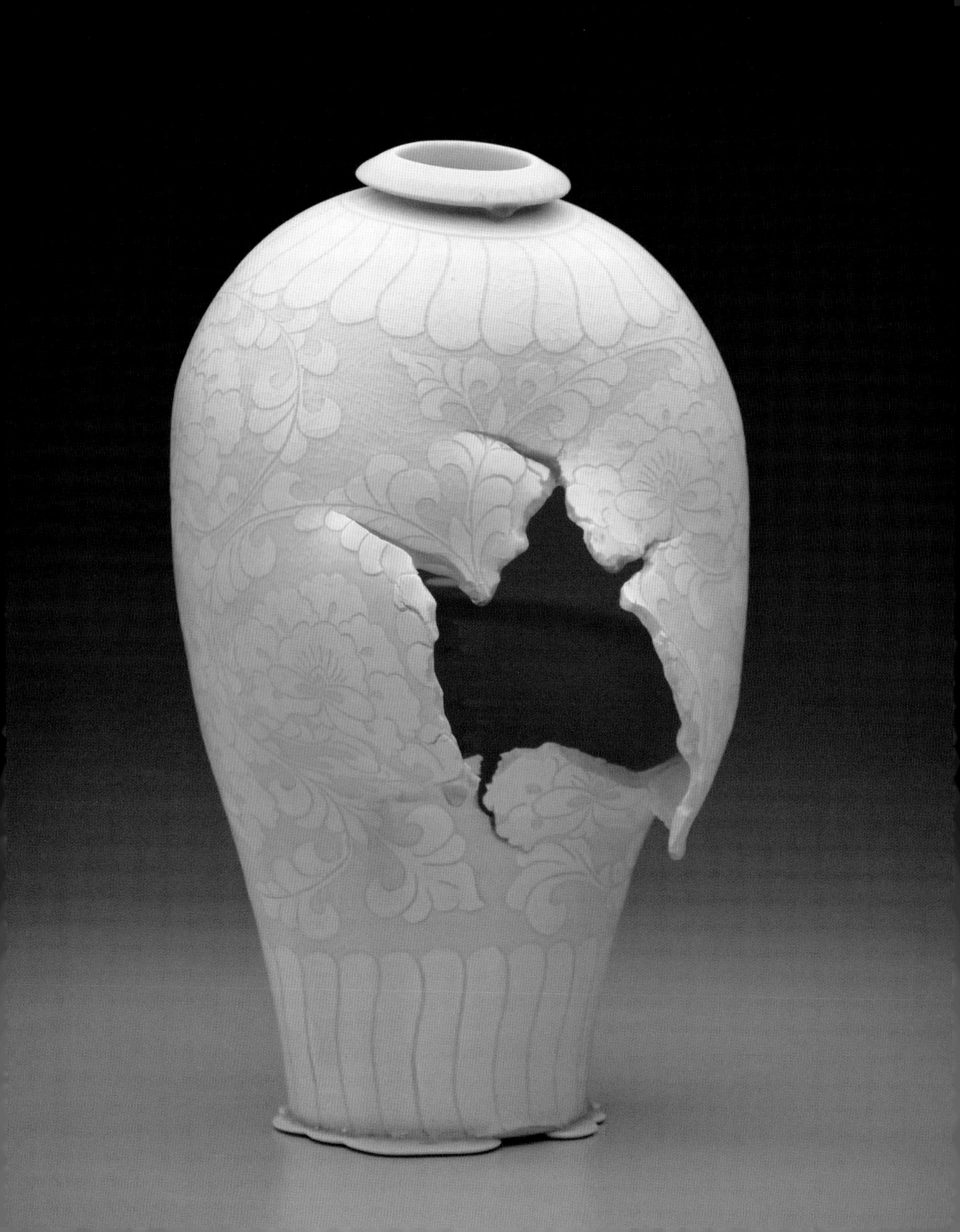

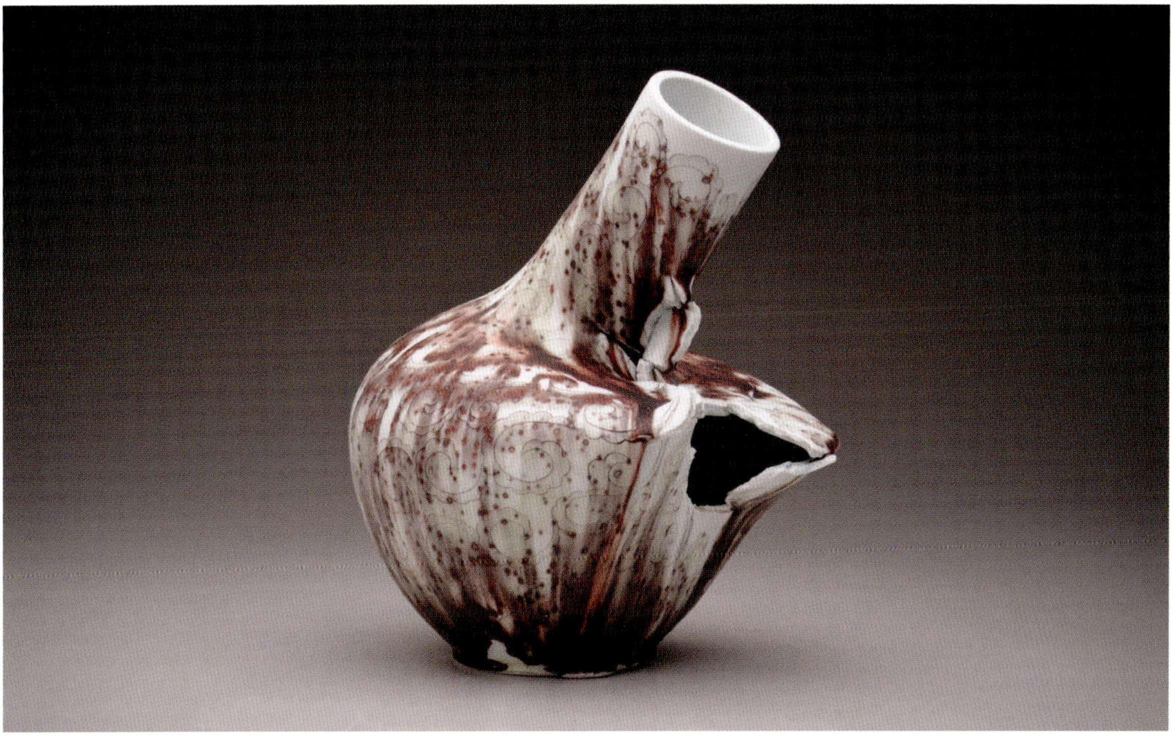

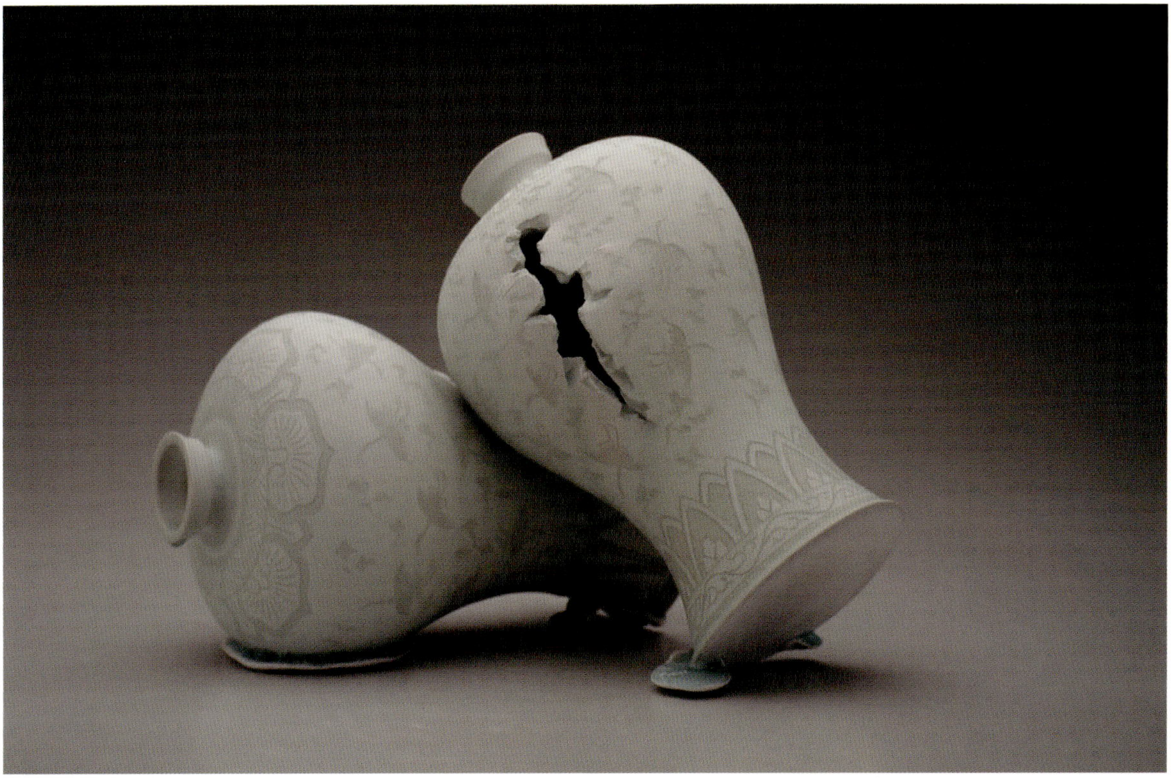

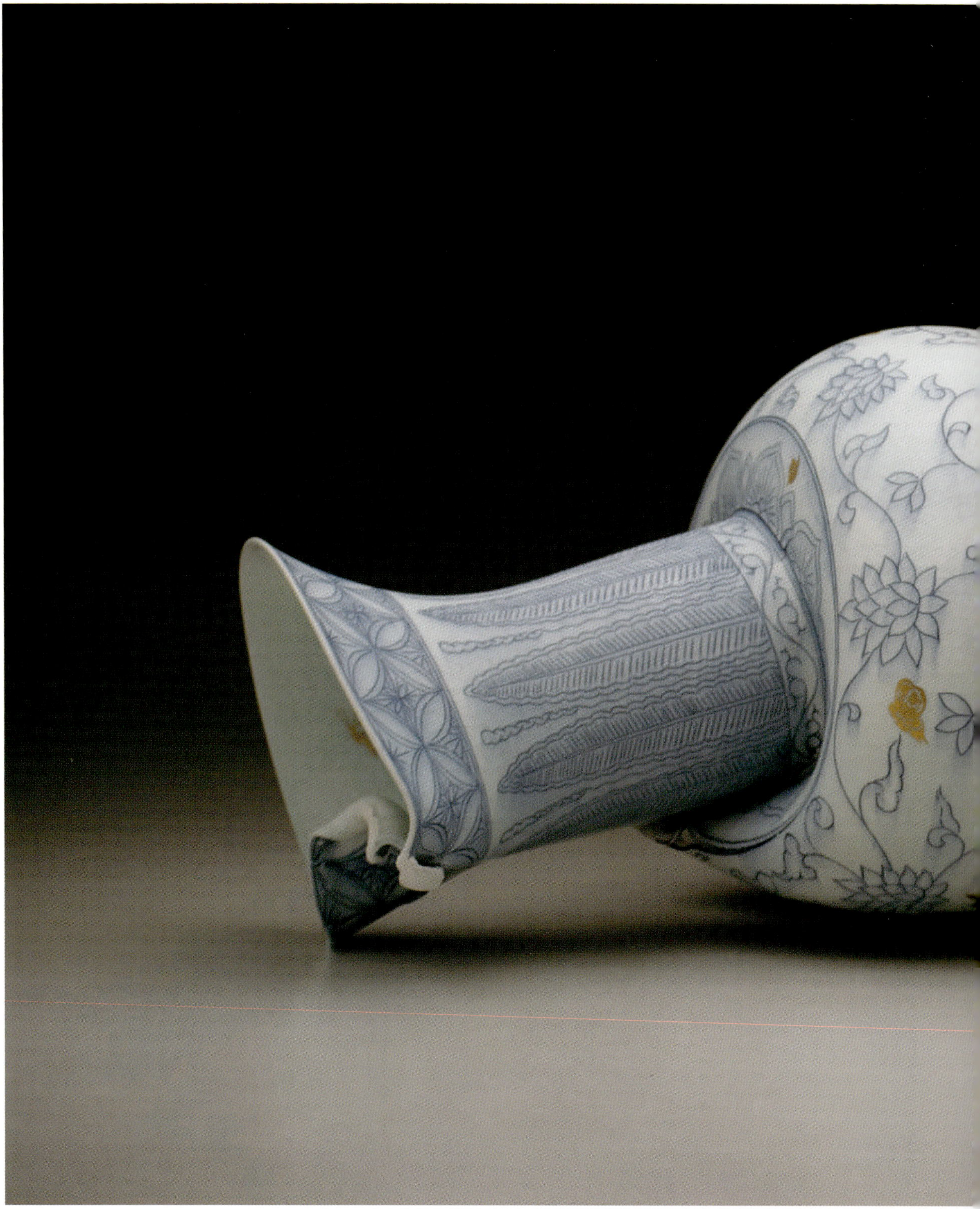

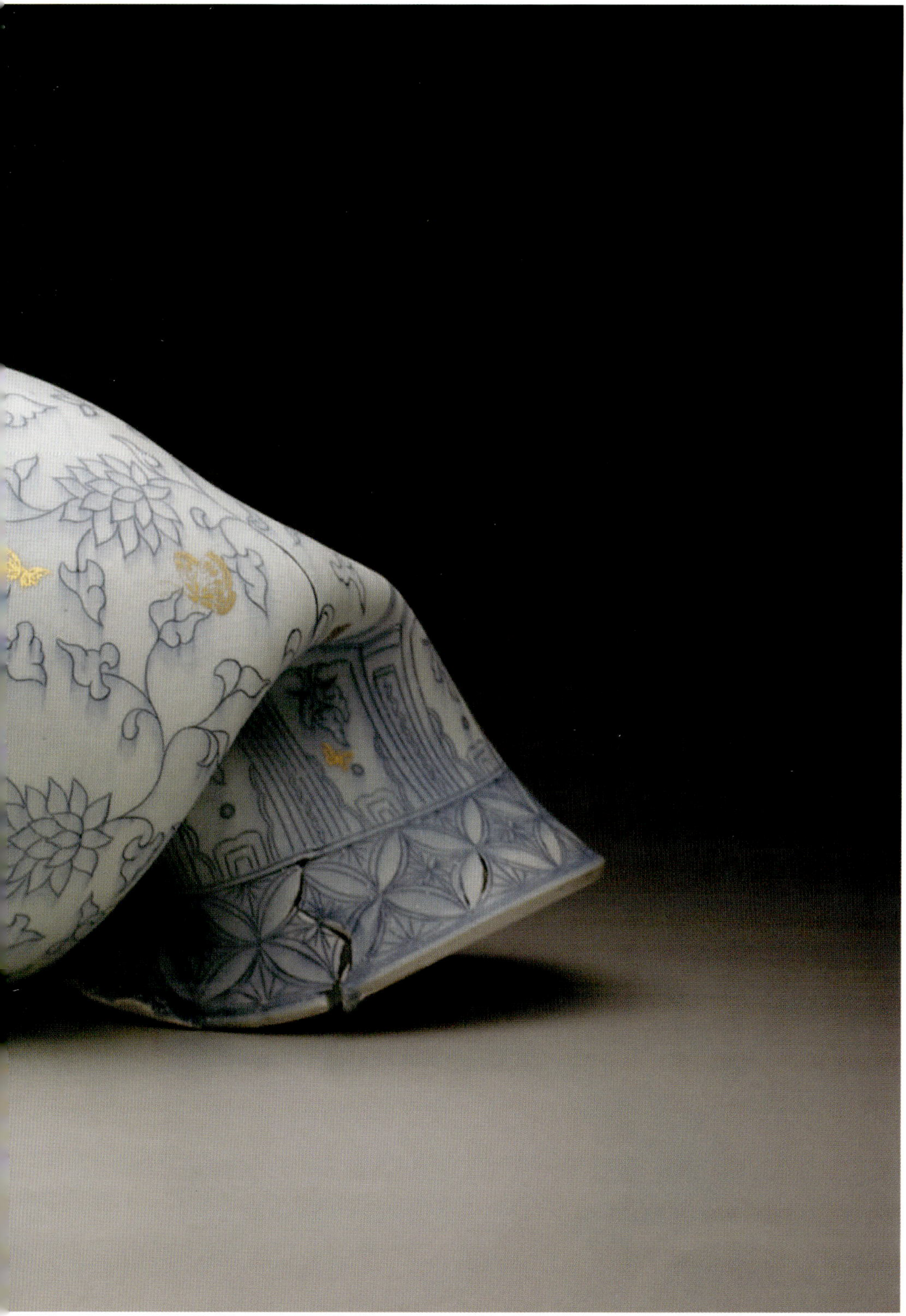

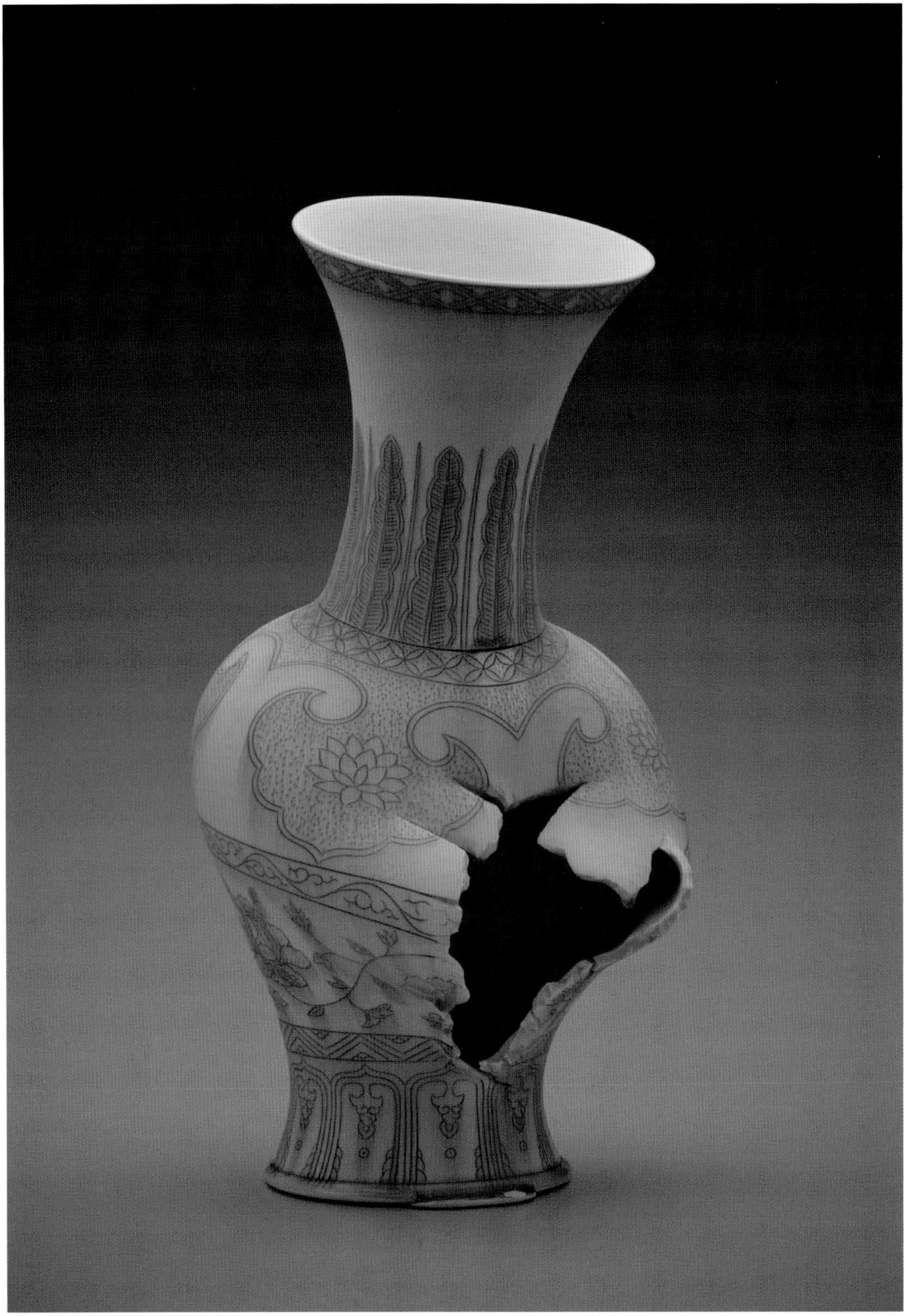

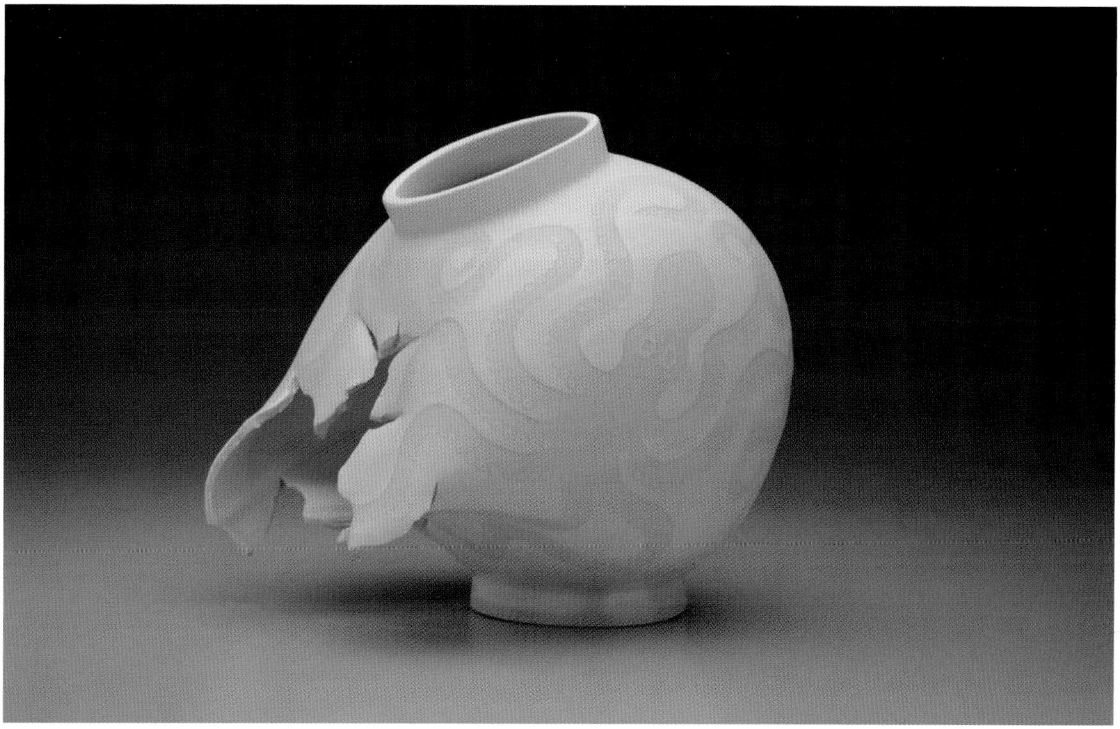

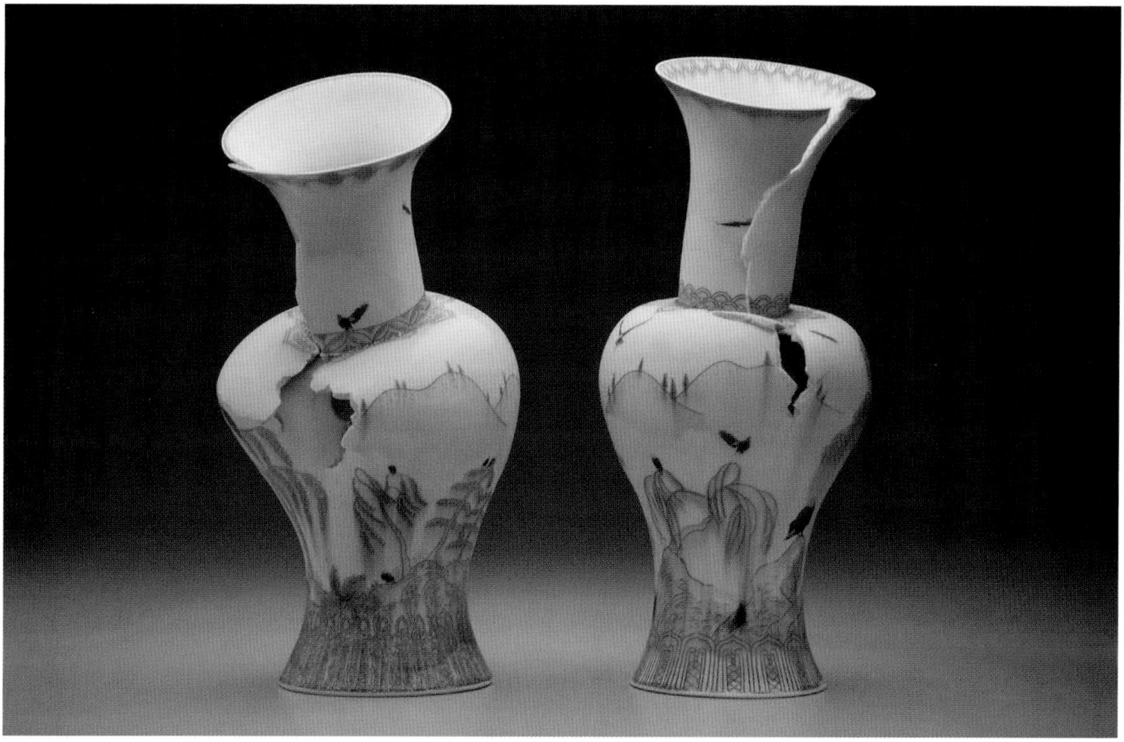

LIVIA MARIN

LOCATION London, UK
YEAR OF BIRTH 1973
MATERIALS Ceramic sculpture
EXPERIENCE 6 Years

Livia Marin is a Chilean-born artist based in London. Marin's work is broadly characterized by large-scale installations and the appropriation of objects of mass-production and consumption. Her work was initially informed by her experiences in Chile during the 1990s as the country transitioned from a right-wing dictatorship to a neo-liberal government. Marin employs everyday objects to explore the nature of our relationship to material objects in an era dominated by standardization and global distribution. The work reflects on the relationship we develop with those often unseen objects that meet our daily needs. Central to the artist's output is a theme of estrangement that seeks to reverse the innate familiarity central to everyday life and the dictates of the marketplace. Livia Marin has exhibited widely, both in her native Chile and internationally.

'I transform that which has already been made: second-hand objects, discarded slip cast moulds and pieces that are somehow lost within the anonymity of the mass-produced. I either employ fragments of ready-made objects, or recreate forms on porcelain.'

Photos courtesy of Sachiyo Nishimura, Futura Tittaferrante and the artist

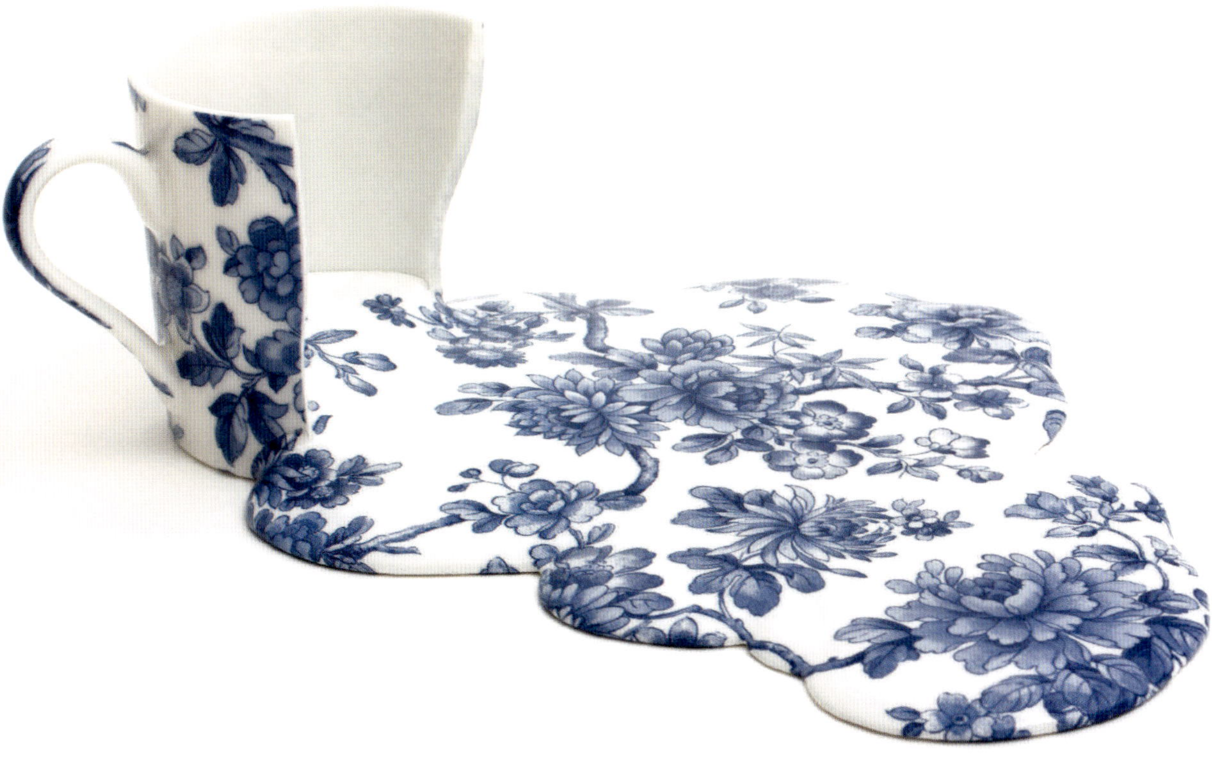

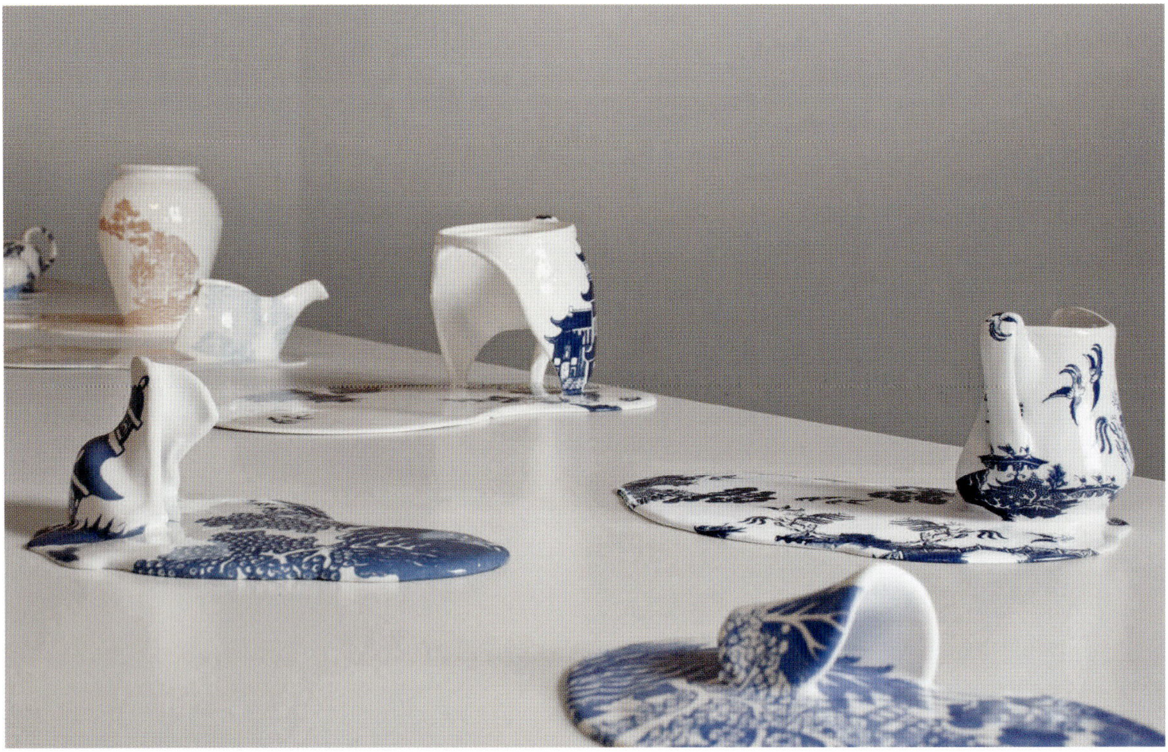

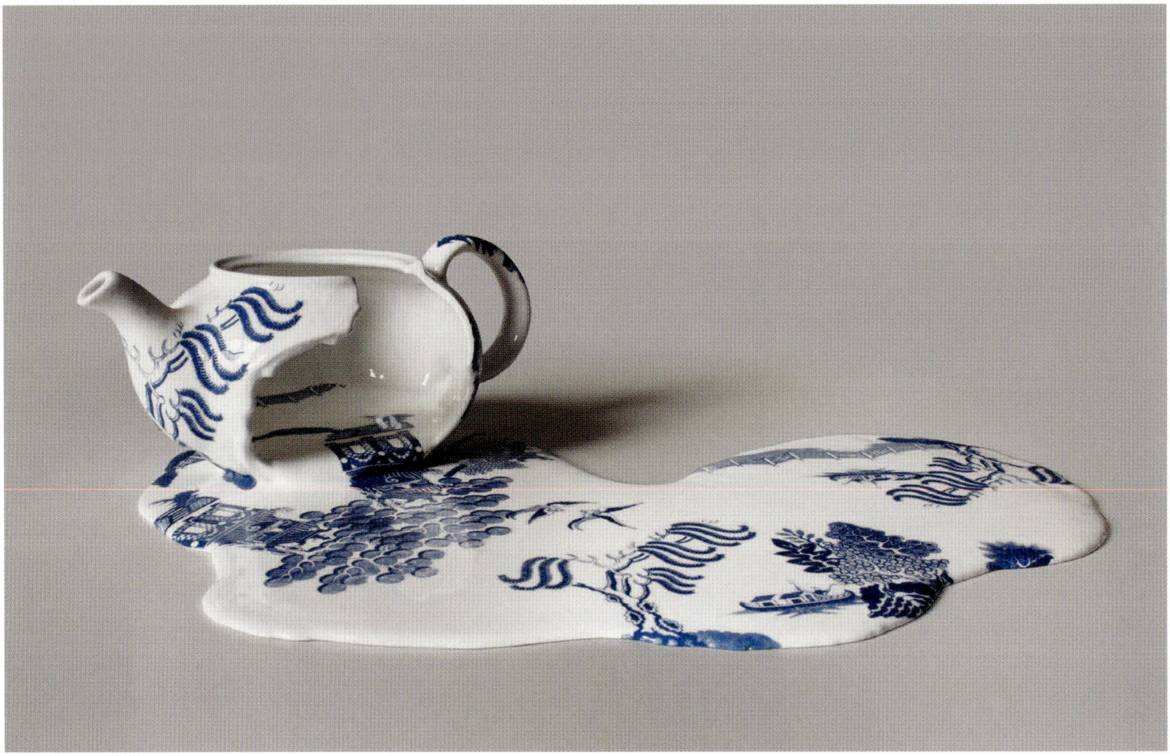

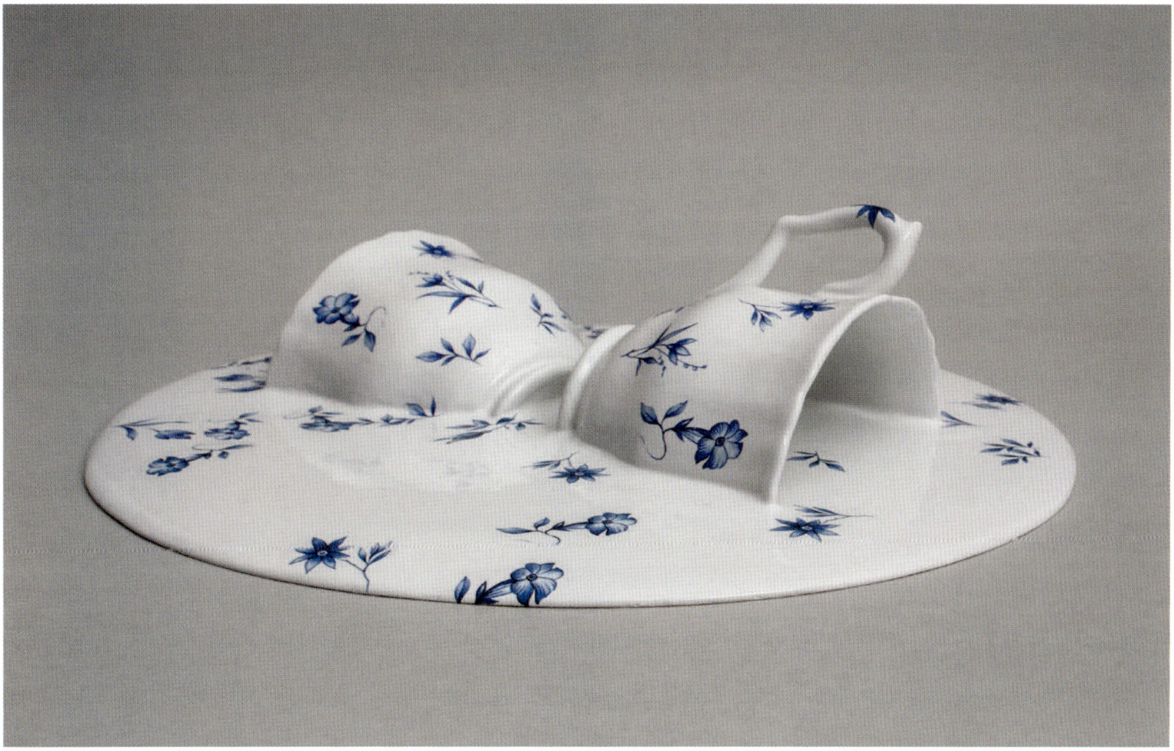

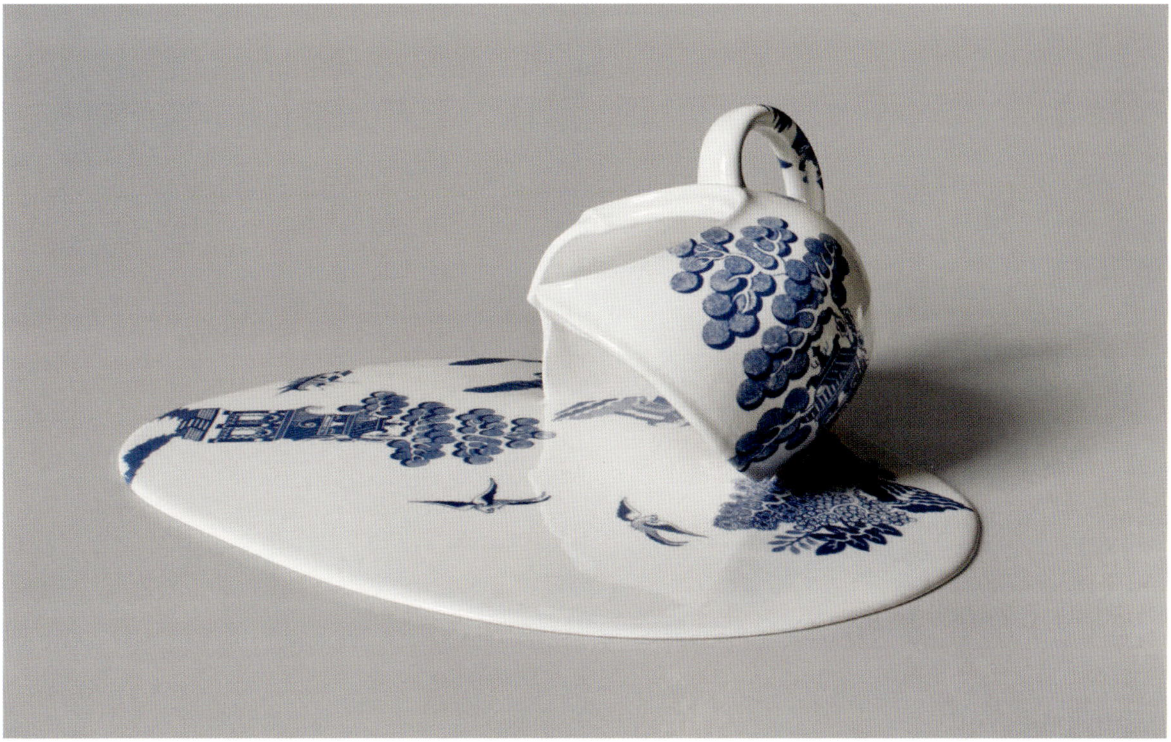

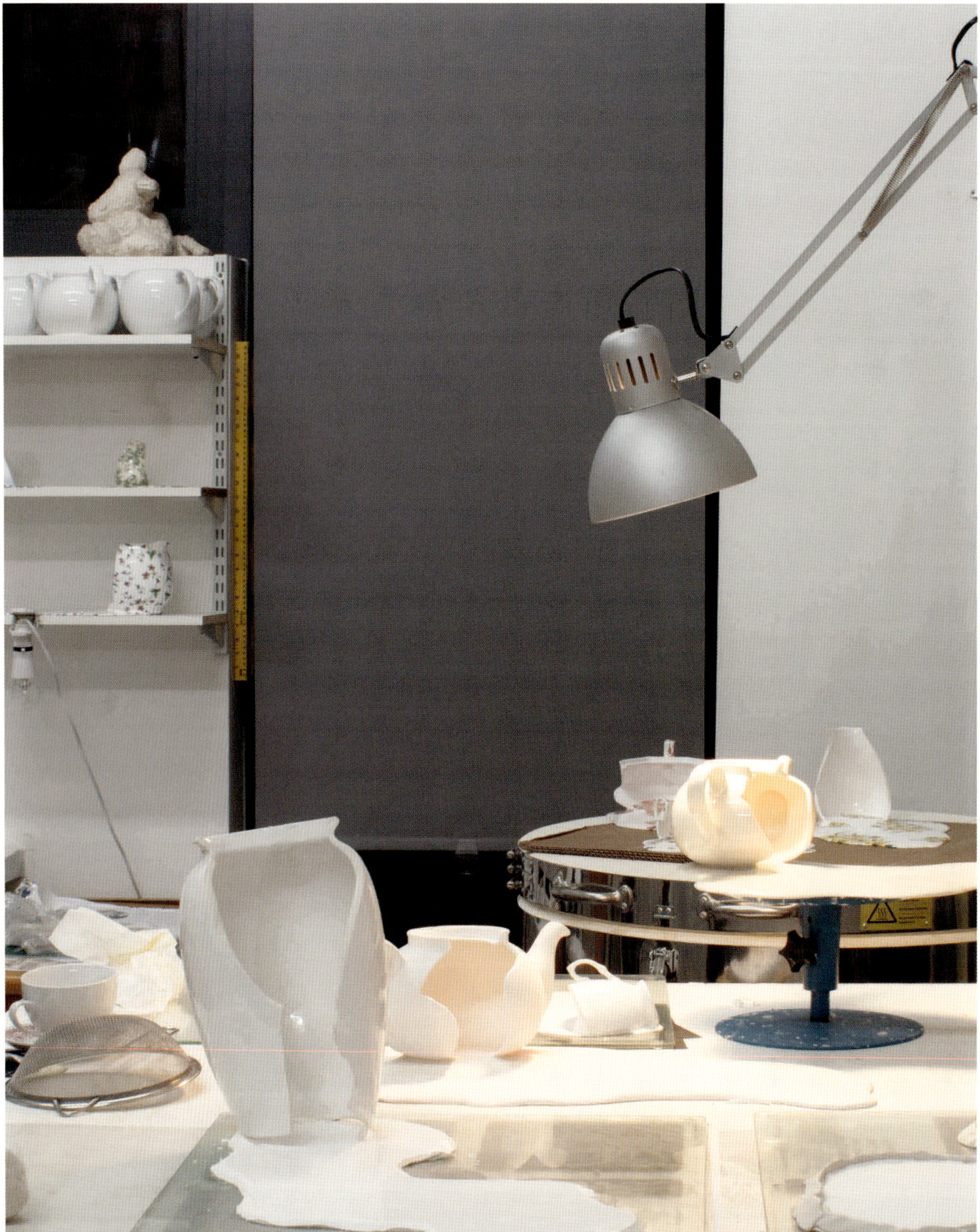

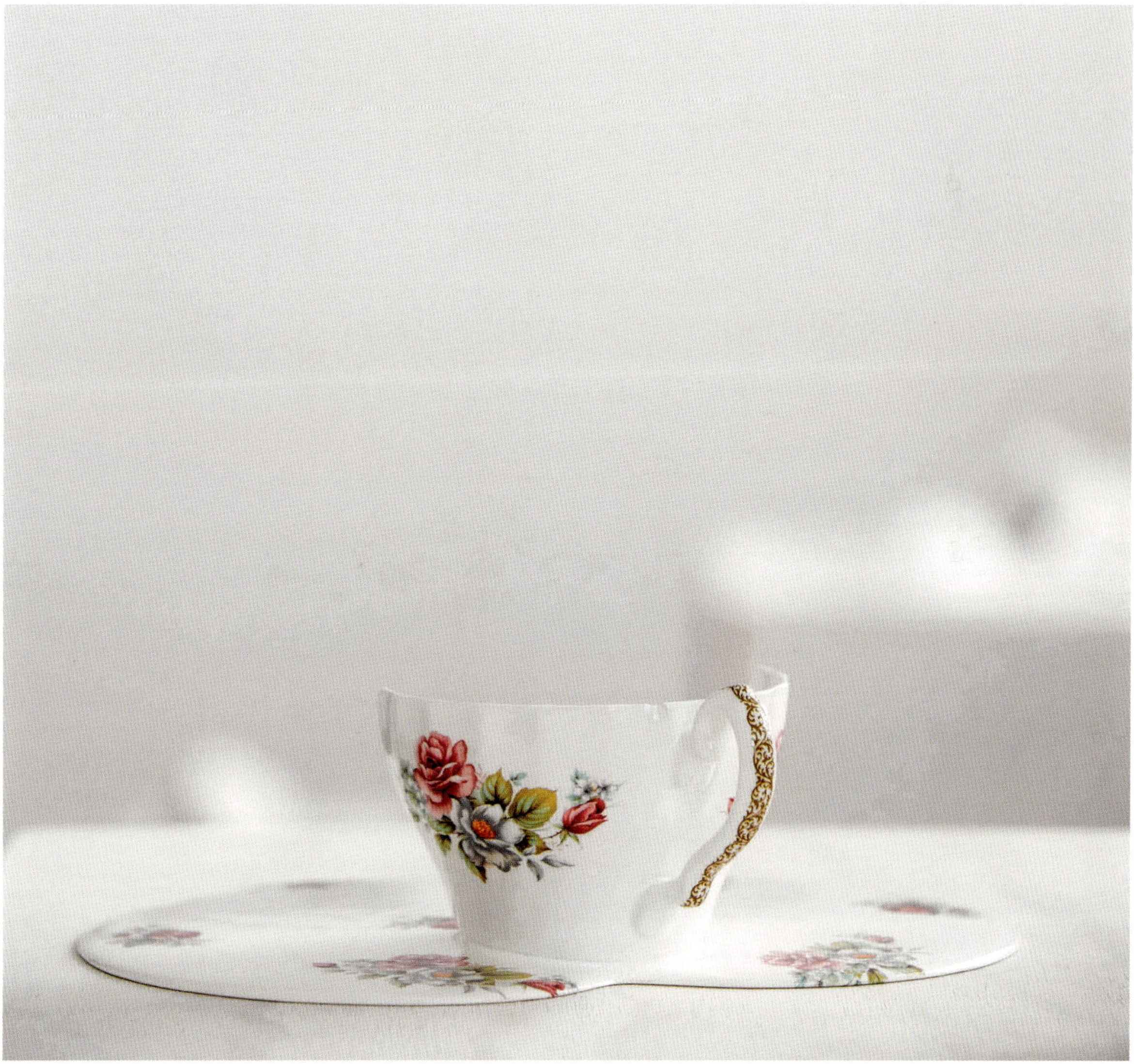

LINDSAY SCYPTA

LOCATION Rossford, OH
YEAR OF BIRTH 1986
MATERIALS Porcelain
EXPERIENCE 8 years

'I am currently an adjunct instructor at Owens Community College in Perrysburg, Ohio and Lourdes University, where I divide my time between teaching and making. Working strictly with porcelain clay, my work is thrown, trimmed, altered and decorated, then fired to cone six in an electric oxidation kiln. I am continuing to push my ideas and am excited to incorporate new research into my studio practice thanks to a generous grant from the Toledo Museum of Art. In my studio practice these towers replace the historical footmen and the frantic modern host, commenting subtly on such social implications through their utilitarian attributes. I am using food as a way of seeing the world, the tableware to create rituals through decorum, and the table to build camaraderie.'

Lindsay Scypta began her ceramics investigation in high school and continued into college, completing a BFA in Art & Design from the New York State College of Ceramics at Alfred University. She then arrived at The Ohio State University where she completed her MFA in ceramic art, taking inspiration from the softness of tufted Victorian sitting room chairs, the architectural motifs and quatrefoils of European Gothic cathedrals, and the construction lines of high fashion wear and textiles. Clay Art Center in Port Chester, New York welcomed her with open arms as an aritst-in-residece upon the completion of her MFA.

Photos courtesy of the artist

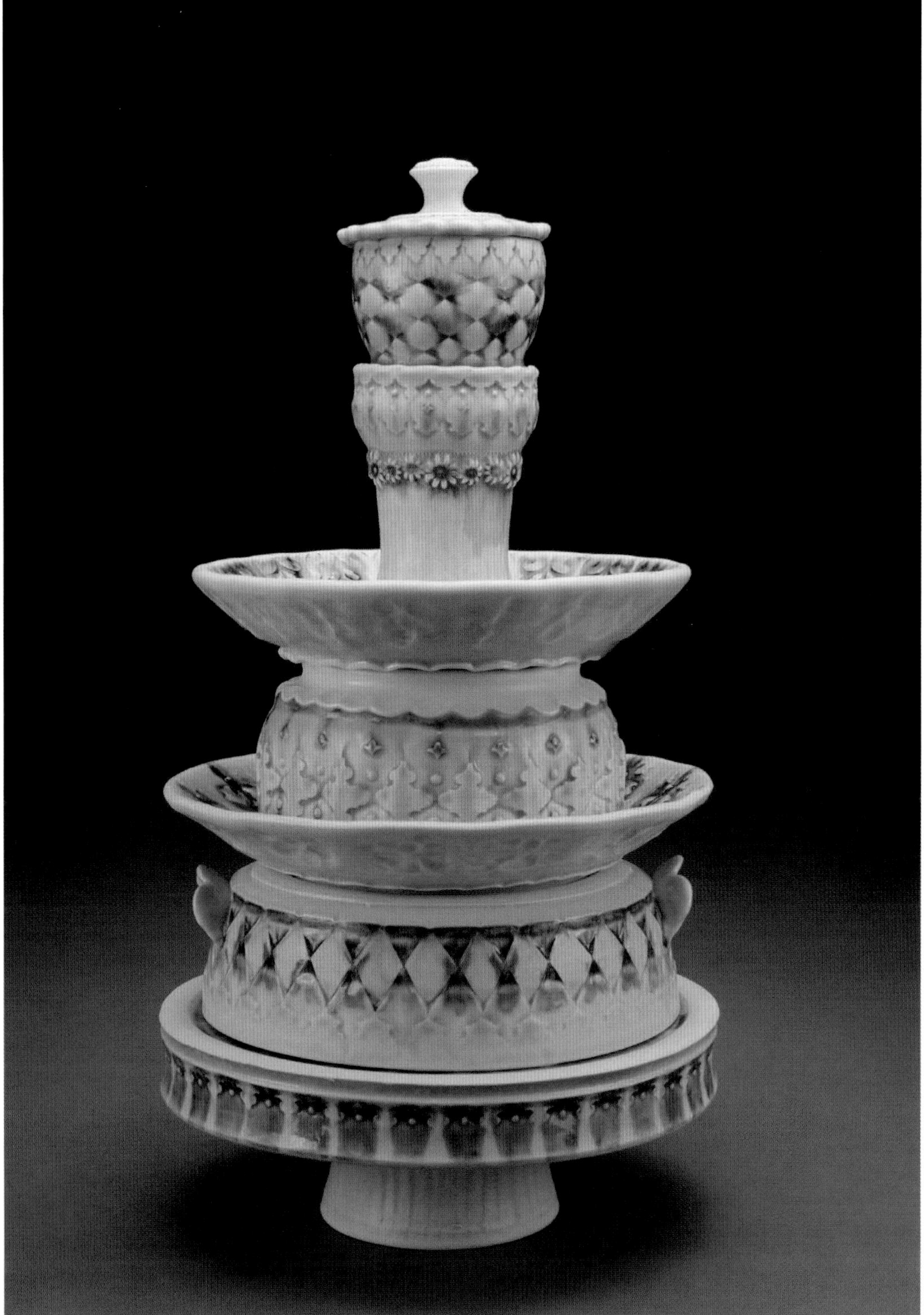

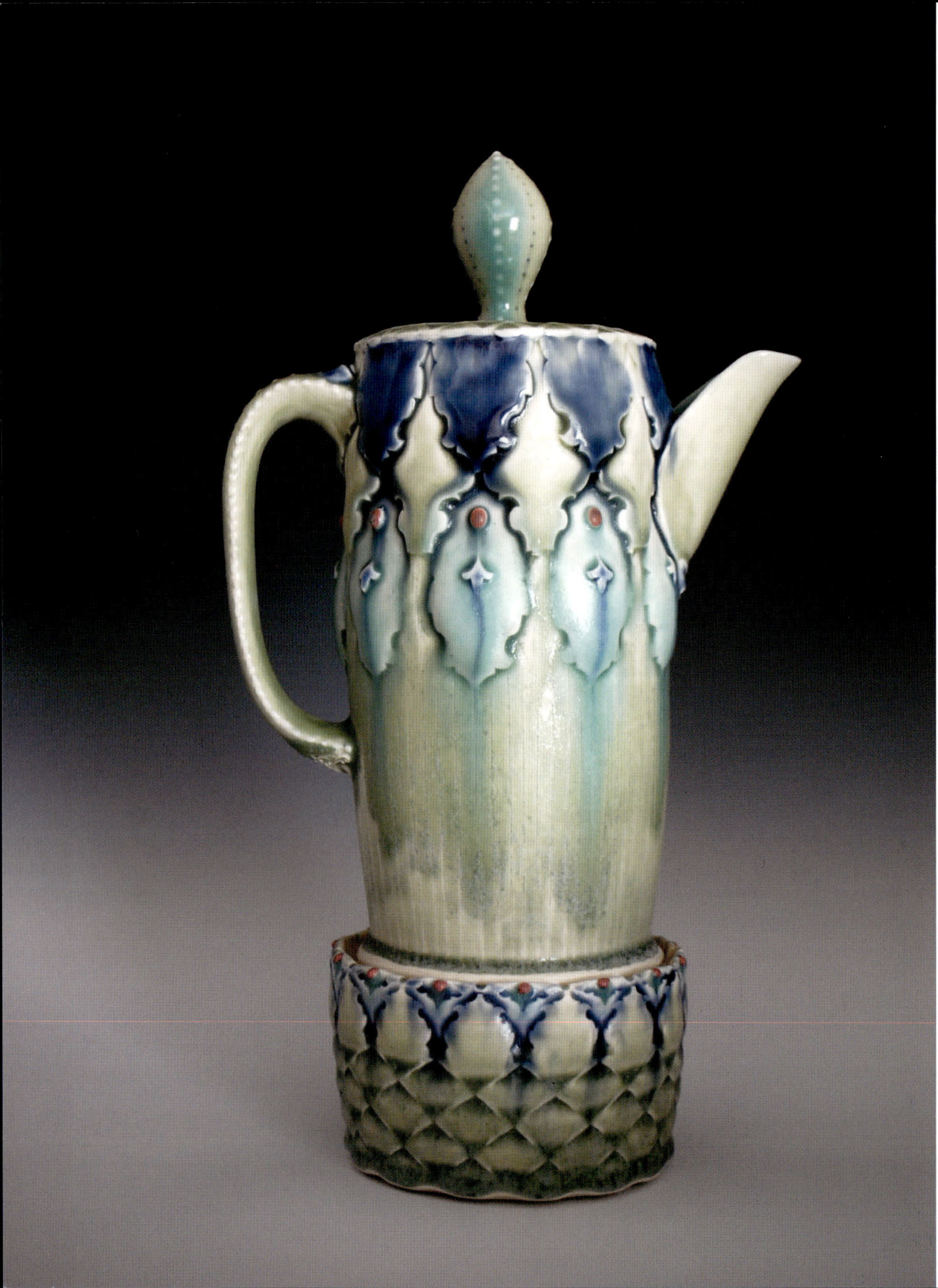

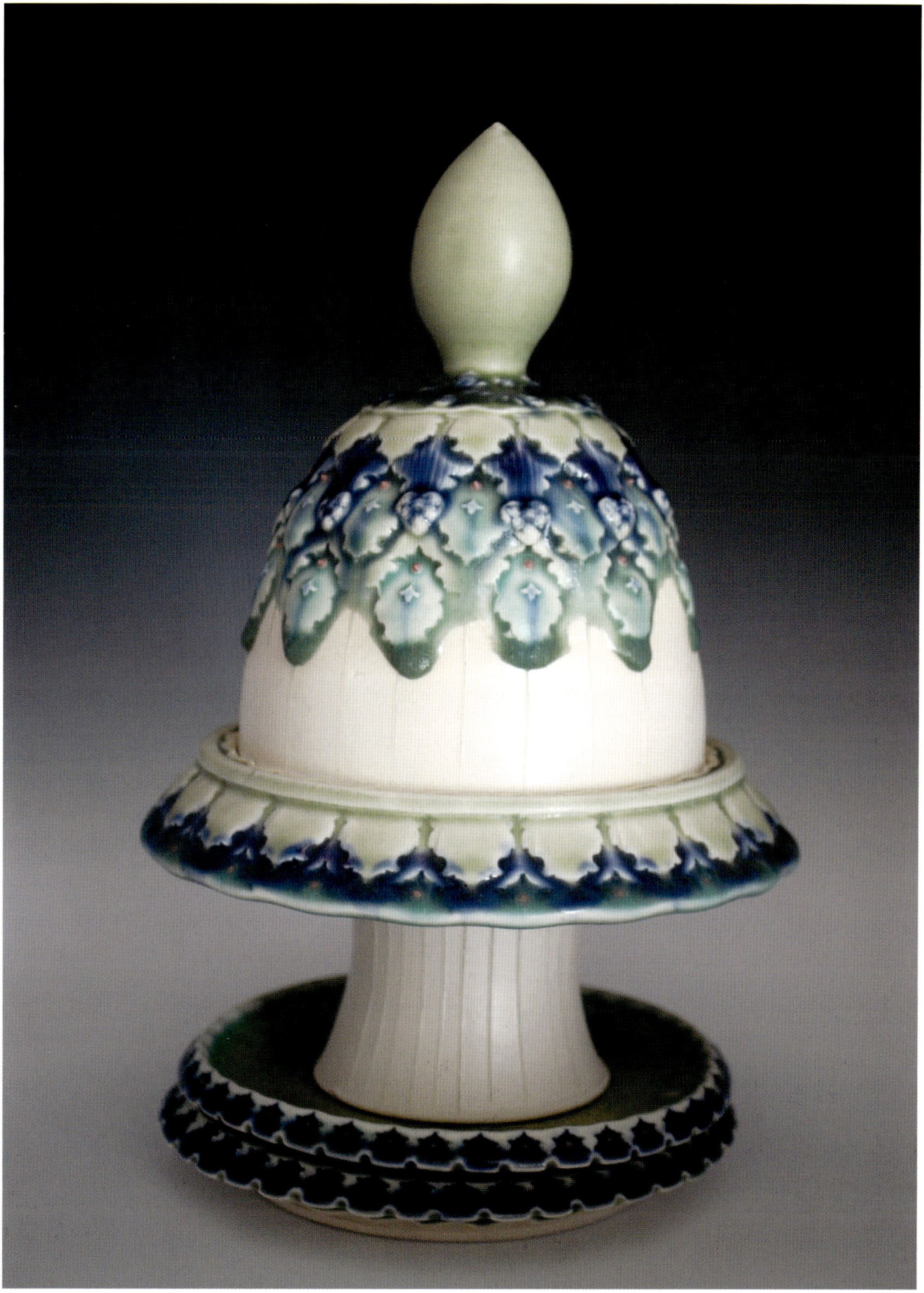

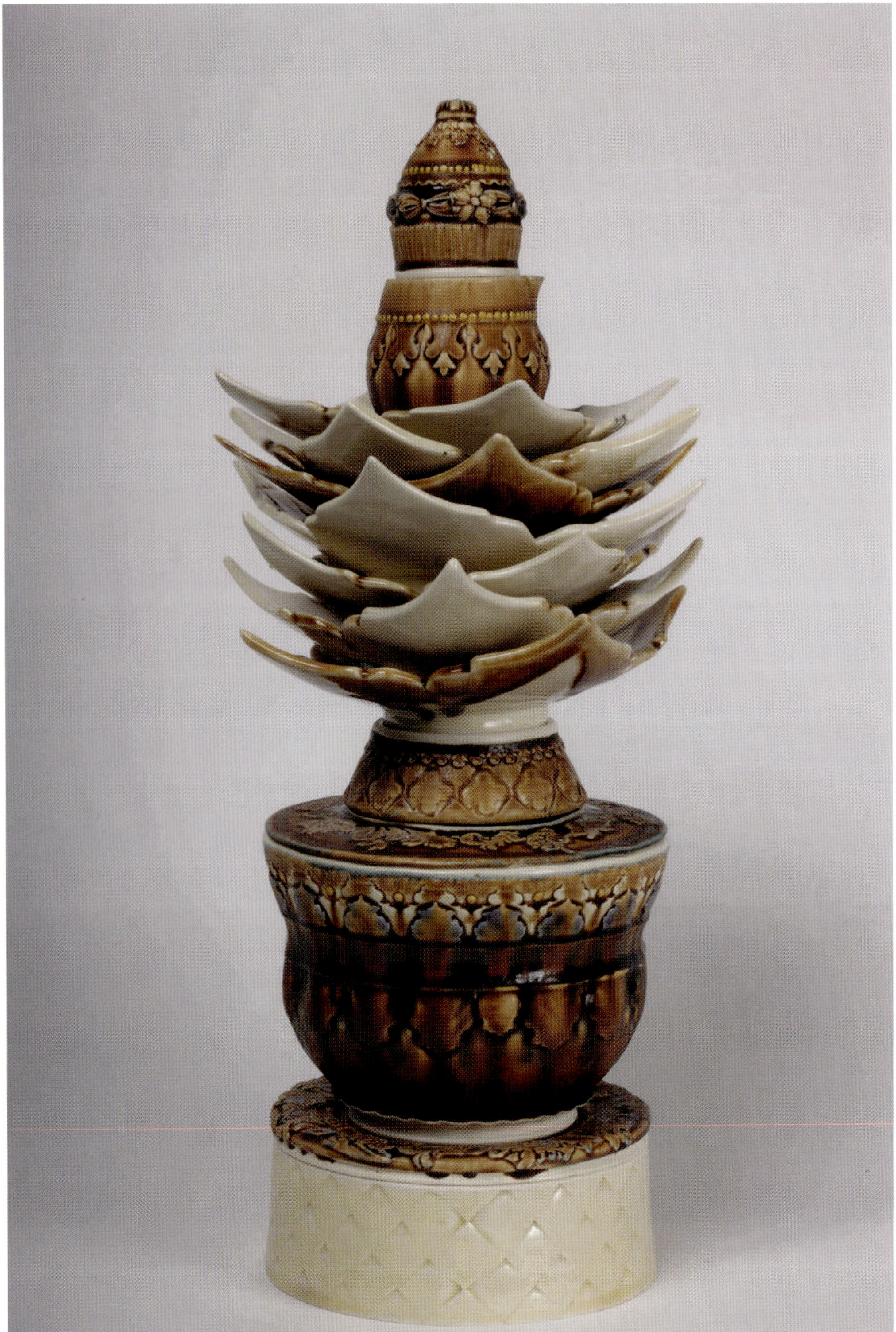

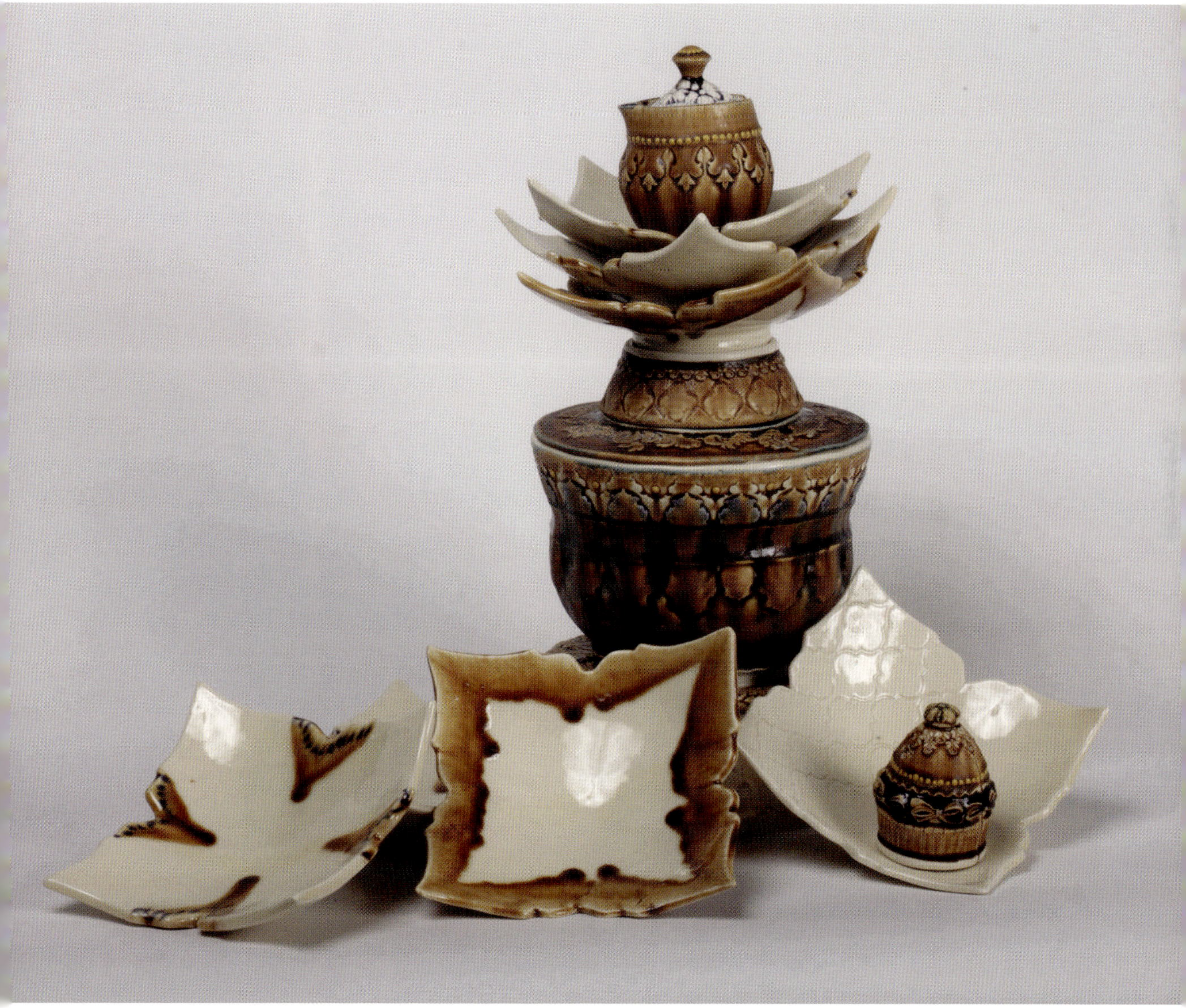

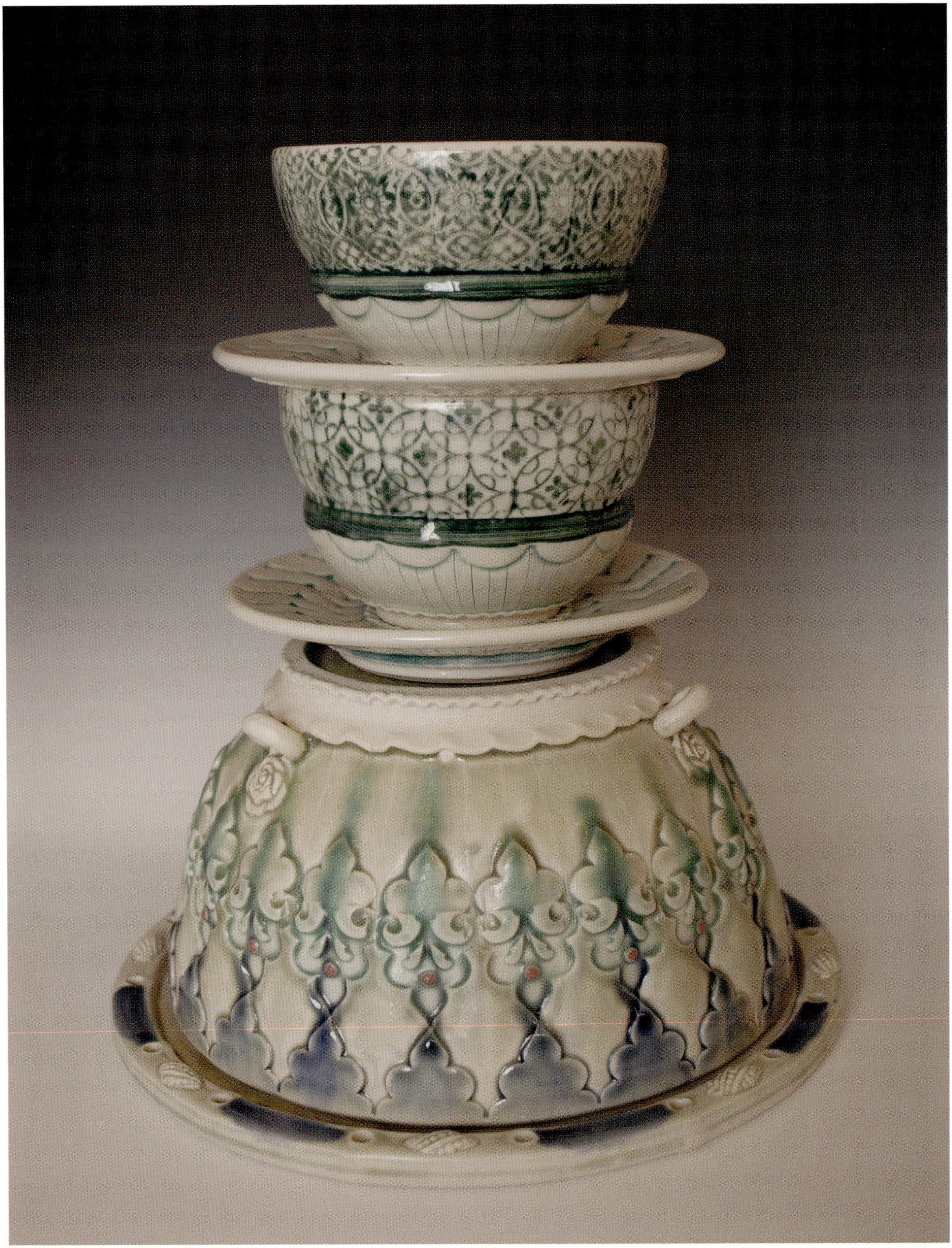

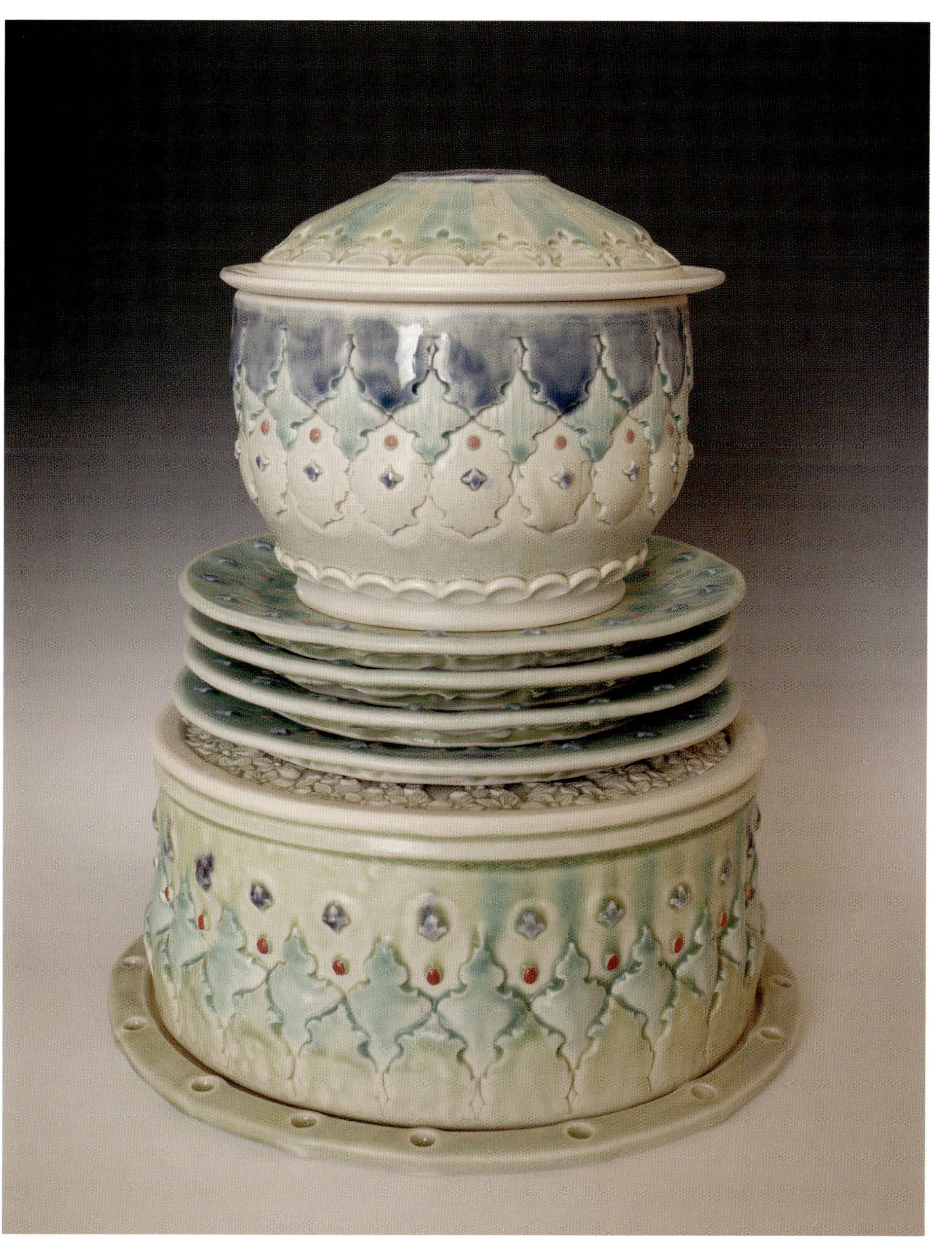

PAUL WOOD

LOCATION Eltham, Victoria Australia
YEAR OF BIRTH 1978
MATERIALS Re-fired ceramics and glass
EXPERIENCE 20 Years

Paul Wood graduated from the Victorian College of the Arts in 1998 with a Bachelor of Fine Art and again in 2002 with a Graduate Diploma of Visual Art.

'I collect mass-produced ceramic objects and glassware gleaned from eBay, op shops, second-hand and variety shops. I spend time stacking and balancing these objects together to create various tableaux. I try to preempt what effects the heat of the kiln may have on the stack before placing the assemblages in the kiln to be fired. For example, various clay bodies and types of glass have differing melting points, so I look at this in combination with the effects of gravity to have an idea of which objects will droop before others. I generally prefer to use a gas kiln as they are not so automated as electric powered ones and I can therefore control the temperature more easily and keep an eye on how the work is coming together through the little peephole. Their shape and glazes melt, warp and re-fuse together into a new oozing mass that I can anticipate to a degree but not completely. It is always an exciting and simultaneously anxious moment when I open the kiln to see how heat and gravity effected the objects.

I am interested in how the process of collecting, collating and firing the objects in assemblages transforms and almost resurrects them from their previous existence in everyday domestic settings, to being rejected as non-fashionable 'clutter' and into a fresh and new context. In one way by being stacked and fused together they become small monuments to their past circumstances.'

Photos courtesy of This is No Fantasy, Siri Hayes, Matthew Stanton and the artist

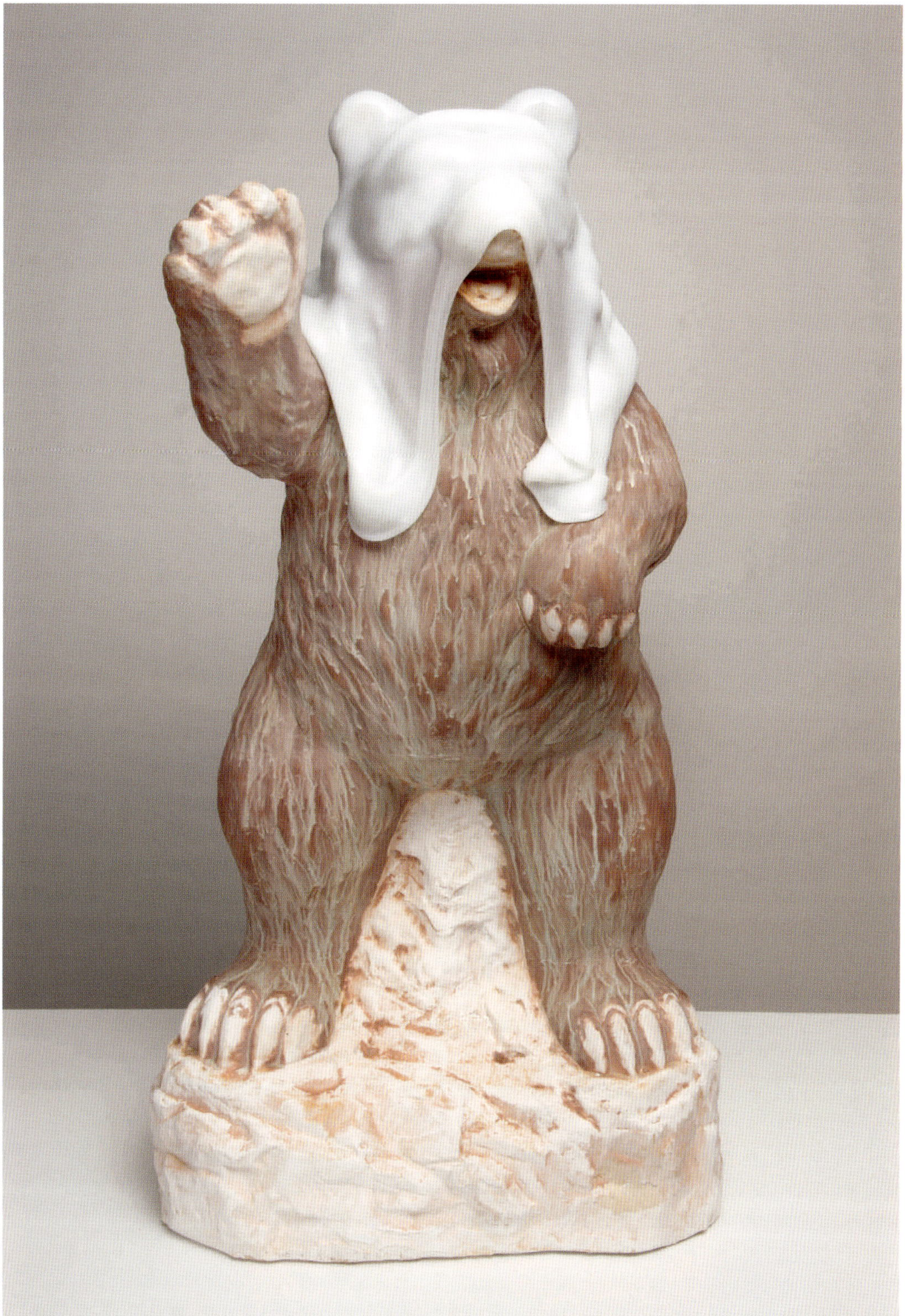

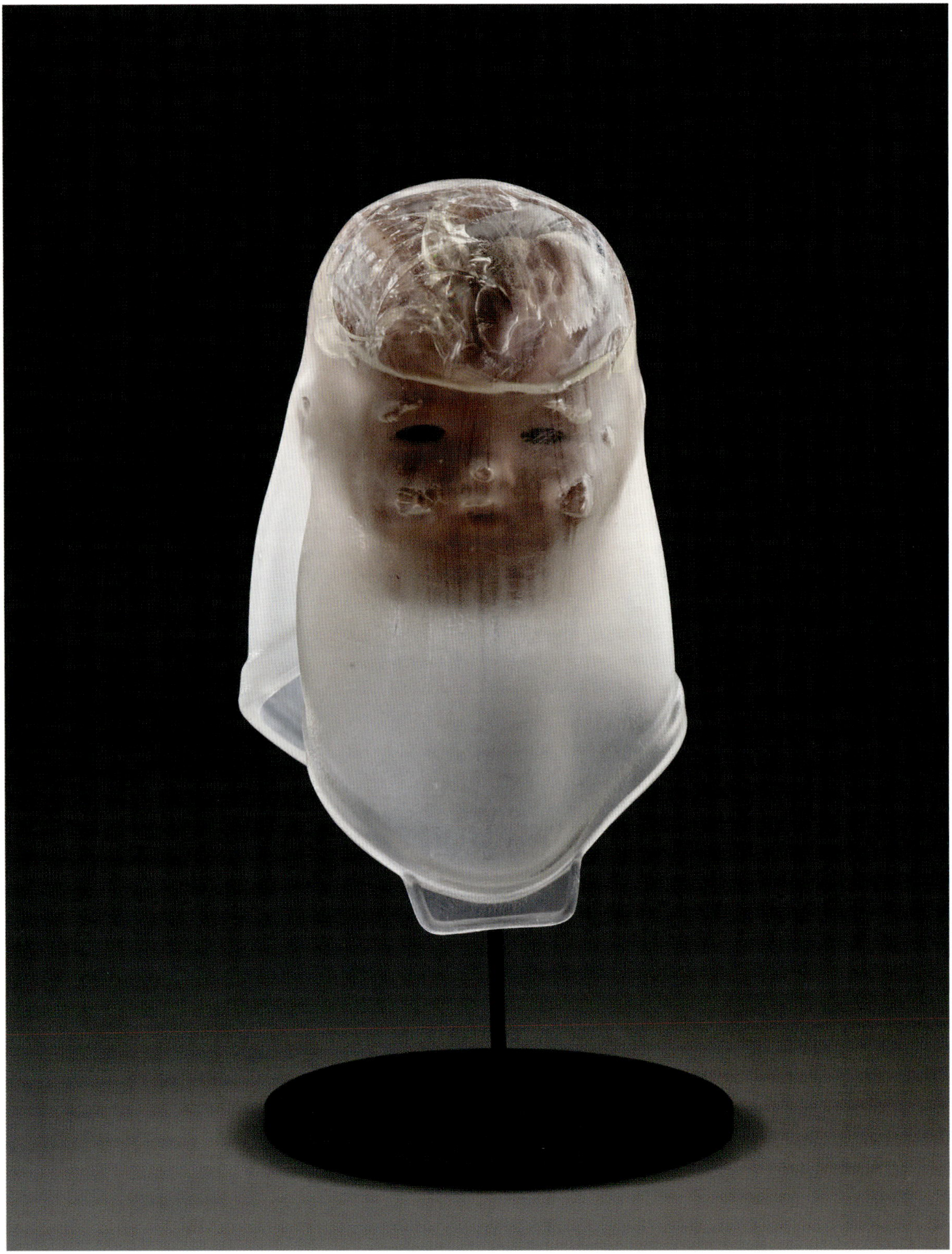

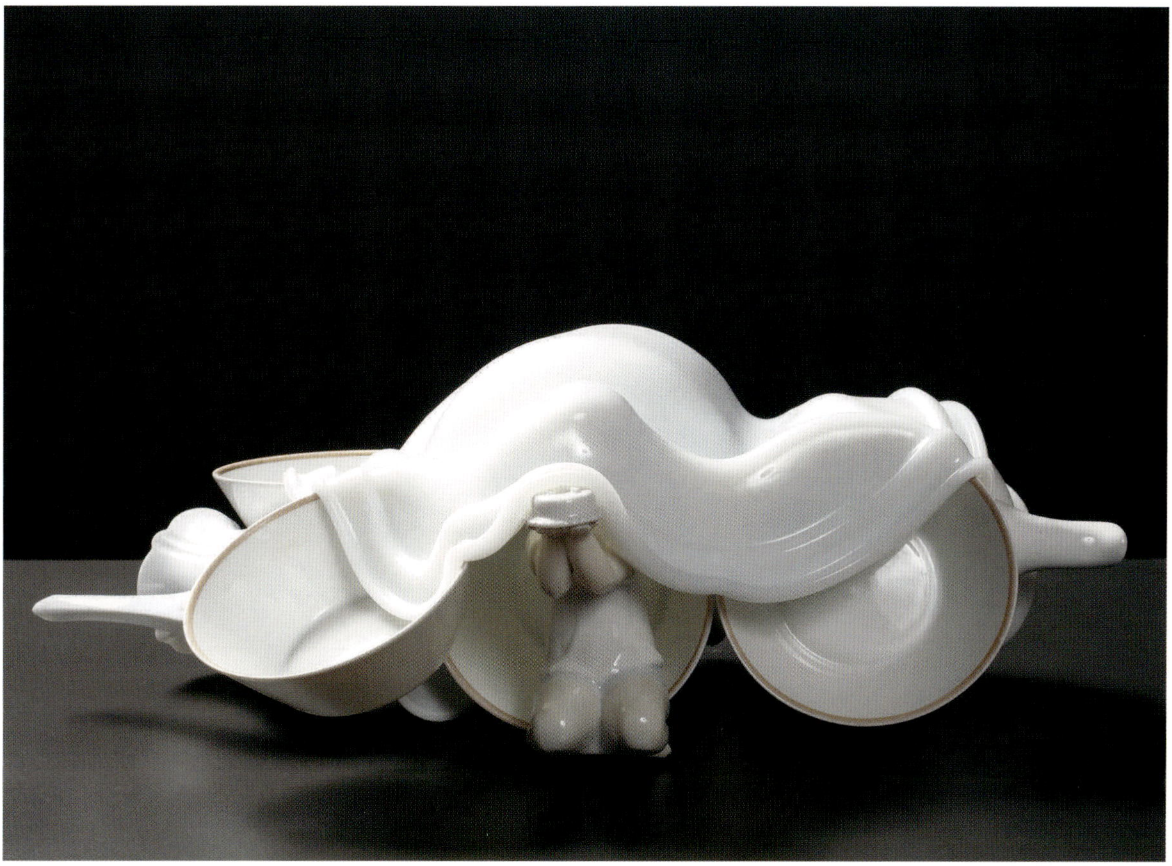

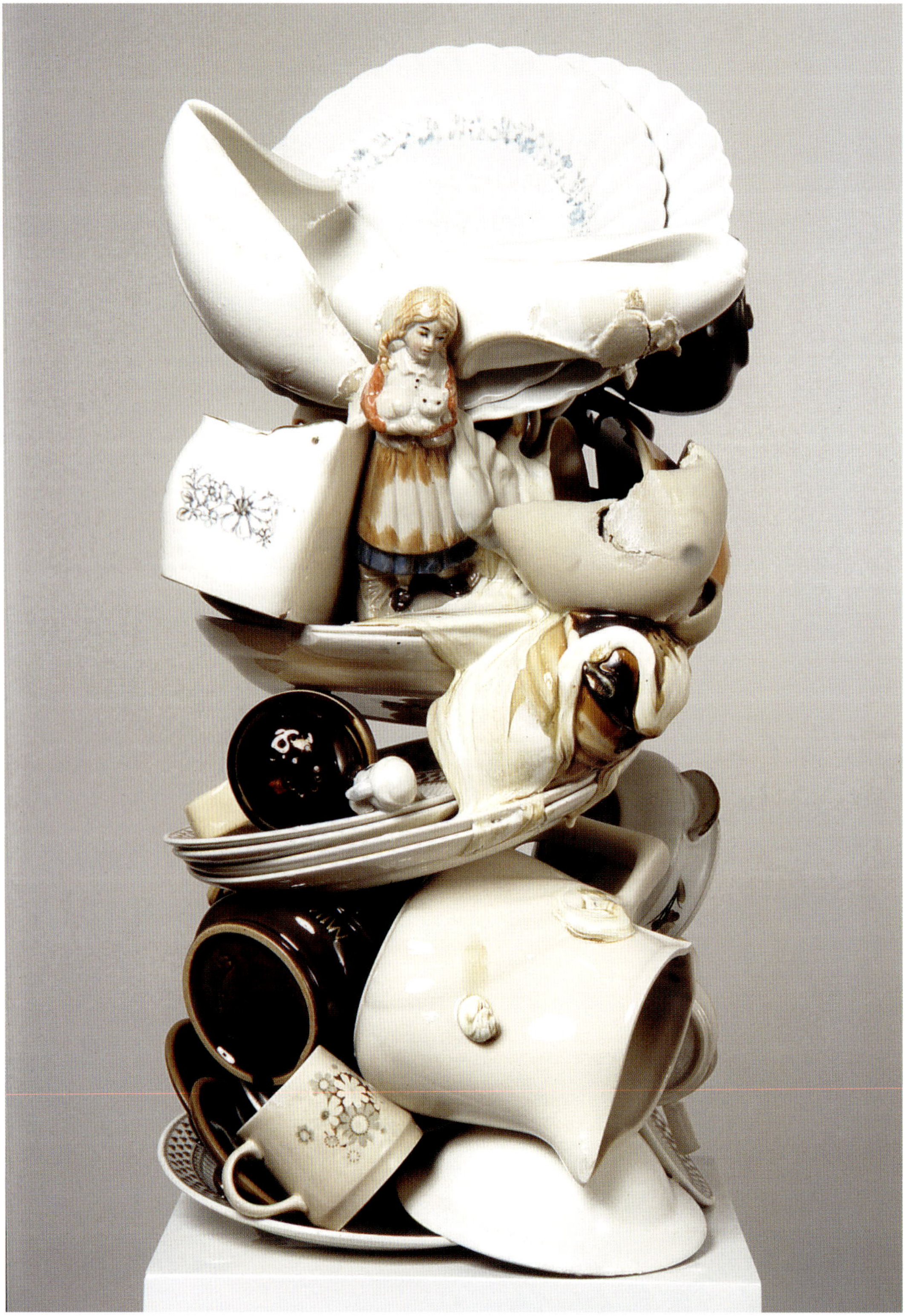

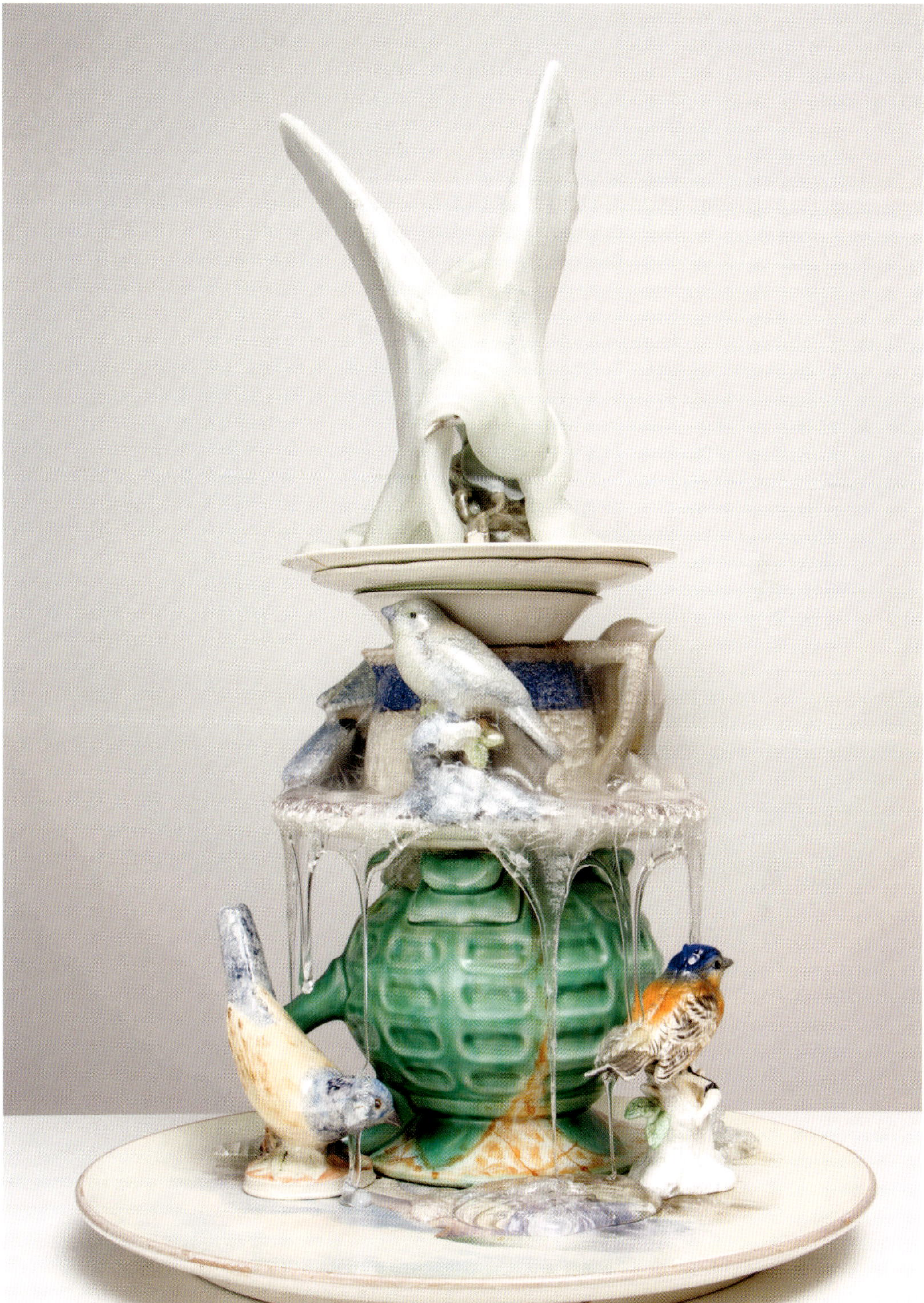

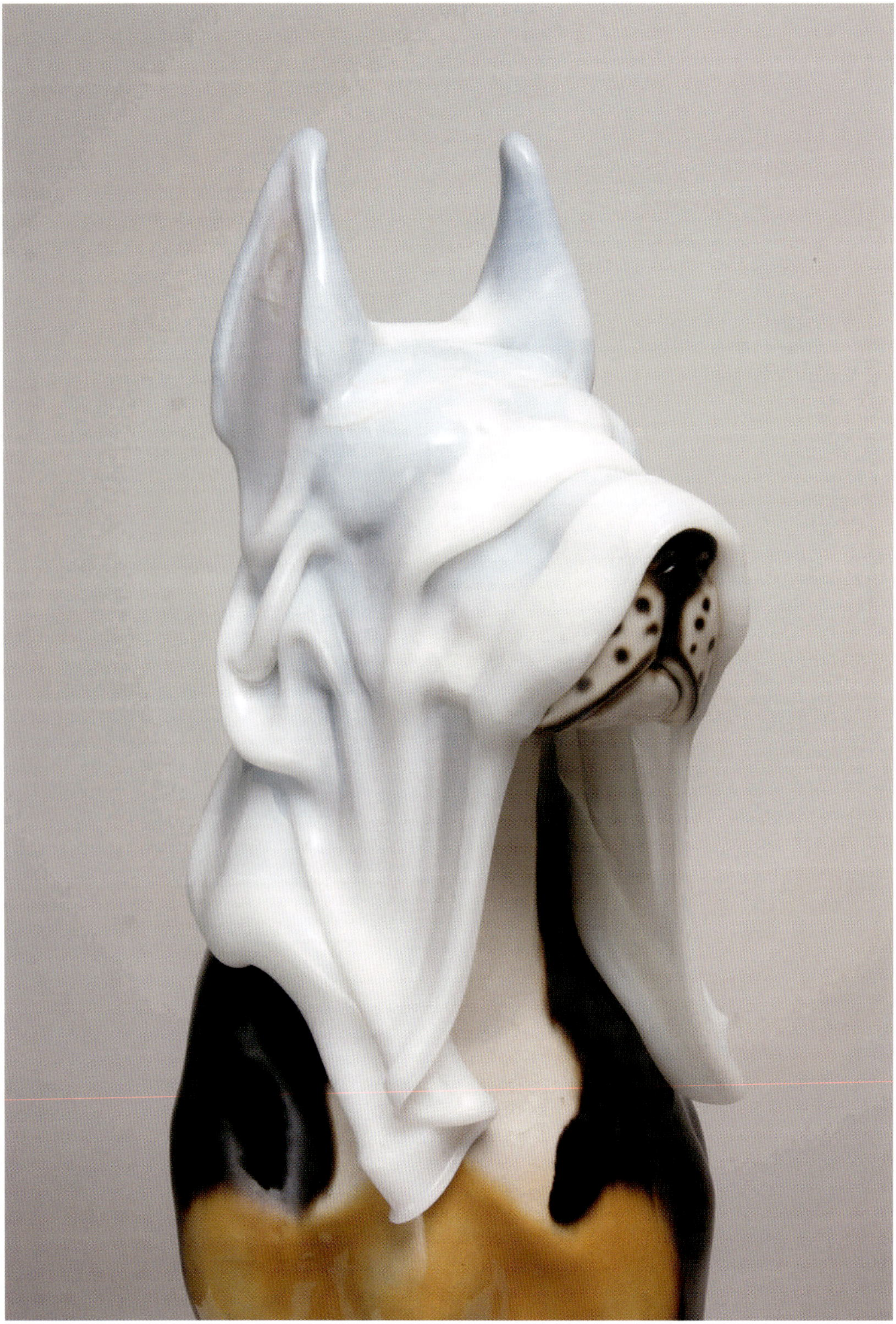

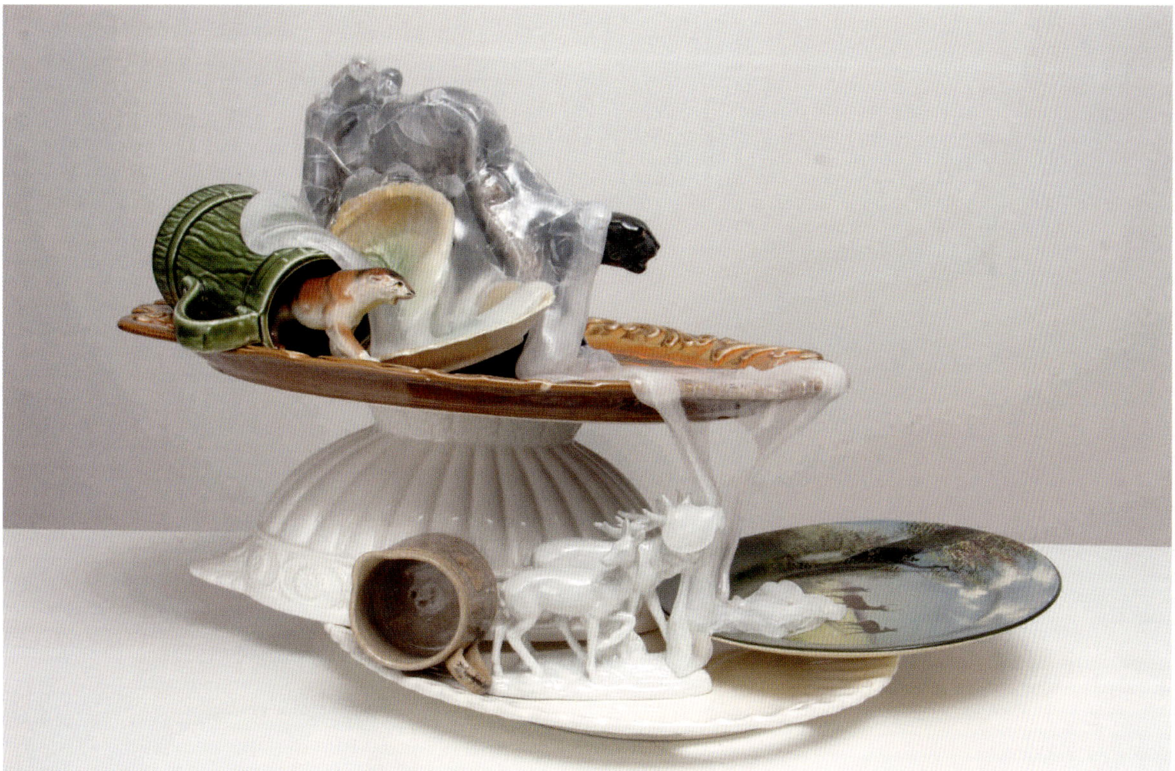

LAURENT CRASTE

LOCATION Montreal, Quebec, Canada
YEAR OF BIRTH 1968
MATERIALS Porcelain
EXPERIENCE 20 years

Of French origin, Laurent Craste has been living and working in Montreal since the early 90s. His practice focuses on the exploration of the multiple layers of meaning of decorative objects: ideologically, aesthetically, and through their sociological and historical dimensions. The porcelain vase, in particular, has for years been the subject of predilection of the artist's work. Laurent Craste appropriates this archetypal figure of decorative arts, using it as matter, support, and playground for his artistic interventions. Considered as instruments of political power, ideological vehicles, and demonstrations of ostentatious luxury and economic power, but also as incarnations of emotions and experiences, the historical archetypes of decorative arts consistently provide him with useful material.

'The different elements of my vases are hand-thrown on the wheel and, after trimming, assembling, and decoration, violently smashed with workers' tools, baseball bats, crucified with nails or darts, torn to pieces or squashed, in a word: vandalized. After a long and delicate drying process, the pieces are bisque-fired, glazed by spraying, and re-fired at high temperature. Then, some works receive a low temperature decoration with China paints, decals, or lusters. Finally, after the last firing, the tools are inserted in the pieces.'

Photos courtesy of David Bishop Noriega, Daniel Roussel and the artist

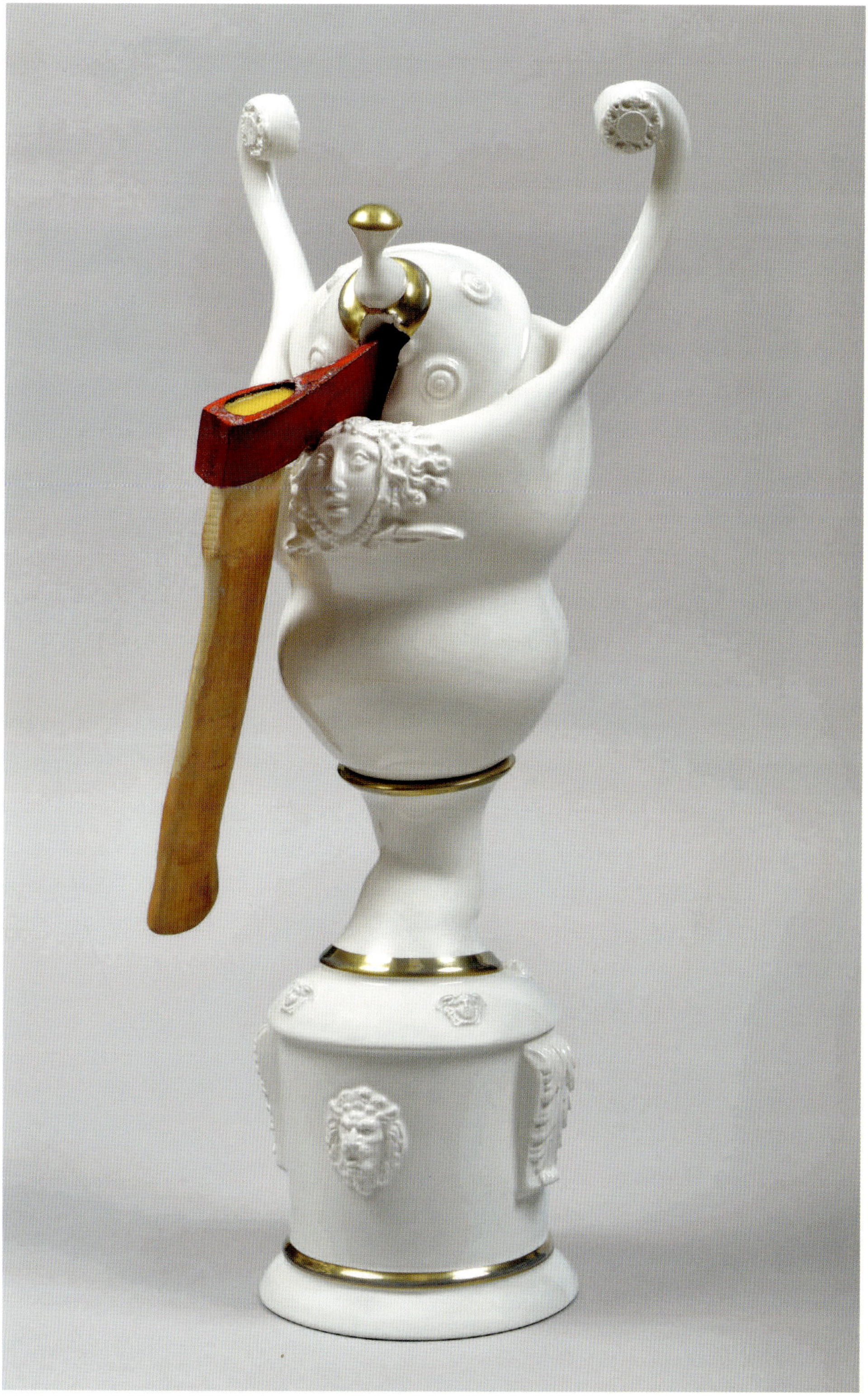

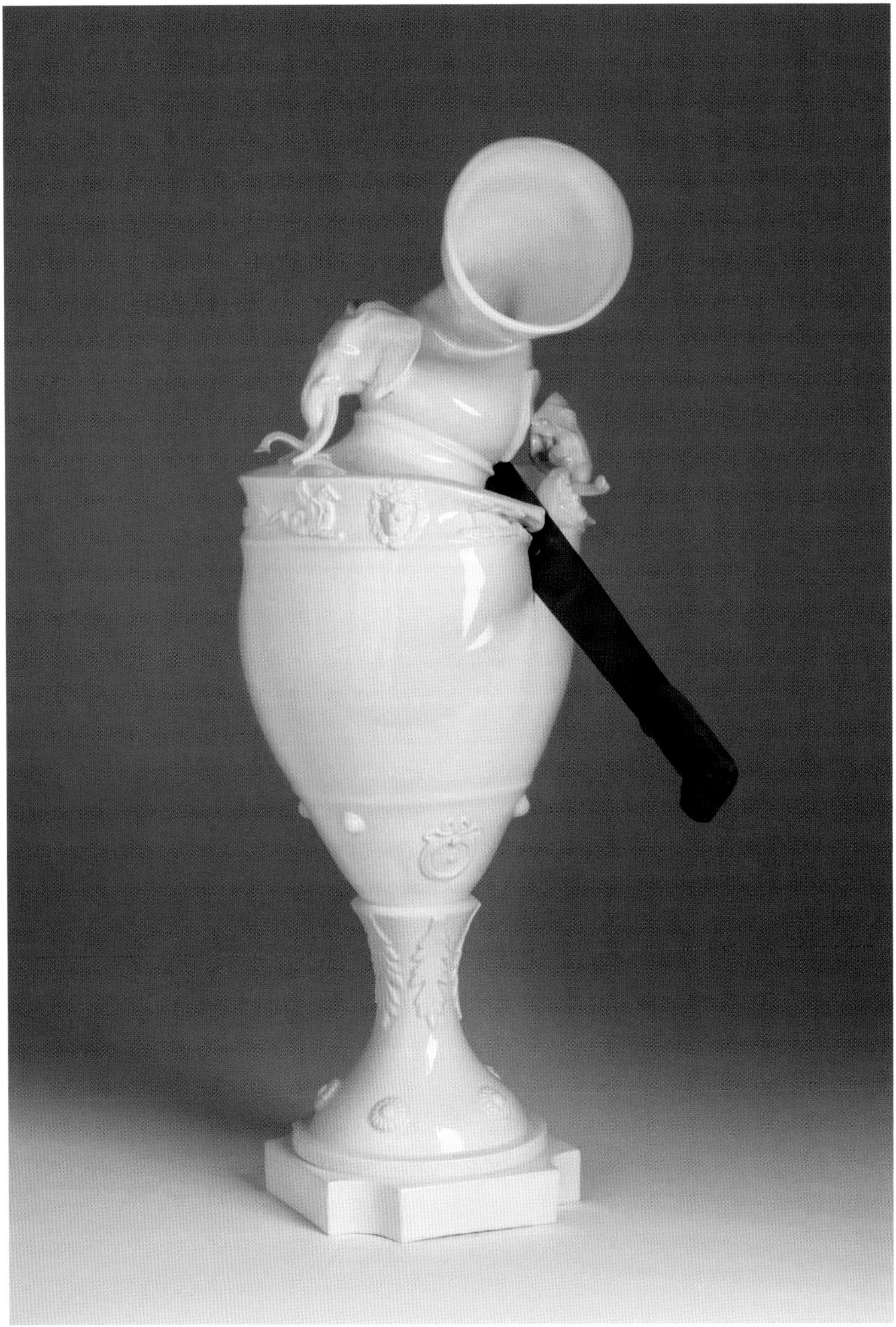

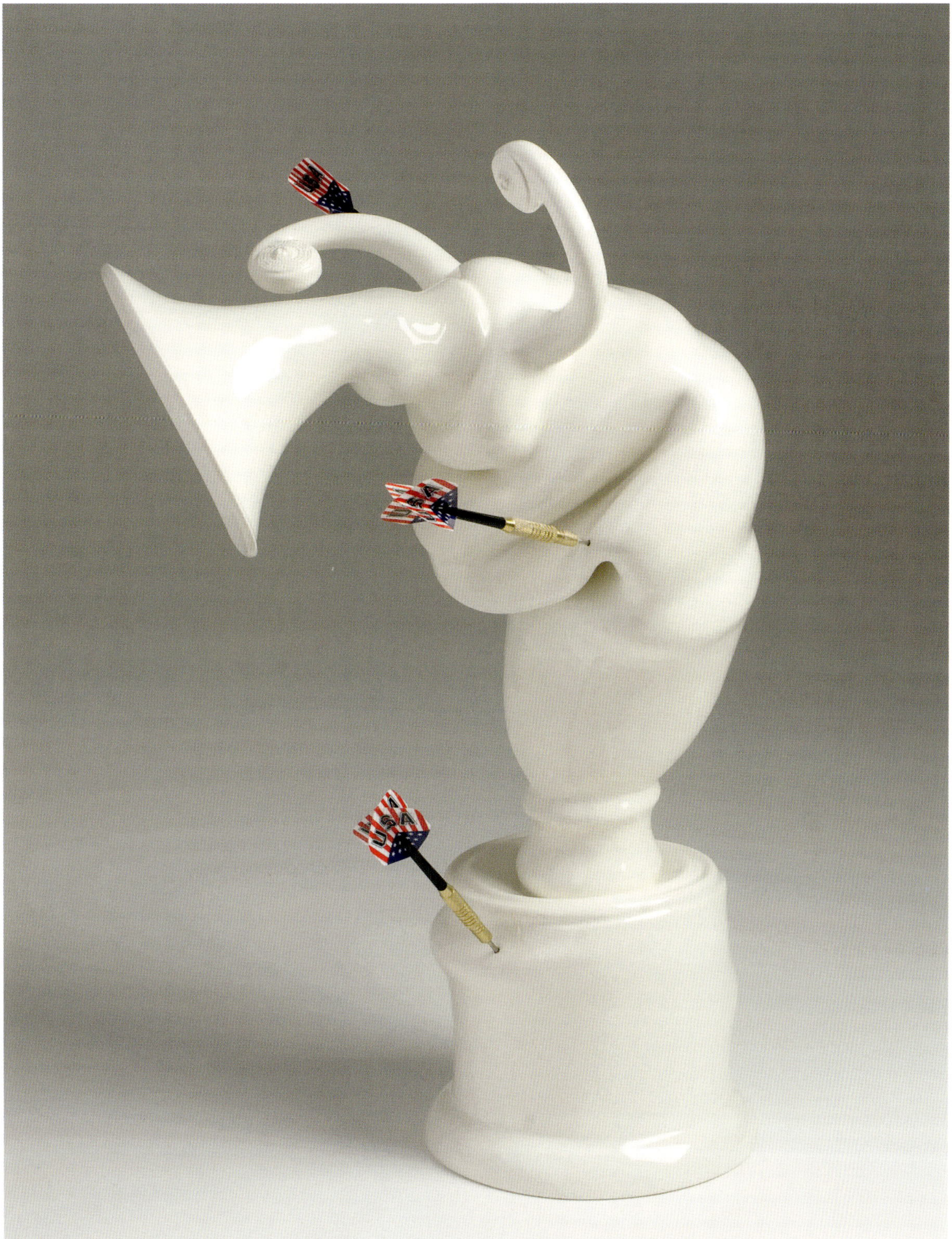

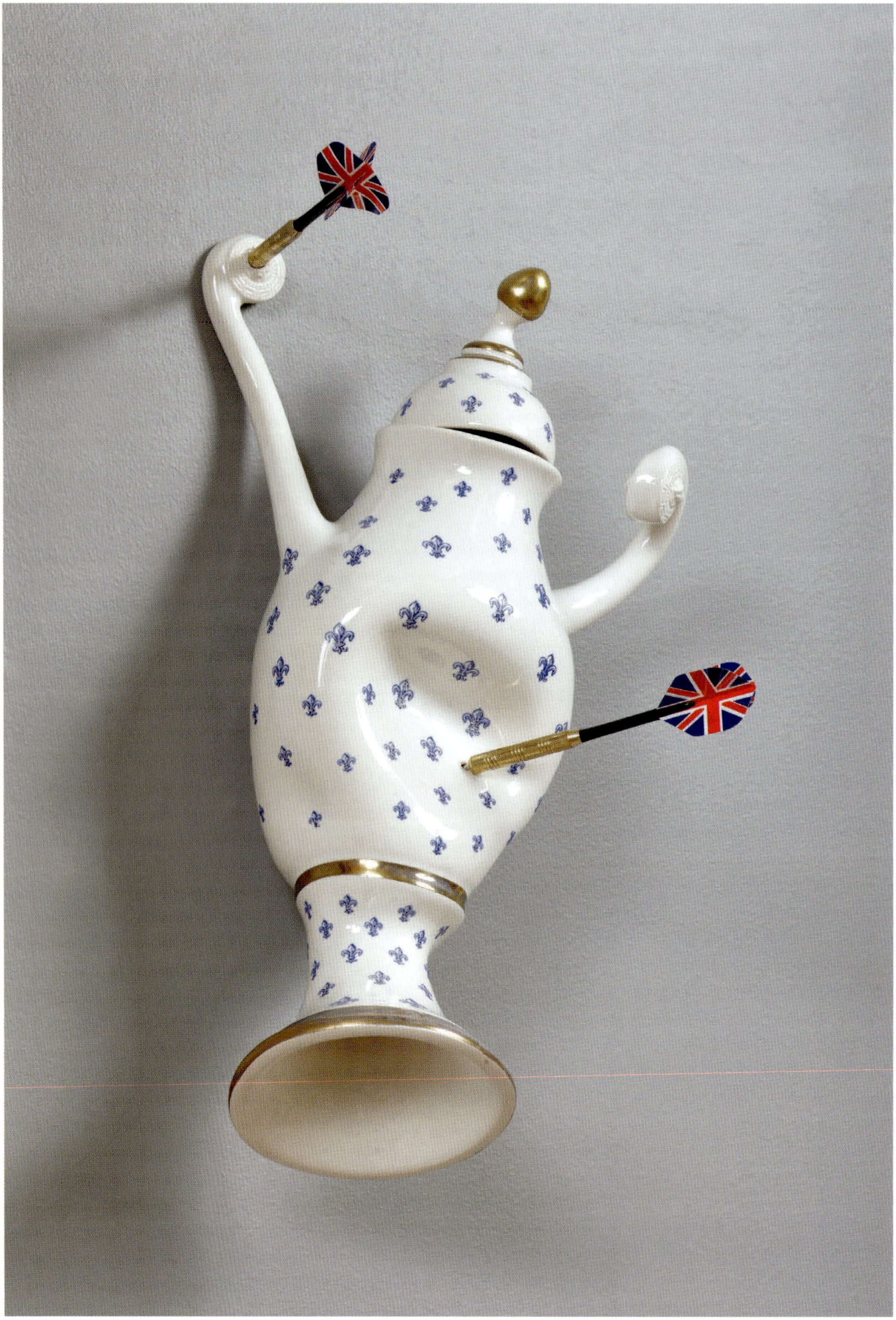

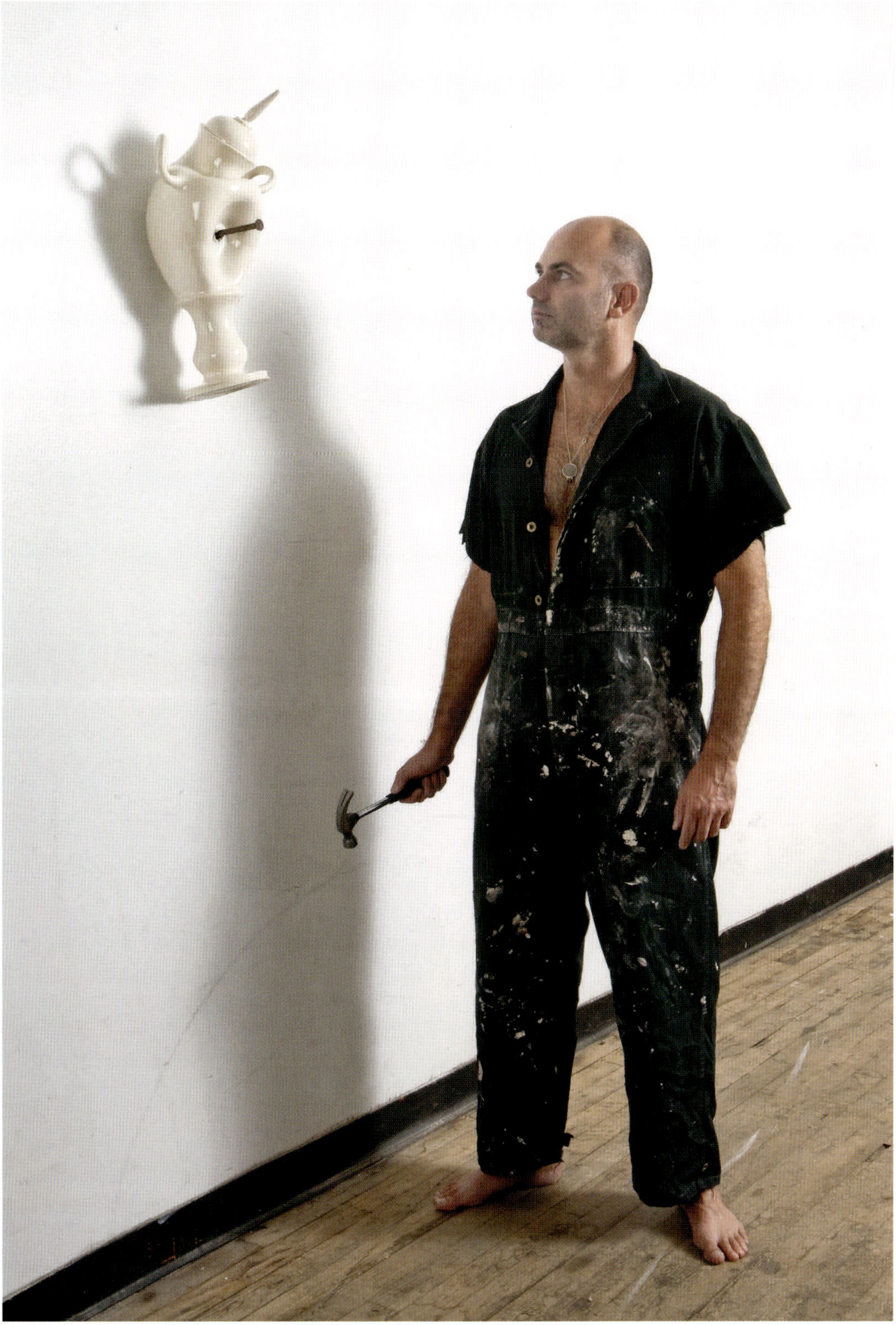

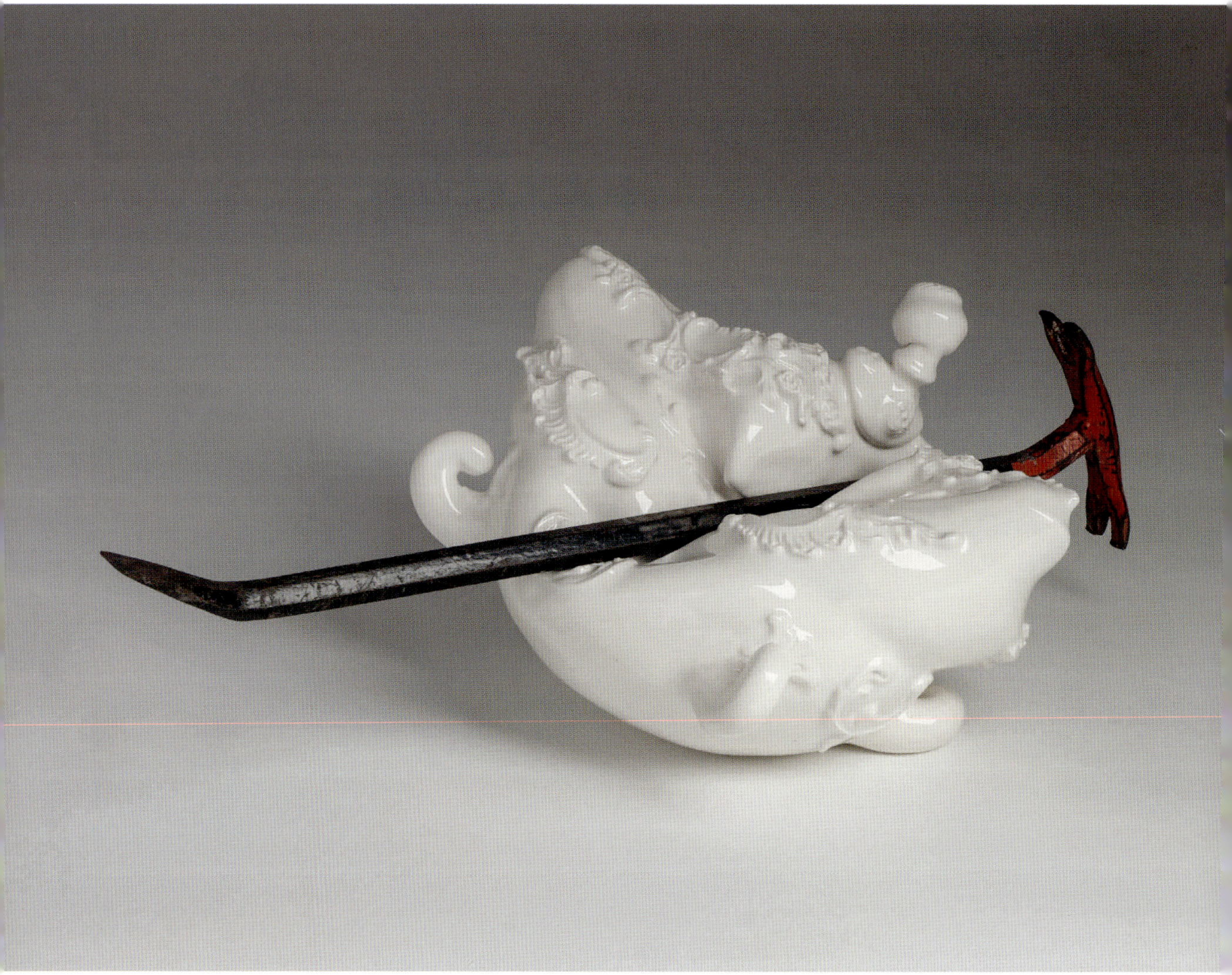

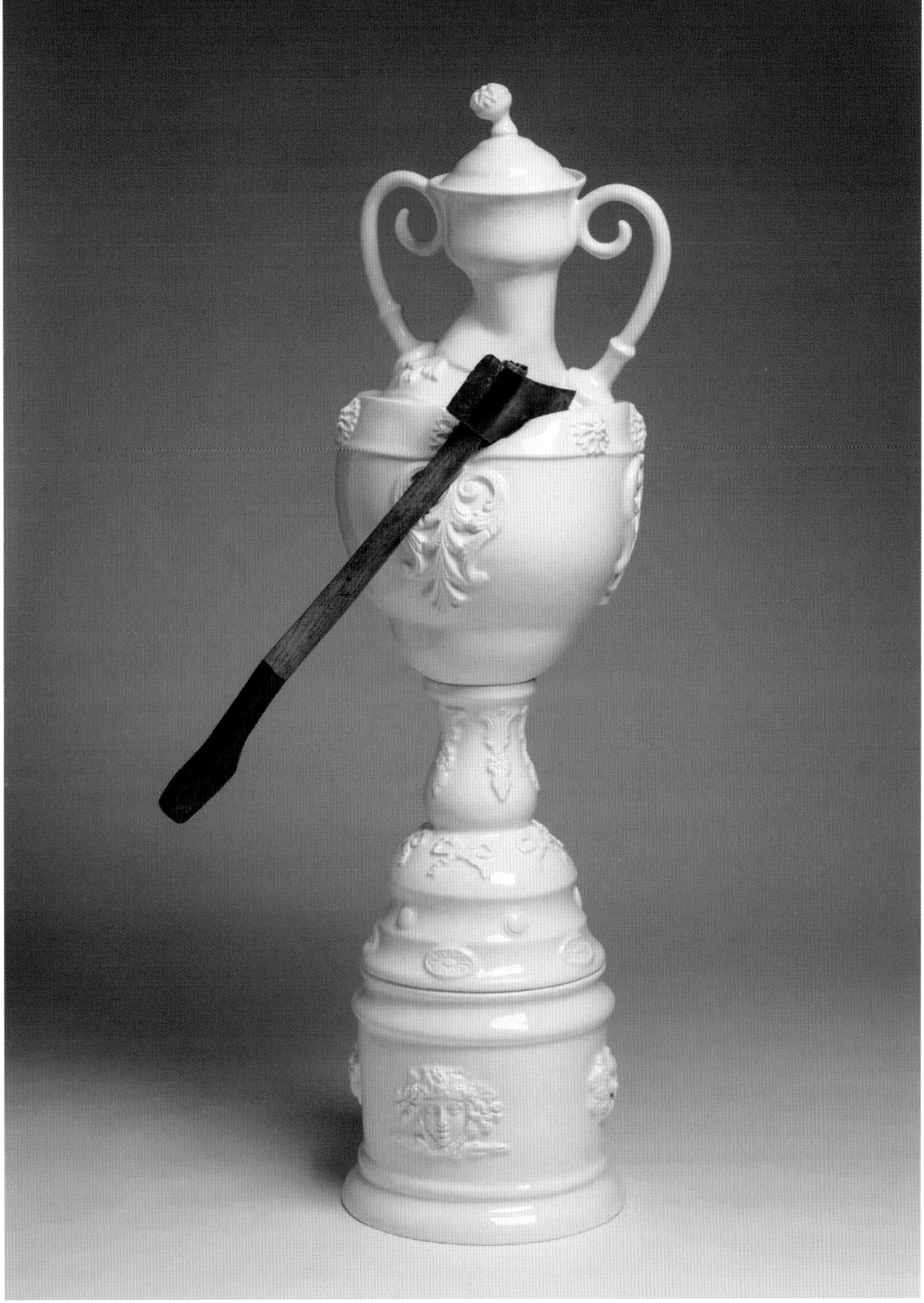

ABSTRACTED WORKS

TAKEUCHI KOUZO

PETER SHIRE

KIMBERLEE JOY ROTH

PATRICIA VOLK

TAKEUCHI KOUZO

LOCATION Hyogo, Japan
YEAR OF BIRTH 1977
MATERIALS Ceramics
EXPERIENCE 20 years

'I don't know why, but since I was a child I have been attracted to geometric shapes, especially aggregated geometric shapes. When I first saw the photos of ancient ruins in South America, my creativity was inspired, and I seriously wanted to express the distinctive ambience of the decayed buildings in my work. That is when I began creating ceramic sculptures composed of square tubes.
In a stage of trial-and-error, one day I dropped a piece by accident. Shocked, I regretted my terrible mistake seriously. However, next came another shock because this broken piece looked to me more beautiful than the unbroken pieces, and it appealed to me very much. Then, with a hammer, I broke all of the other finished tubes in my studio and exhibited them in a solo exhibition in 2006. This is how my Modern Remains Series was born.

When they see the ruins of decayed buildings, many people have similar nostalgic and sentimental feelings, but if those ruins possess an exquisite balance between the contemporary existence and the deterioration from the past, I believe it is possible to enhance the ephemeral and create dynamic beauty. When I break my pieces to create a sculpture I try to be meticulous in my effort to find the perfect balance between the original forms and the deterioration.'

Photos courtesy of the artist

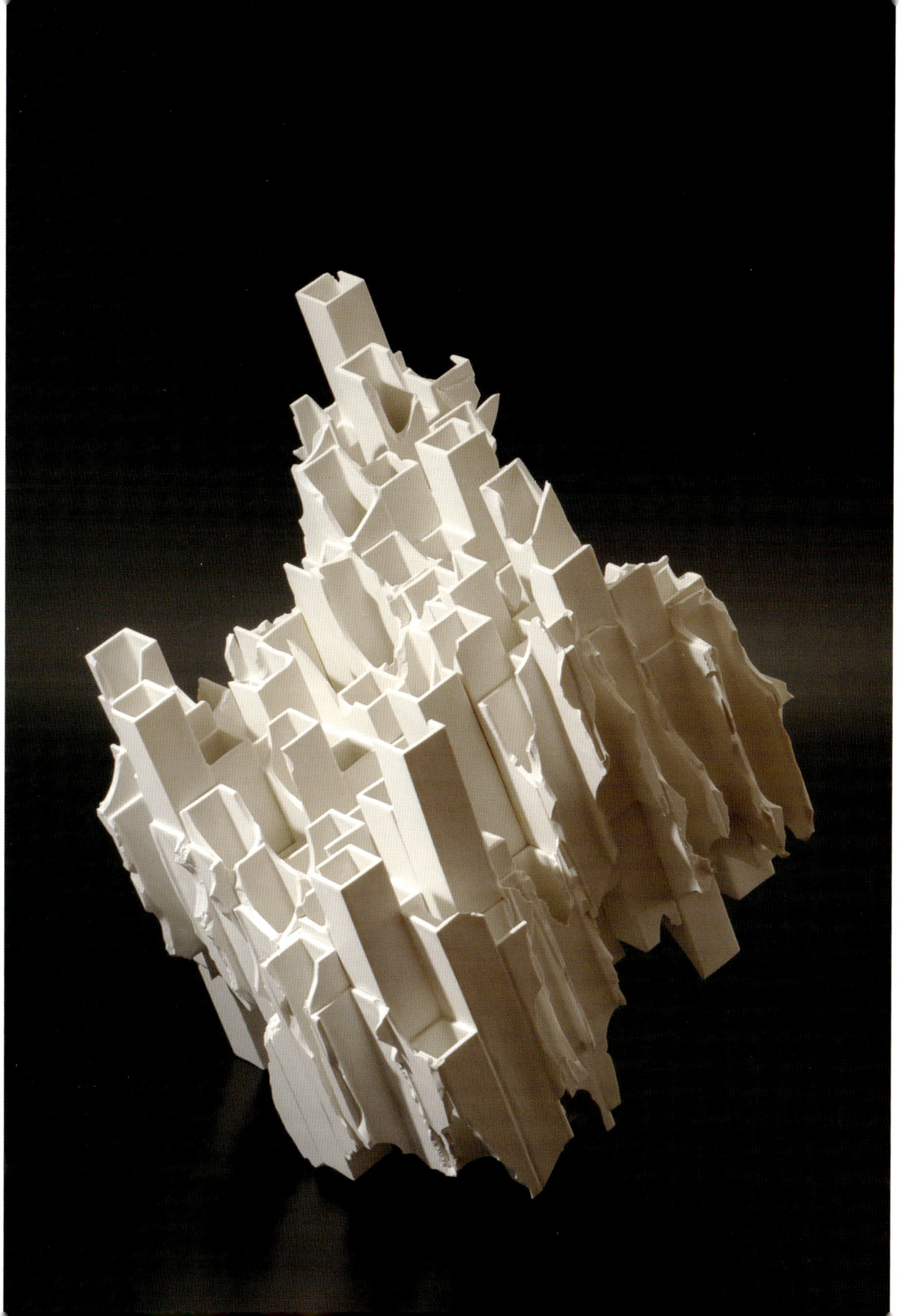

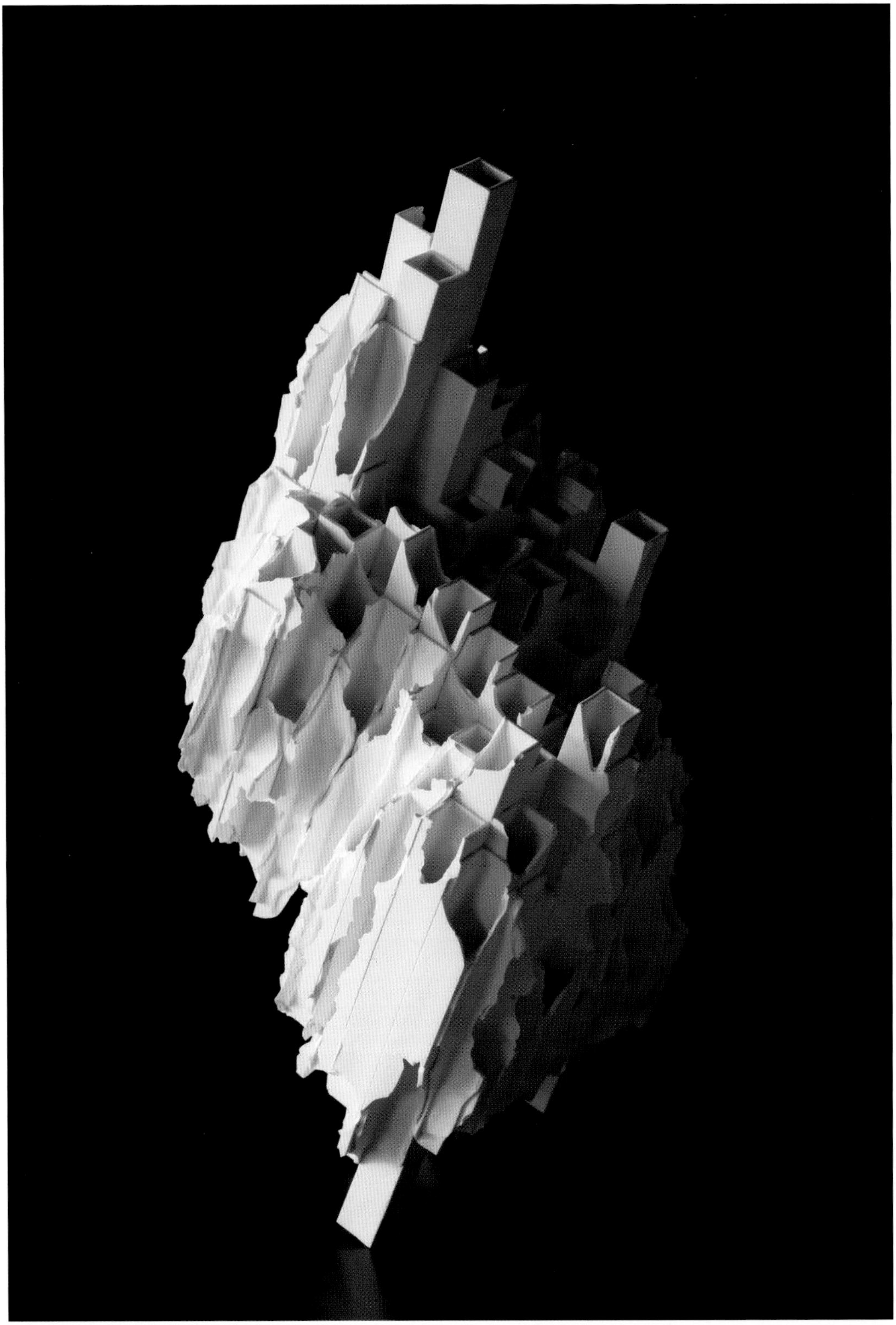

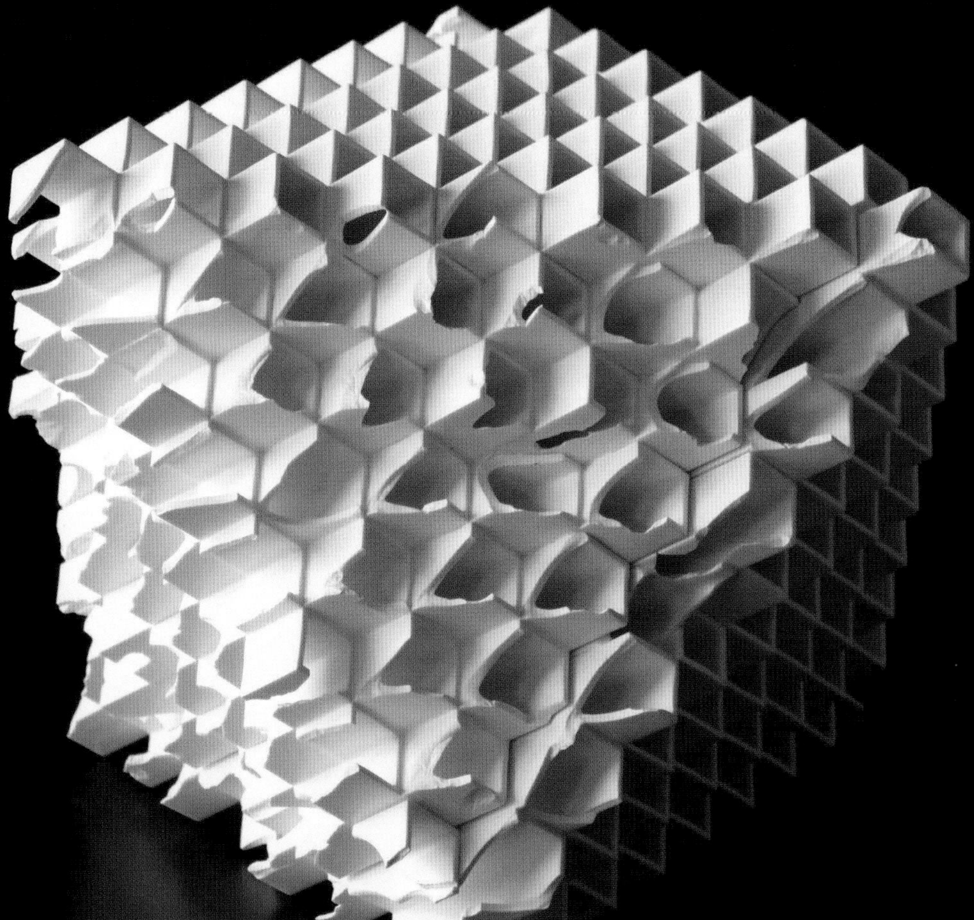

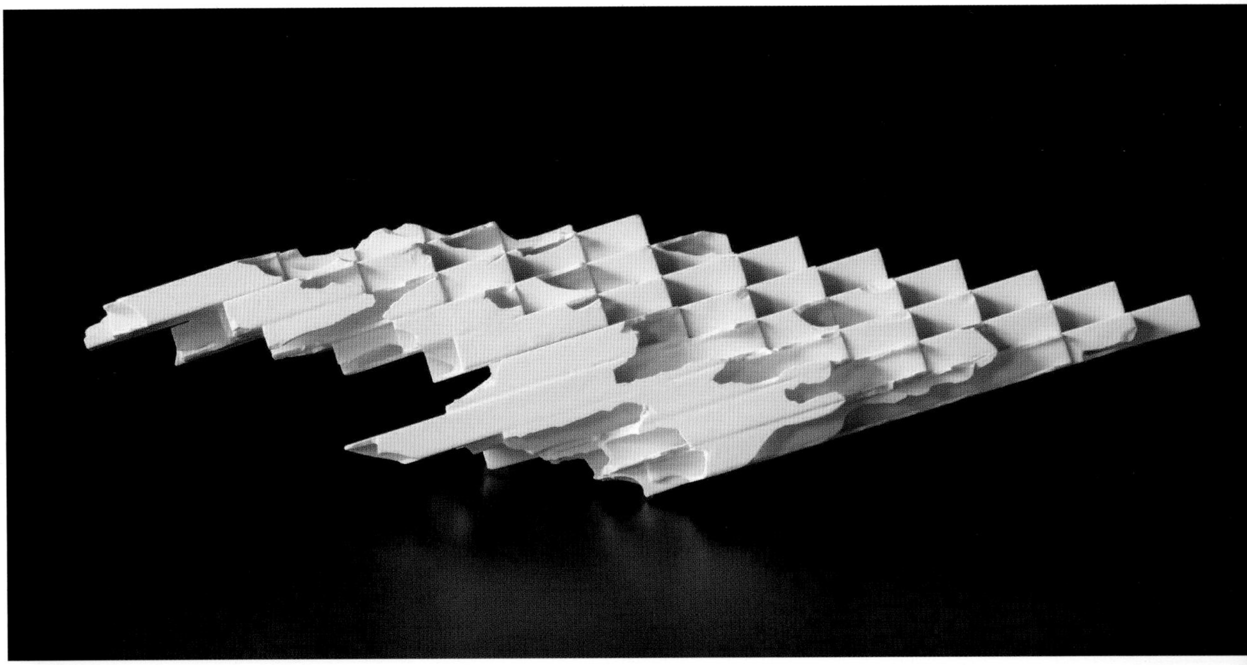

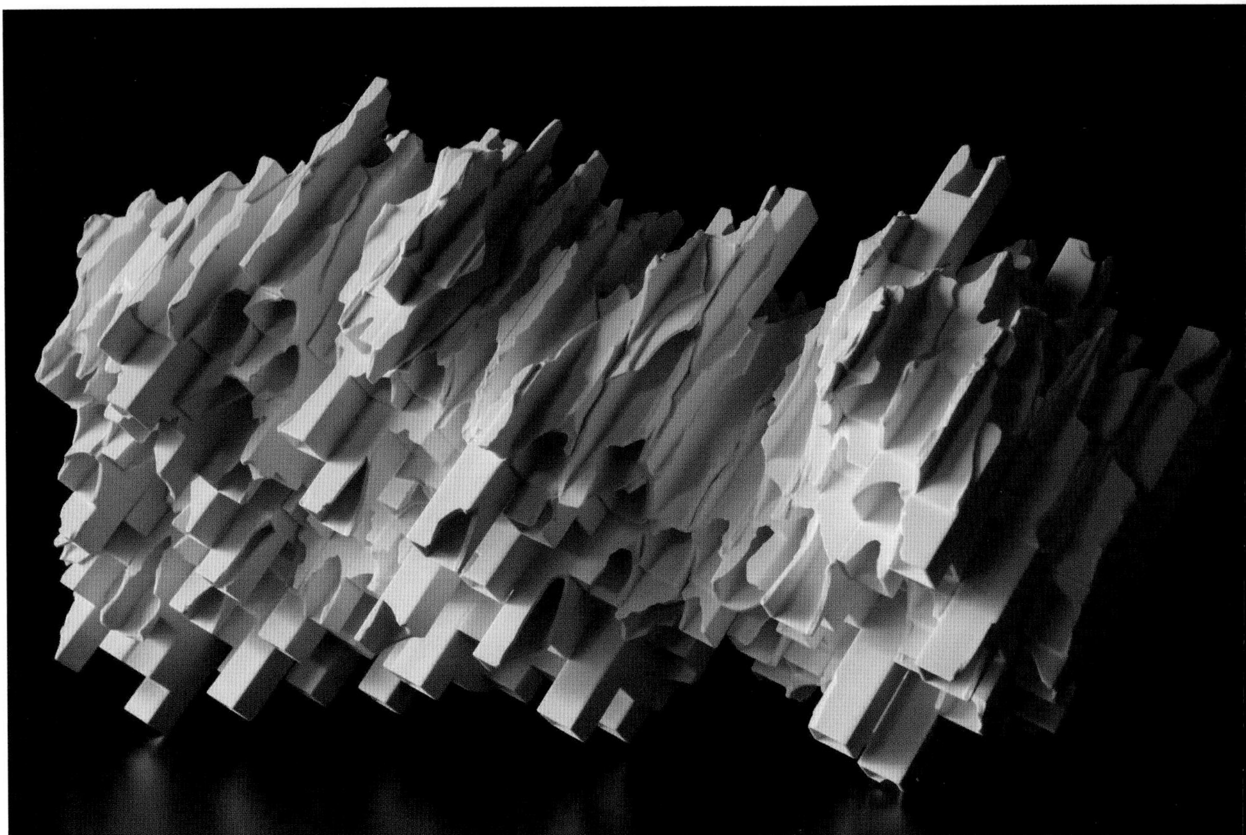

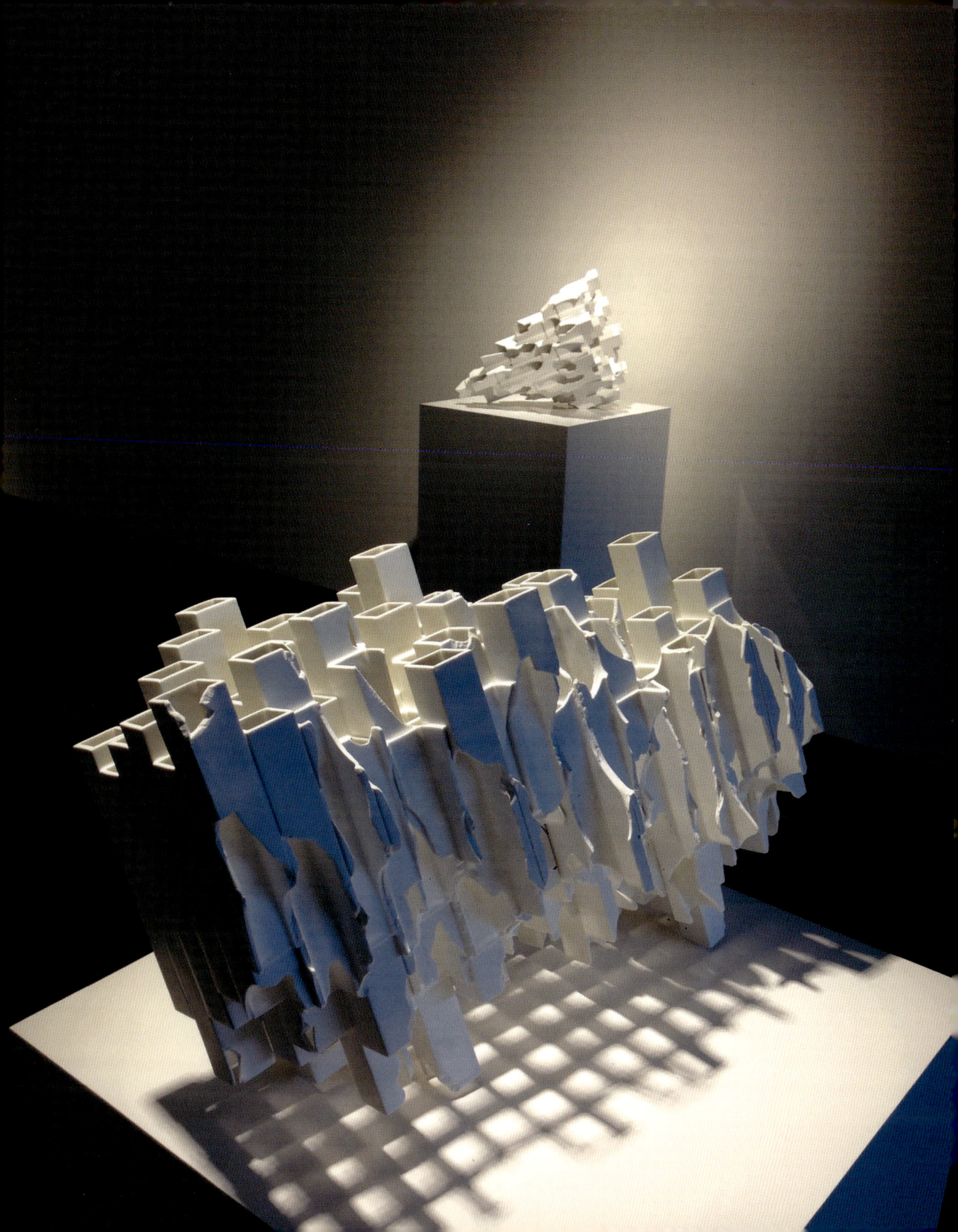

PETER SHIRE

LOCATION Echo Park Los Angeles, CA
YEAR OF BIRTH 1947
MATERIALS Ceramics, slab construction
EXPERIENCE 52 years

Peter Shire has crossed the borderlines of traditional painting and sculpture and has delved into fields such as ceramics, furniture, and toys. His work contains eclectic playfulness and subtle irony. Establishing his own studio in 1972 positioned Shire to be part of what has come to make Southern California unique and important in international art. As one of the original members of the Memphis Group, Shire has forged a new language defined by the dialogue between shapes and surfaces and between technology and aesthetics. Shire is concerned with revealing new artistic possibilities. His teapots (he's made over 4,000) deal with the notion of functionality in an industrial artistic world. His furniture reaffirms his commitment to the human condition. Cones, cylinders and grids are combined in his furniture in defiance of typical modernist style in a way that swings between the usual ideas of sculpture and a functional object, creating a third category that goes beyond both of them that he calls Stanzaroto, or Ol' Squint Eye. The use of different materials reinforces the artist's concern with experimenting beyond traditional techniques and recognizable styles.

Photos courtesy of Hannah Stouffer and the artist

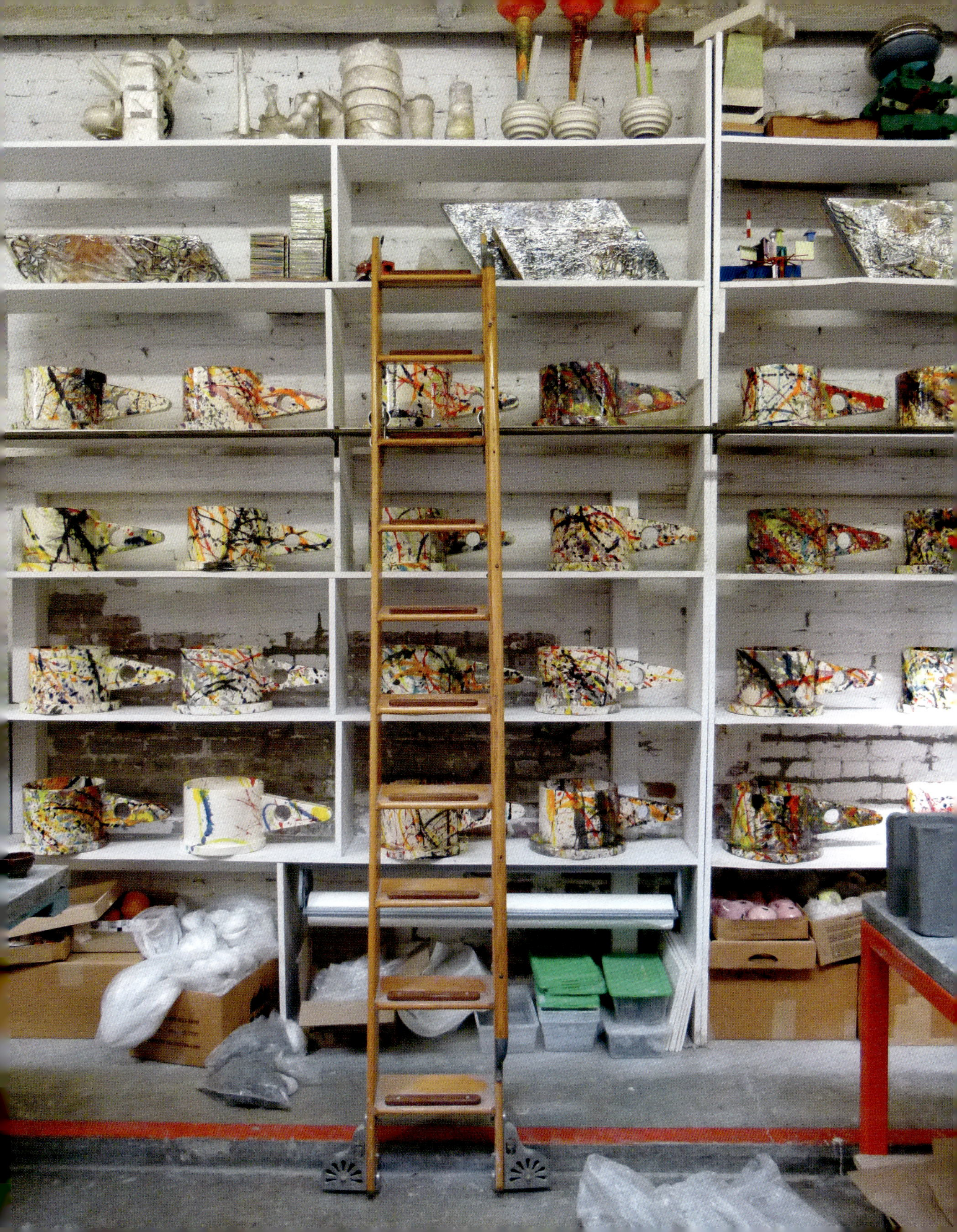

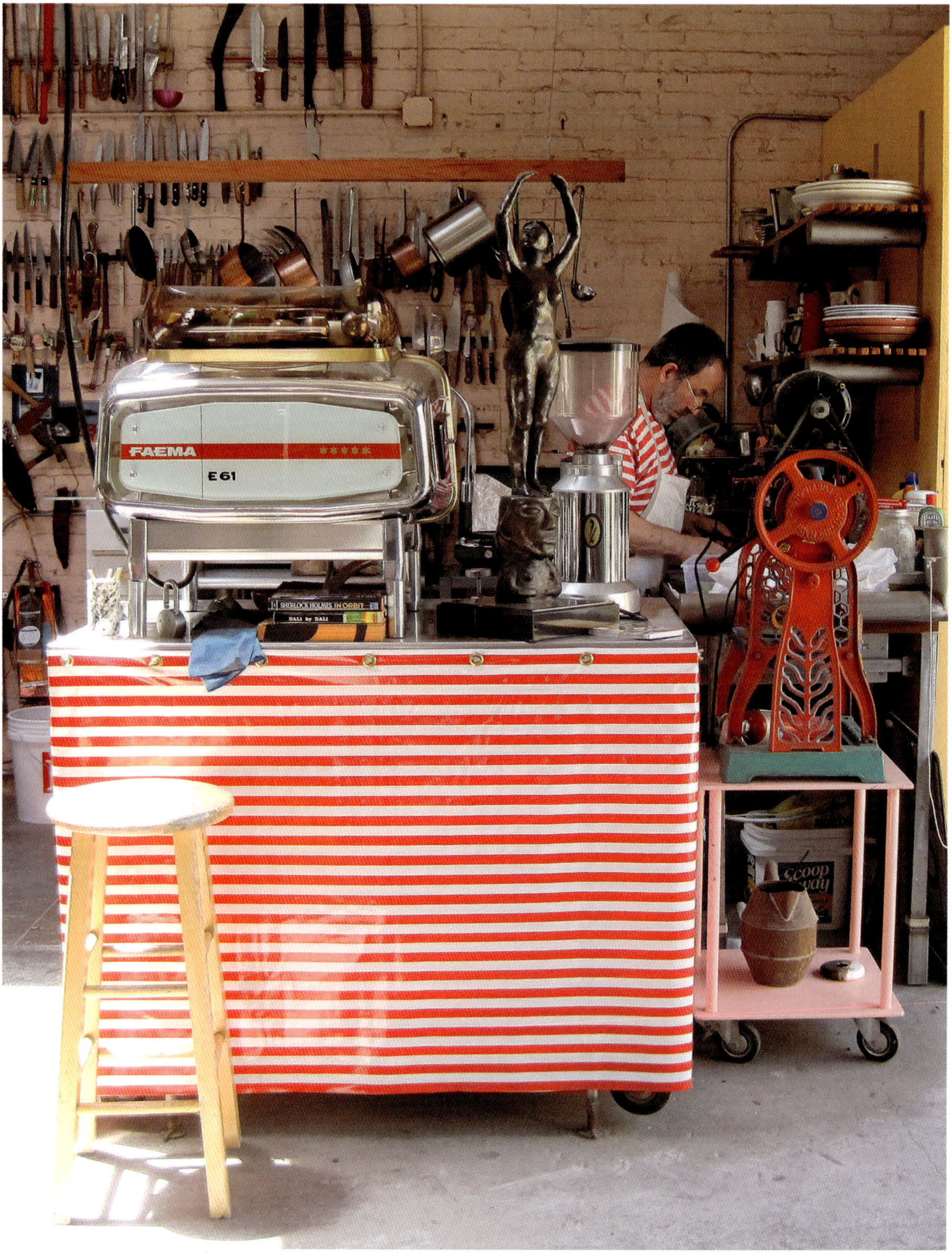
FAEMA
E 61
SHERLOCK HOLMES IN ORBIT
DALI by DALI

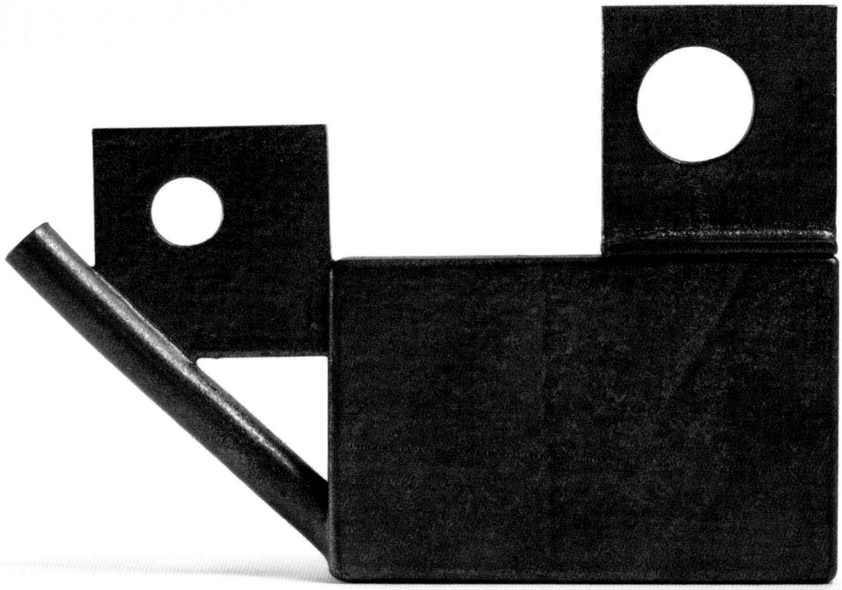

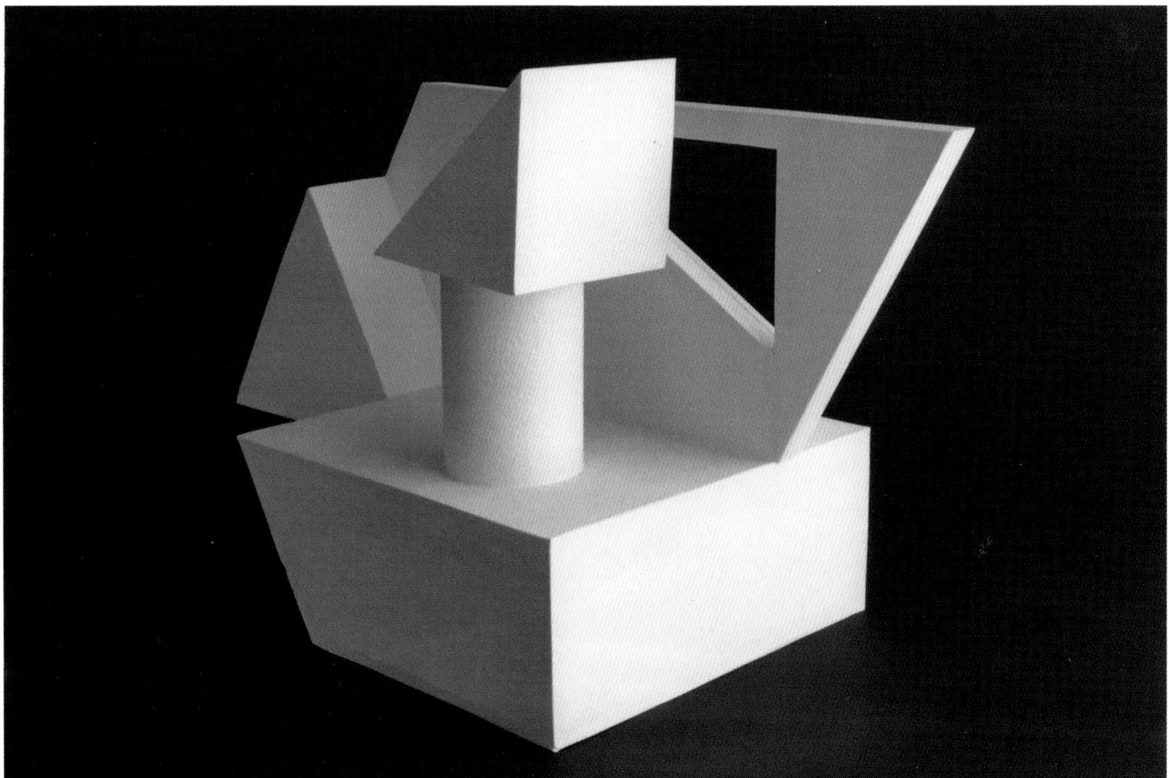

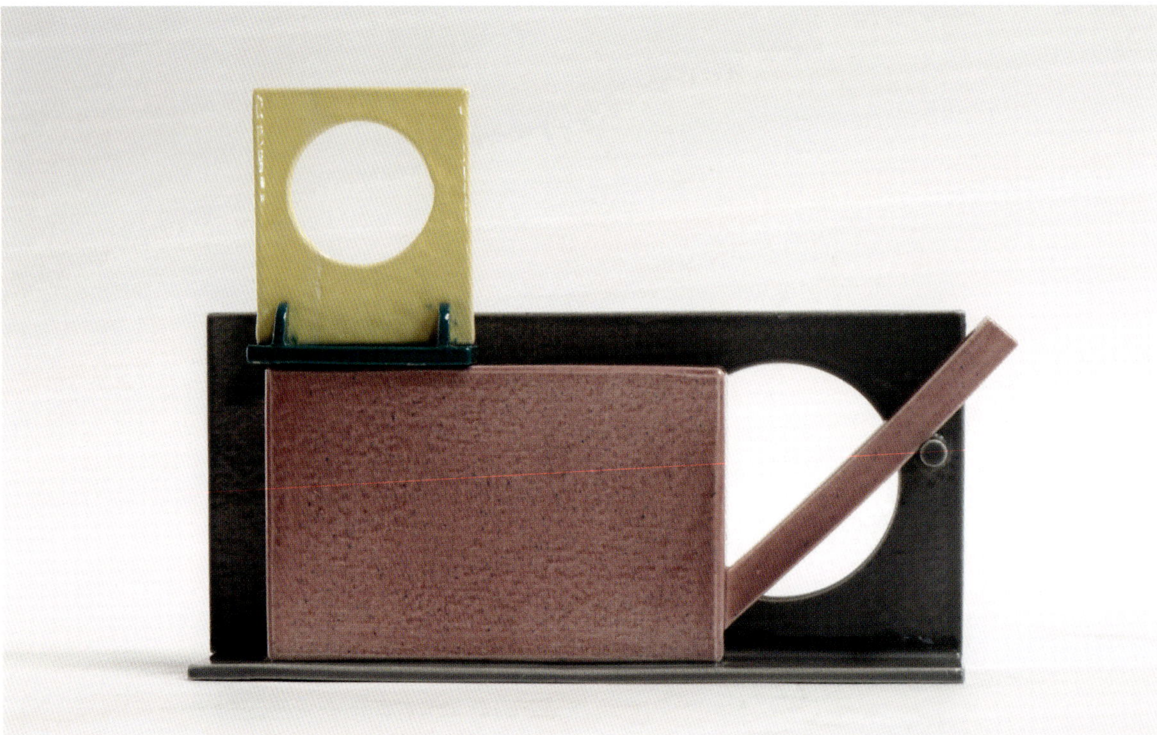

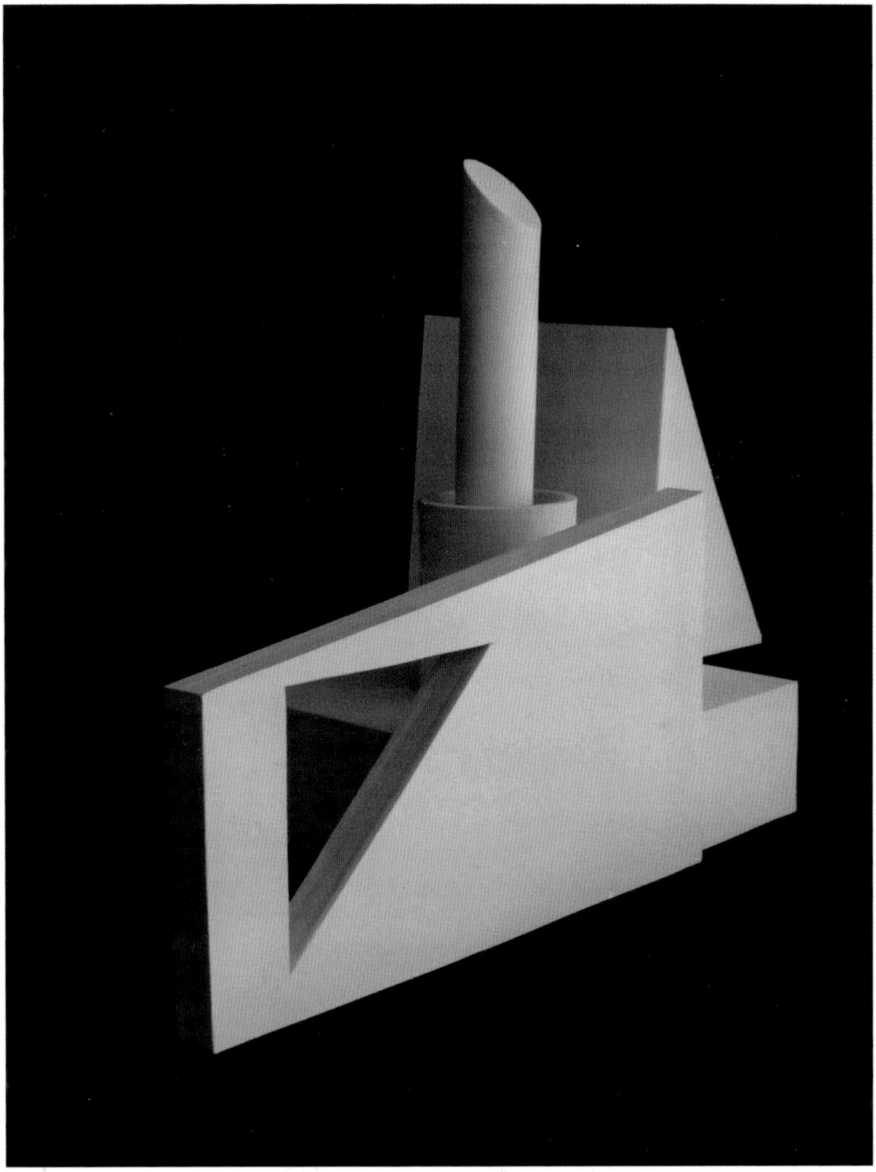

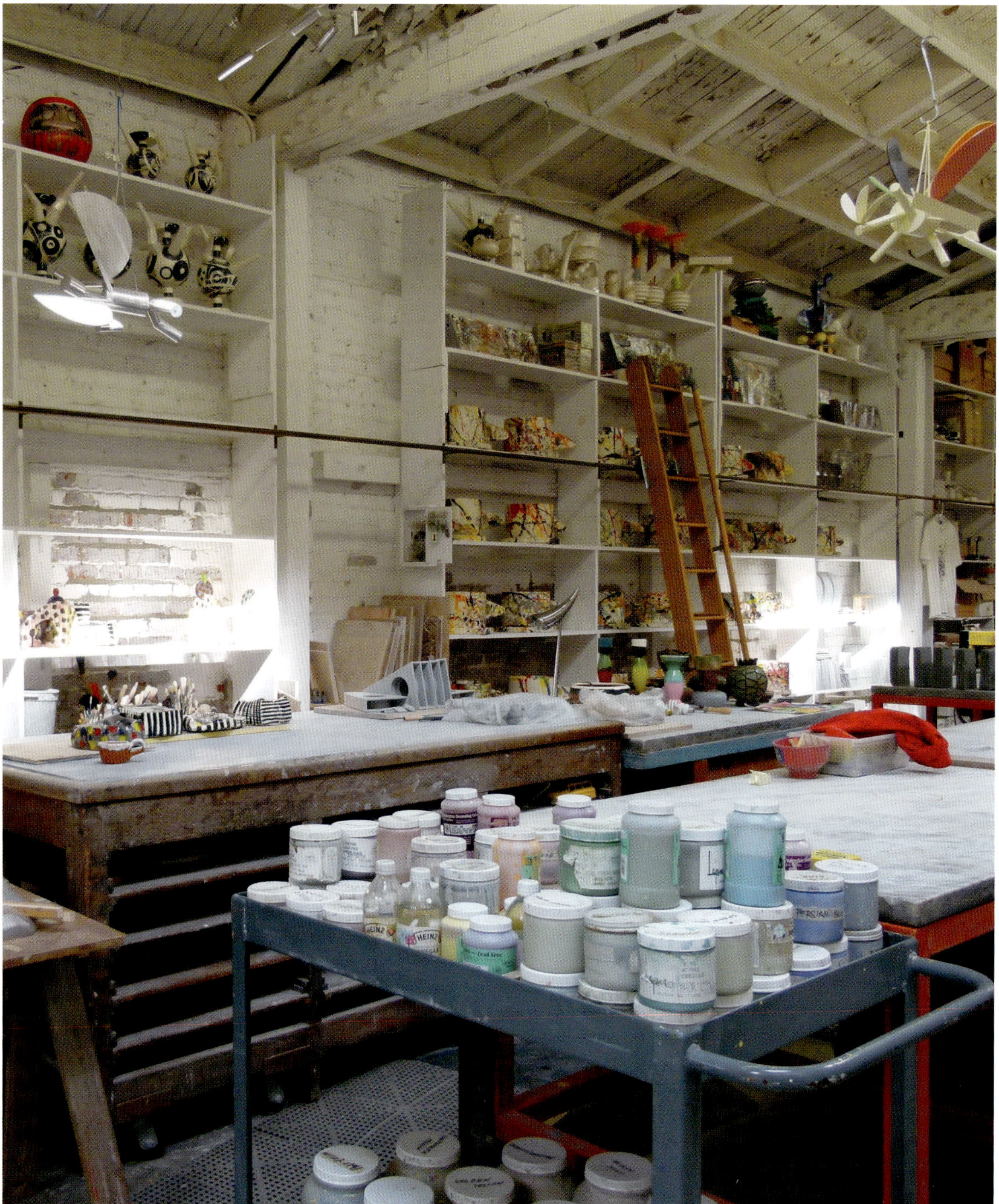
HEINZ
HEINZ

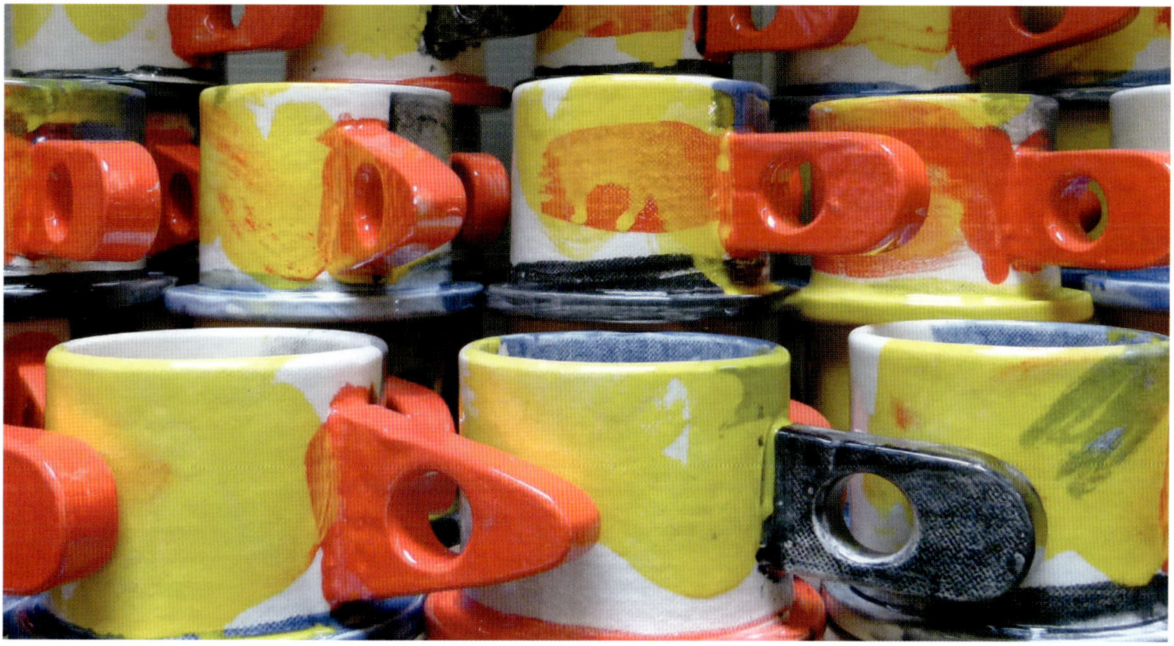

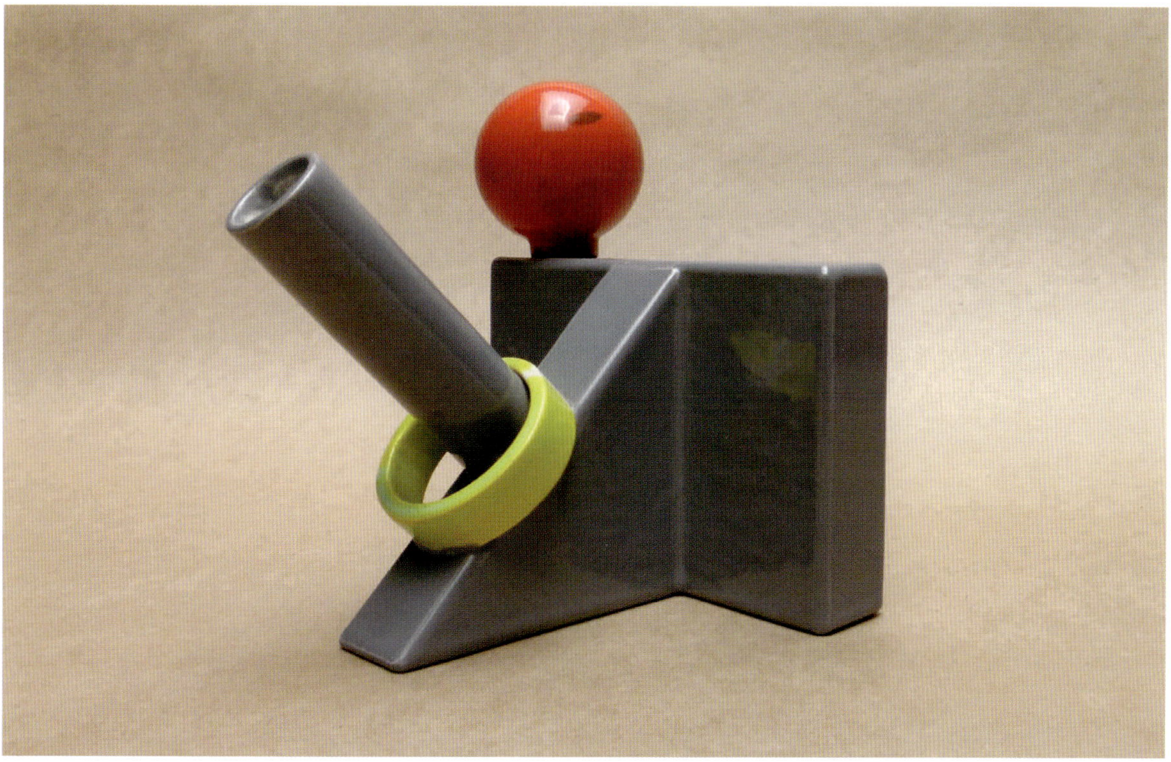

KIMBERLEE JOY ROTH

LOCATION Minneapolis, MN
YEAR OF BIRTH 1968
MATERIALS Porcelain
EXPERIENCE 9 years

Roth's work is wall sculpture that is still functional; it can be used to contain and serve food. She sees it as a bridge between decorative tile and utilitarian ware, and as a vehicle for beginning a communication between these two disparate uses of the same material.

'Currently, my method of production begins with full scale drawings of the topographical structure of a form. It is enlarged and scaled to compensate for shrinkage. This 'footprint' is then used to carve a three dimensional form either from Styrofoam or a low fire white clay body. I then arrange these forms into symmetrical repetitively patterned compositions that subtly allude to Islamic tiled façades. The negative spaces in-between the plate arrangements inform the development of the overall background design. These formal relationships between the objects and background create a cohesive and aesthetically pleasing composition.'

Kimberlee Joy Roth graduated from the University of Minnesota – Twin Cities with an MFA in Ceramics and an Art History minor in 2007. Her ceramic work is in the permanent collection of the Frederick R. Weisman Art Museum in Minneapolis, MN, the Lincoln Arts and Culture Foundation in Lincoln, California, the Mayo Clinic, Rochester, MN, and has been shown nationally in juried ceramic exhibitions.

Photos courtesy of the artist

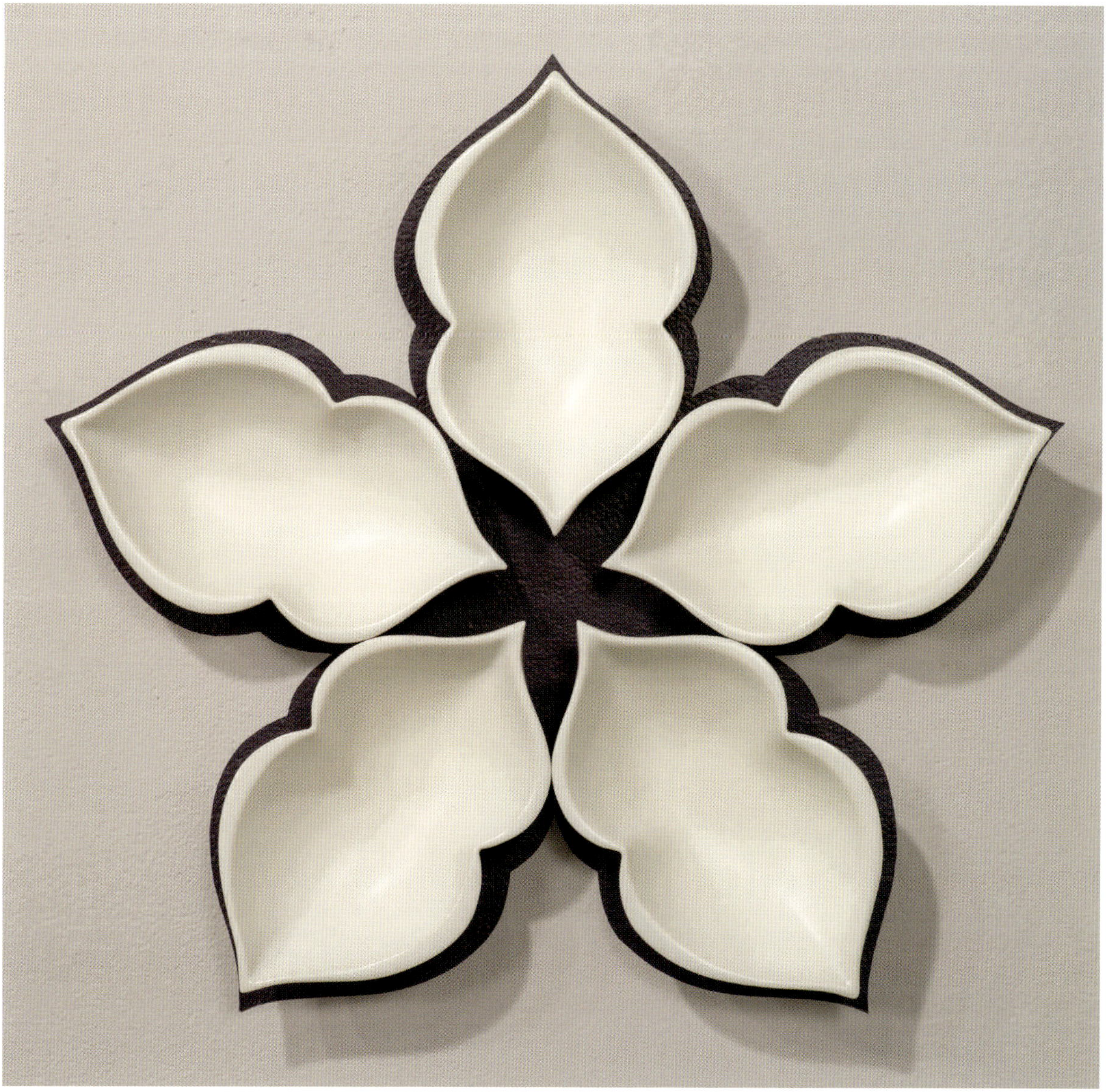

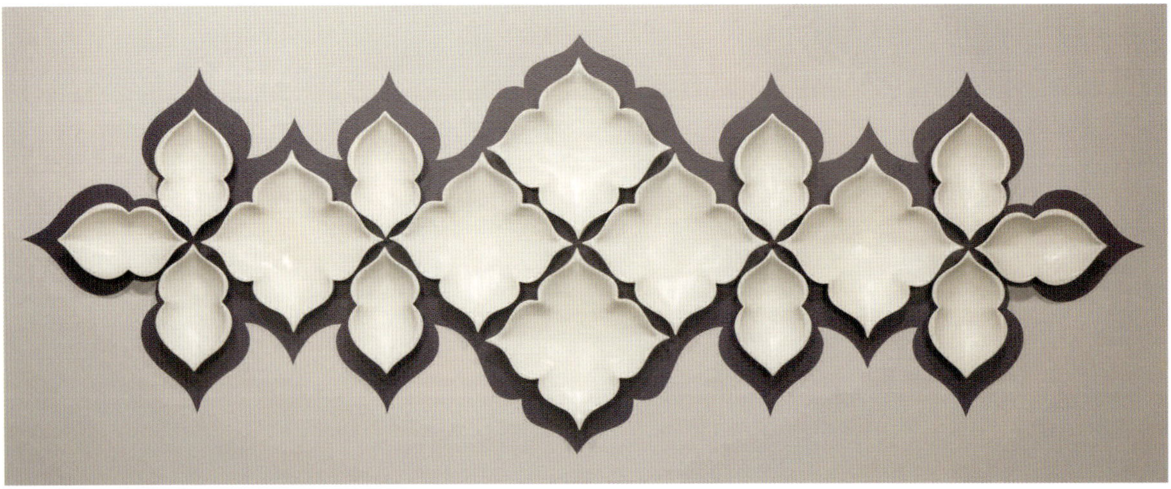

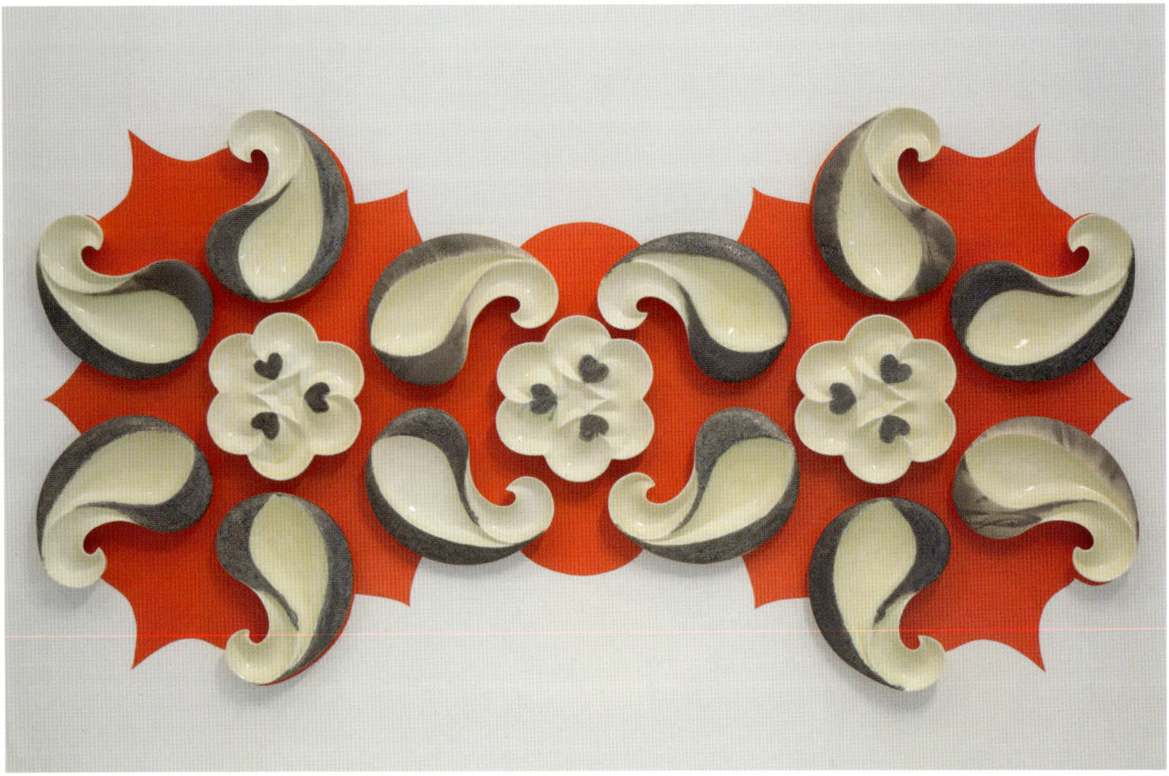

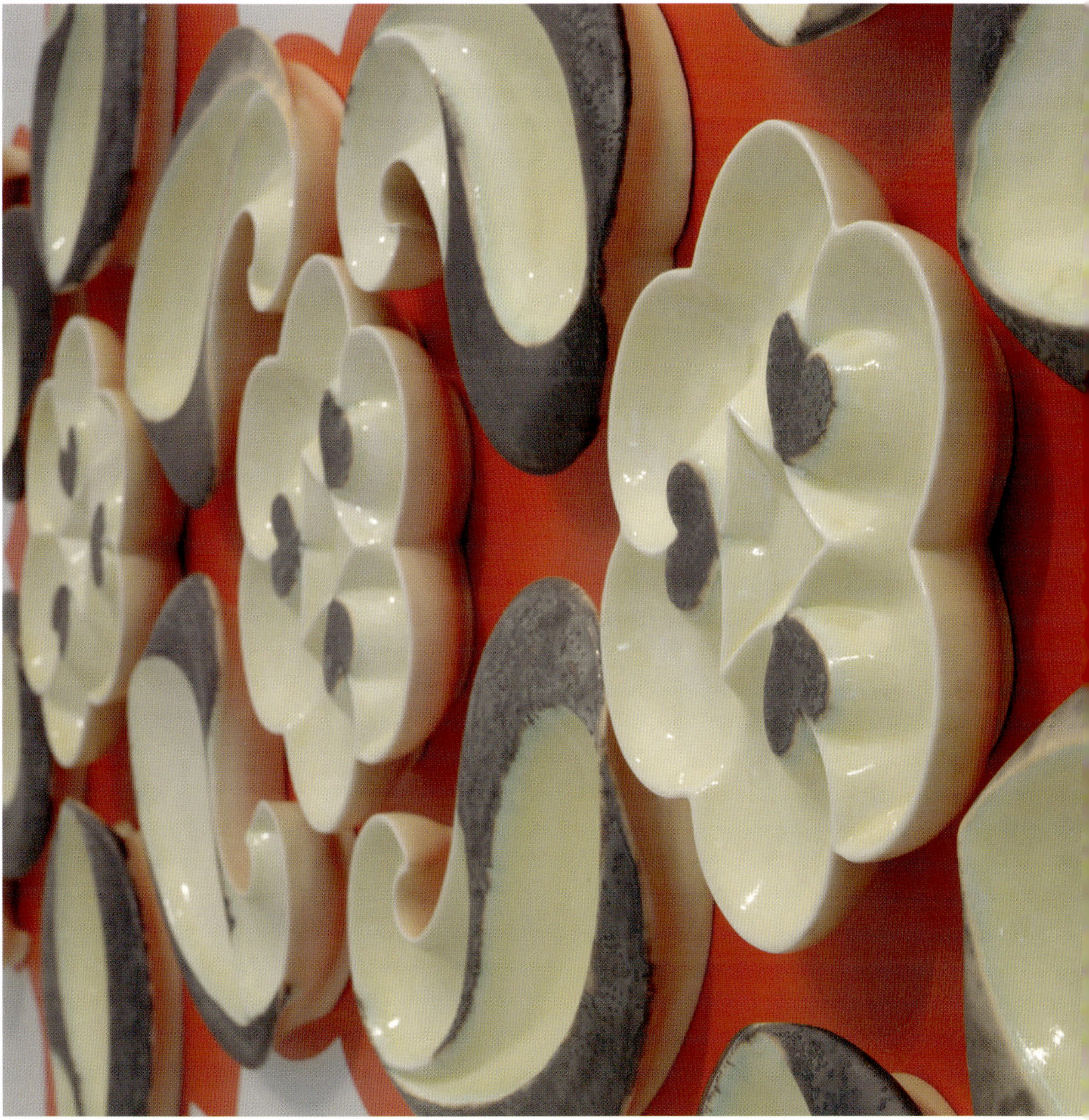

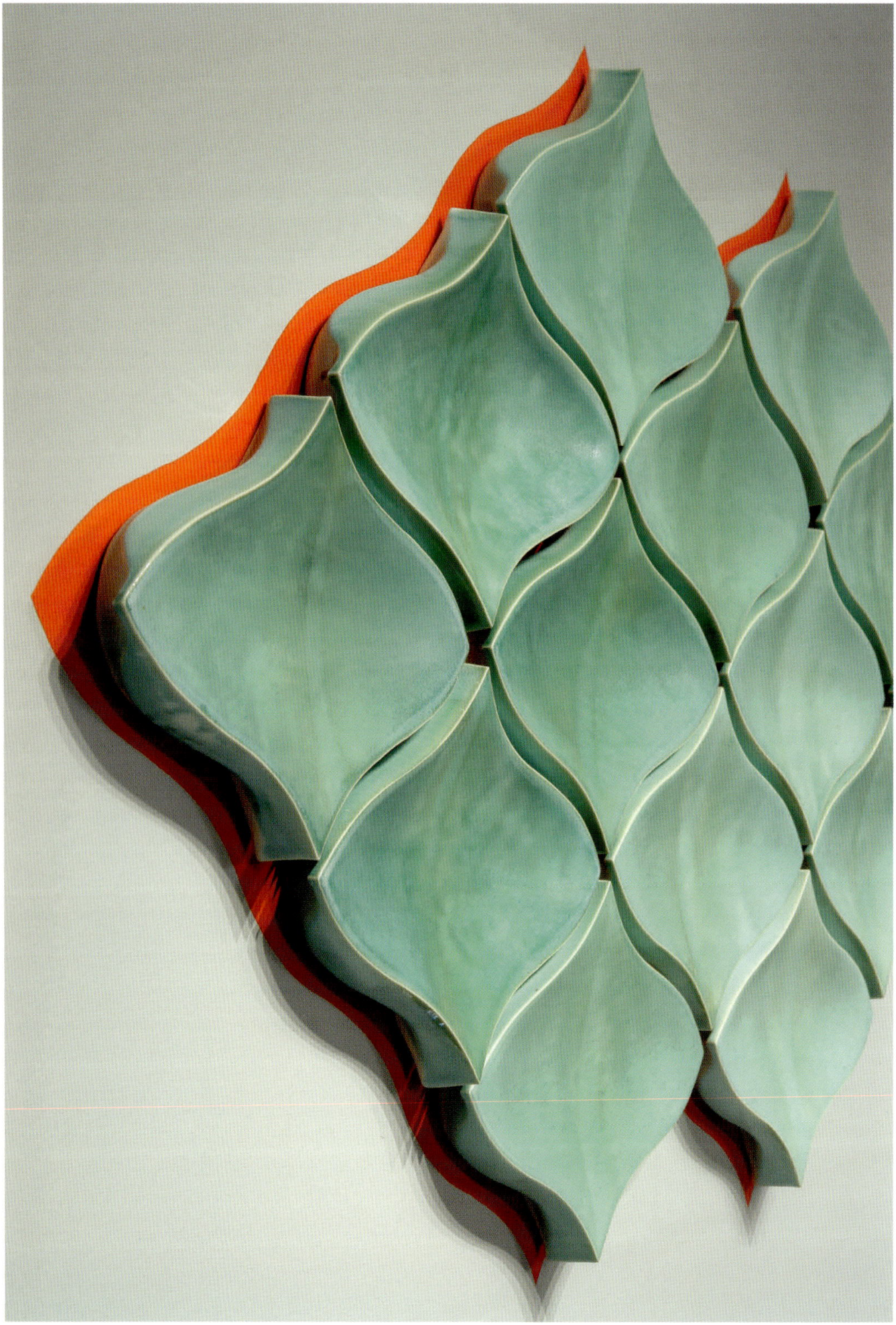

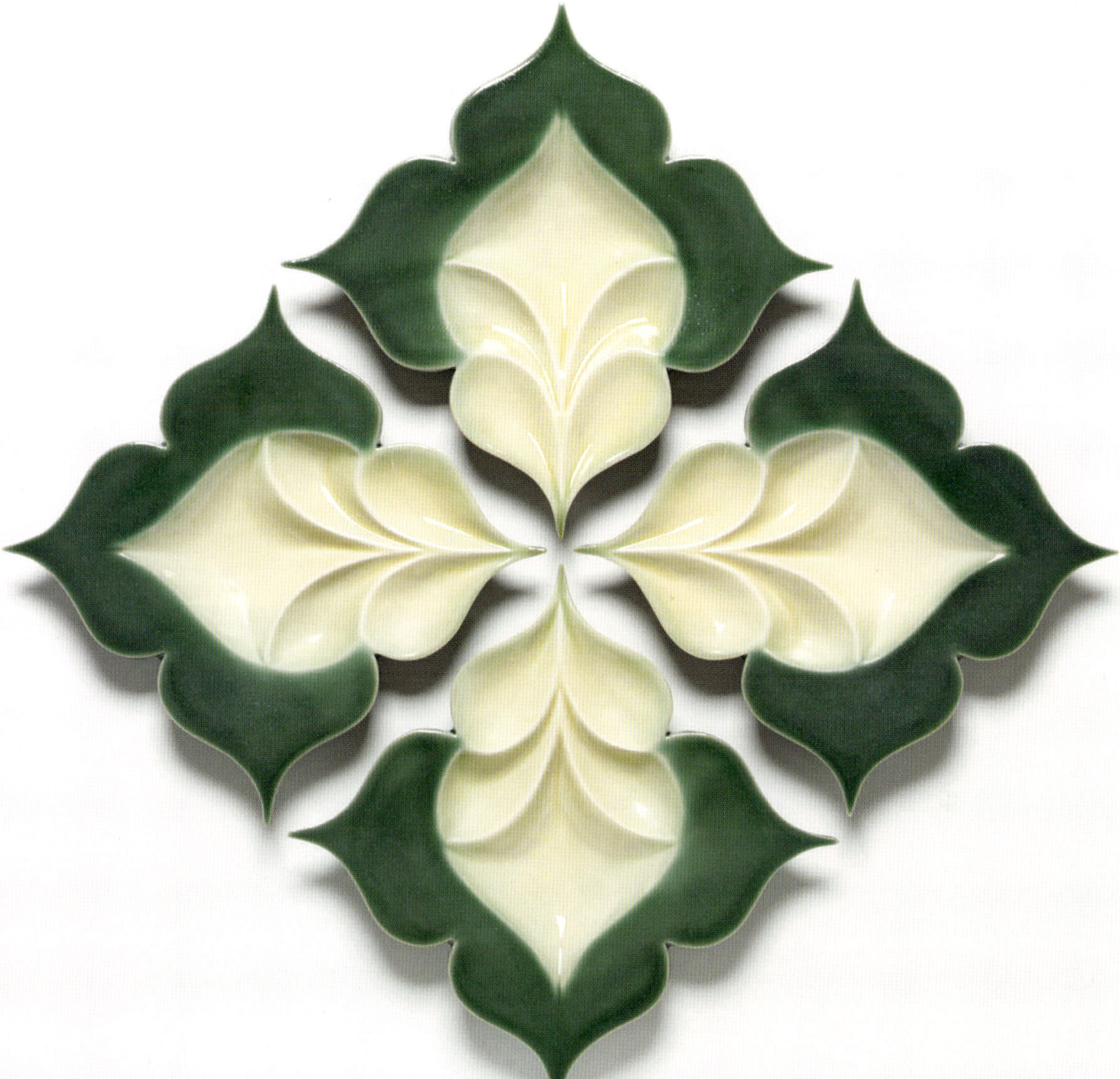

PATRICIA VOLK

LOCATION Bradford on Avon, Wiltshire, UK
YEAR OF BIRTH 1951
MATERIALS Ceramic slab building and coiling with acrylic paint
EXPERIENCE 25 years

'My obsession as an artist is catching a very simple form or line, then enhancing it with colour. Sometimes these are juxtapositions that I hope suggest contradictions of strength and fragility, stability and precariousness — like the relationships between human beings. But also of rest and activity, grace and motion, that give an object a lively presence and a sense of individual character. But really all this is open to the eye of the beholder. I never want to be too prescriptive, and in saying this much I've probably said too much.'

Patricia Volk is a member of the Royal Society of British Sculptors and was born in Belfast, Northern Ireland. Her work has been exhibited nationally and internationally, including at Buekenholf-Phoenix Gallery in Belgium, Chichester Cathedral, and the Royal West of England Academy, and is also held in the collections of Lord Carrington, Simon Relph CBE, the British Consul (Ivory Coast), and Mary Portas. She was Regional Winner of the ING "Discerning Eye" prize in 2007 and has been shortlisted for the prestigious Brian Mercer Residency. One of her pieces was also selected as a Southern Arts prize.

Photos courtesy of the artist

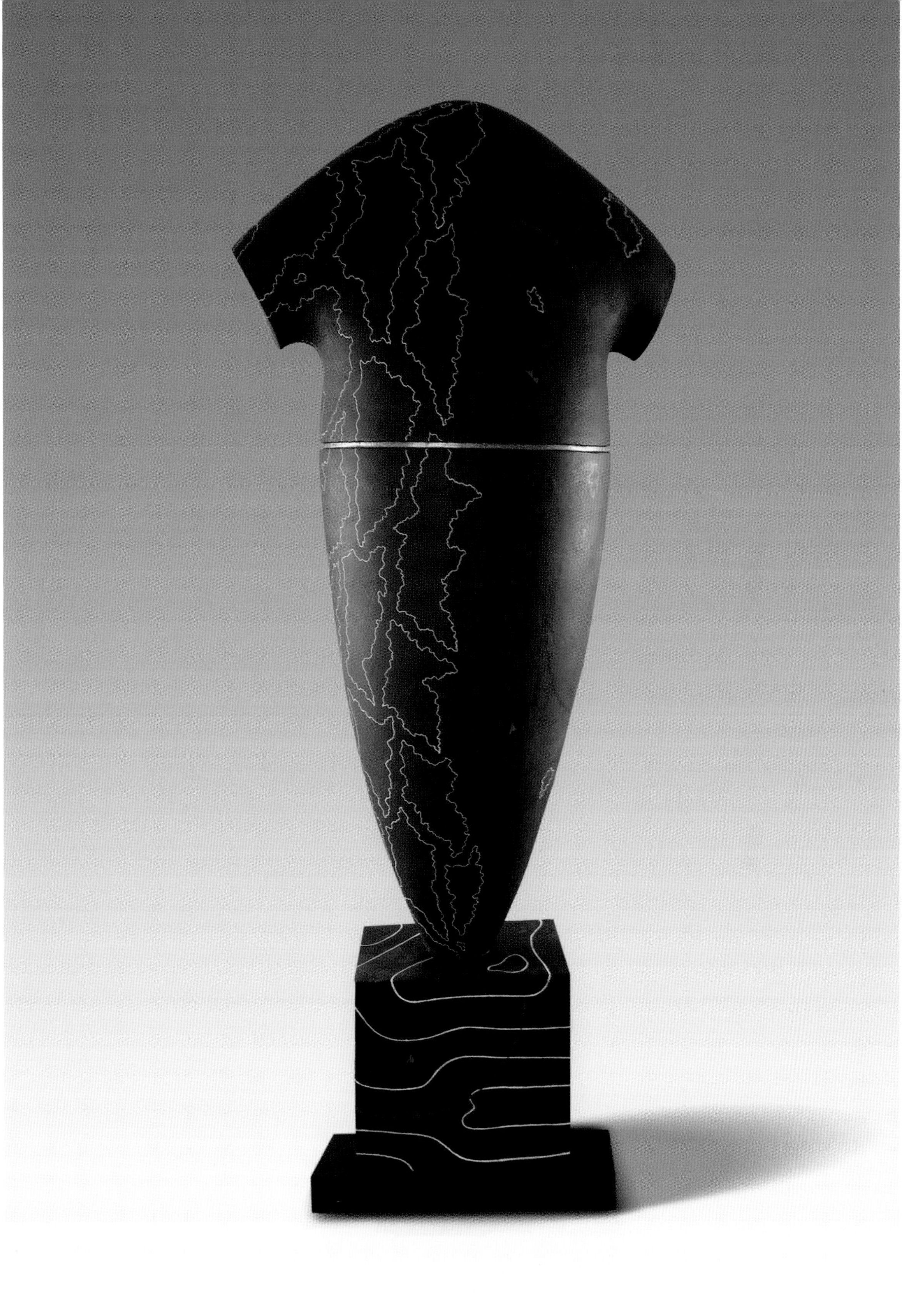

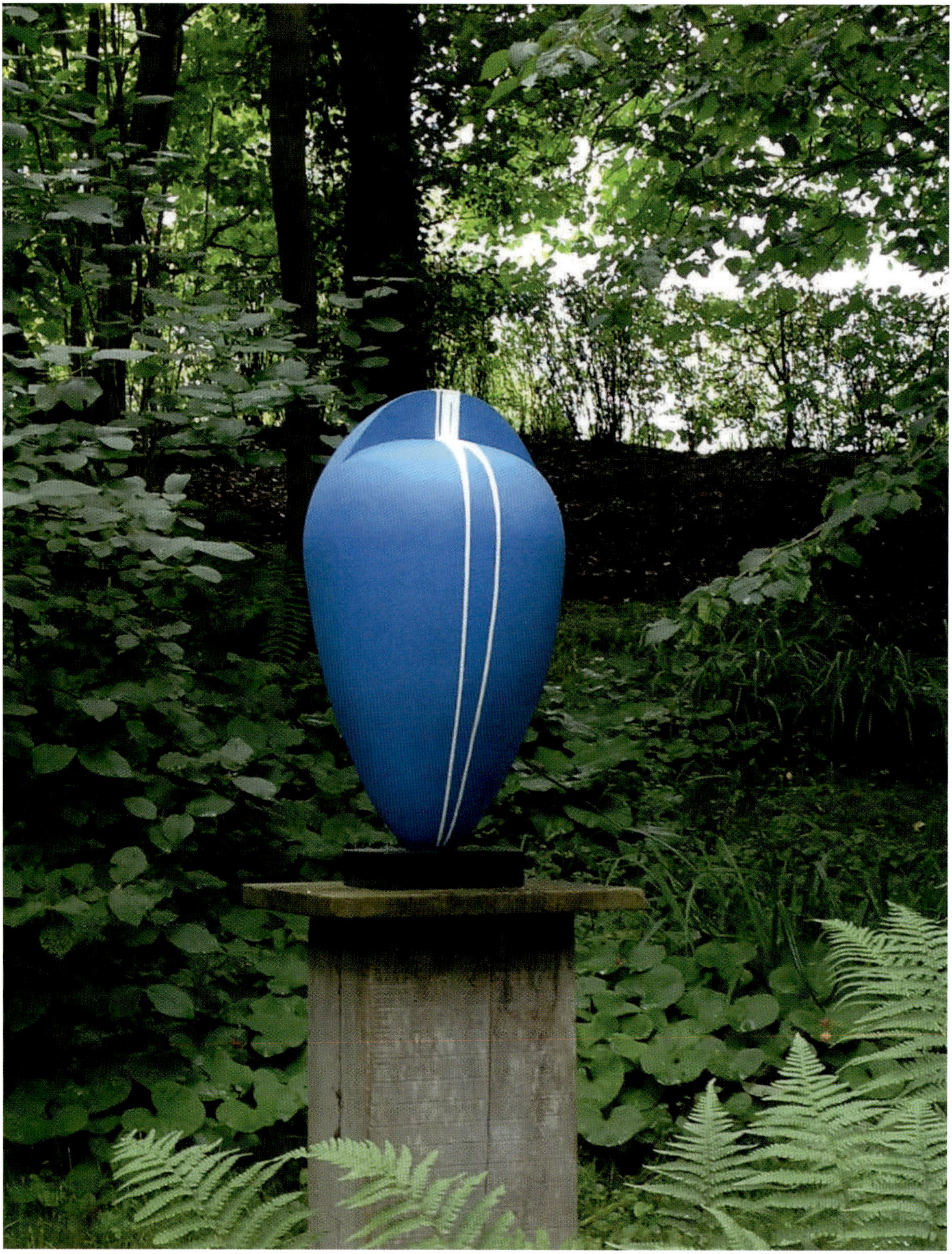

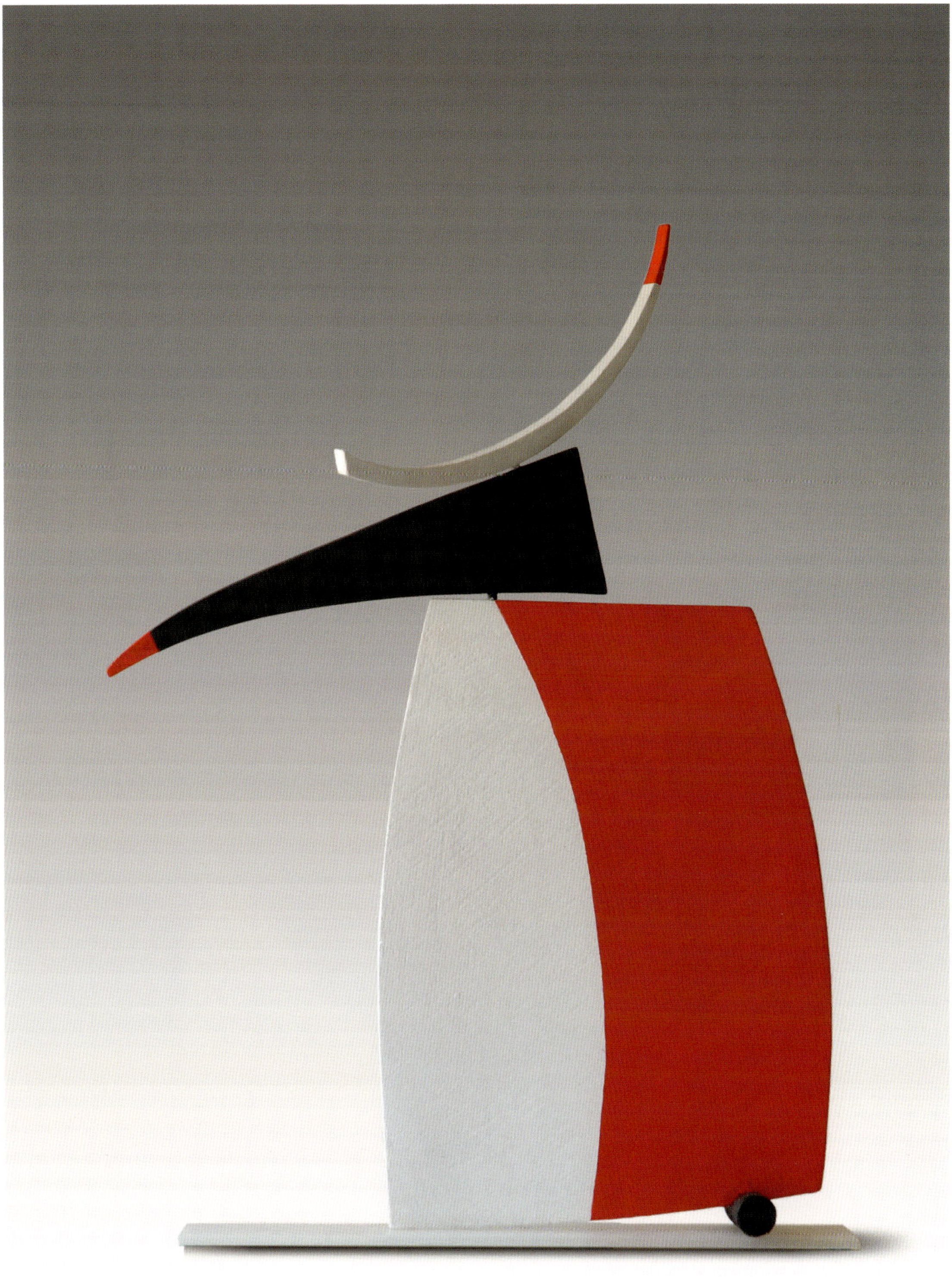

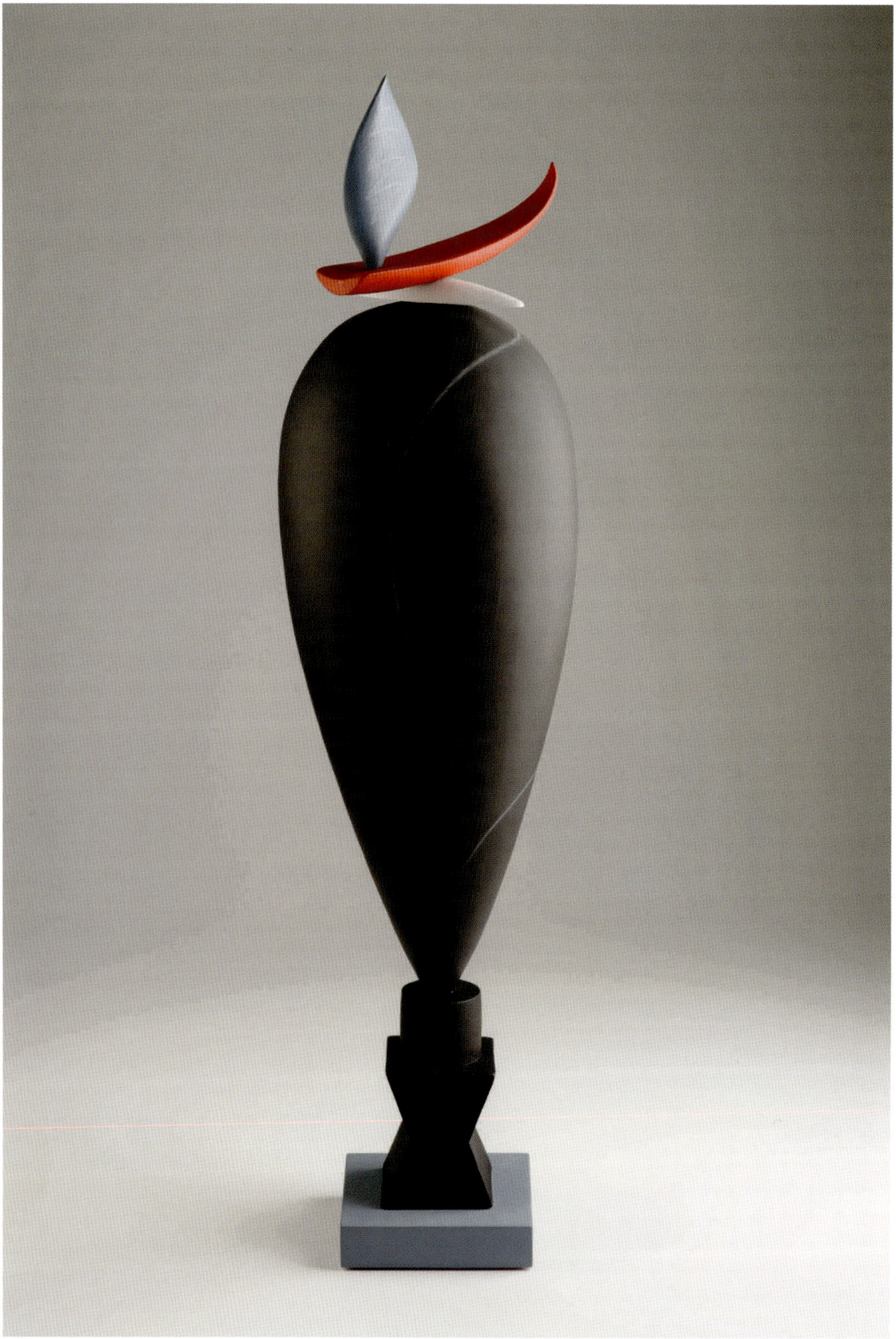

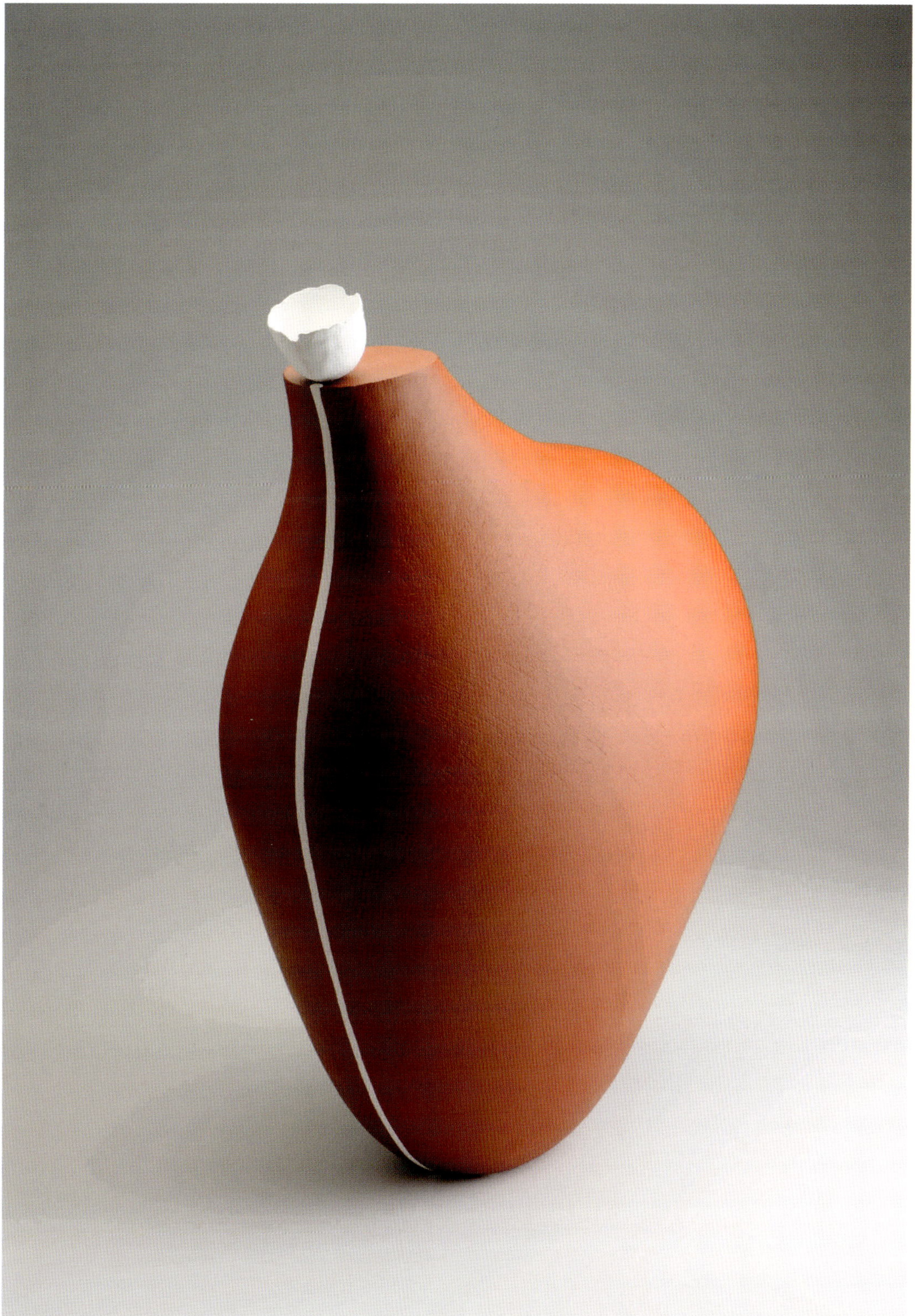

SCULPTING AND CASTING

MARK WHALEN

CASE STUDYO

JEN DWYER

RICHARD SHAW

MATTHEW GROVES

SUMMERLAND CERAMICS

MARK WHALEN

LOCATION Los Angeles, CA
YEAR OF BIRTH 1982
MATERIALS Hand built and slip cast ceramics
EXPERIENCE 3 years

Mark Whalen is an Australian born, Los Angeles based painter. Using a tightly controlled painting technique, Whalen expresses satirical social narratives in seemingly universal situations. Characters caught in the throes of ambiguous actions, shimmering trees, geometric patterns, architectural anomalies and intricate textural beads of color collide in kaleidoscopic intensity, at once immaculately composed and careening out of control. His use of sly oppositions reflects life's puzzling duality, evoking dark humor through luminous color, chaos through controlled detail and the sublime through the specific. The result is a seductive synesthesia that sheds new light on humanity, which, like his works, is both erotic and oppressive, vibrant and reclusive, beautiful and uncanny.

Photos courtesy of Carlos Gonzalez, Bobby Alderman, Julian Berman and the artist. Special thanks to Tyler Durtschi.

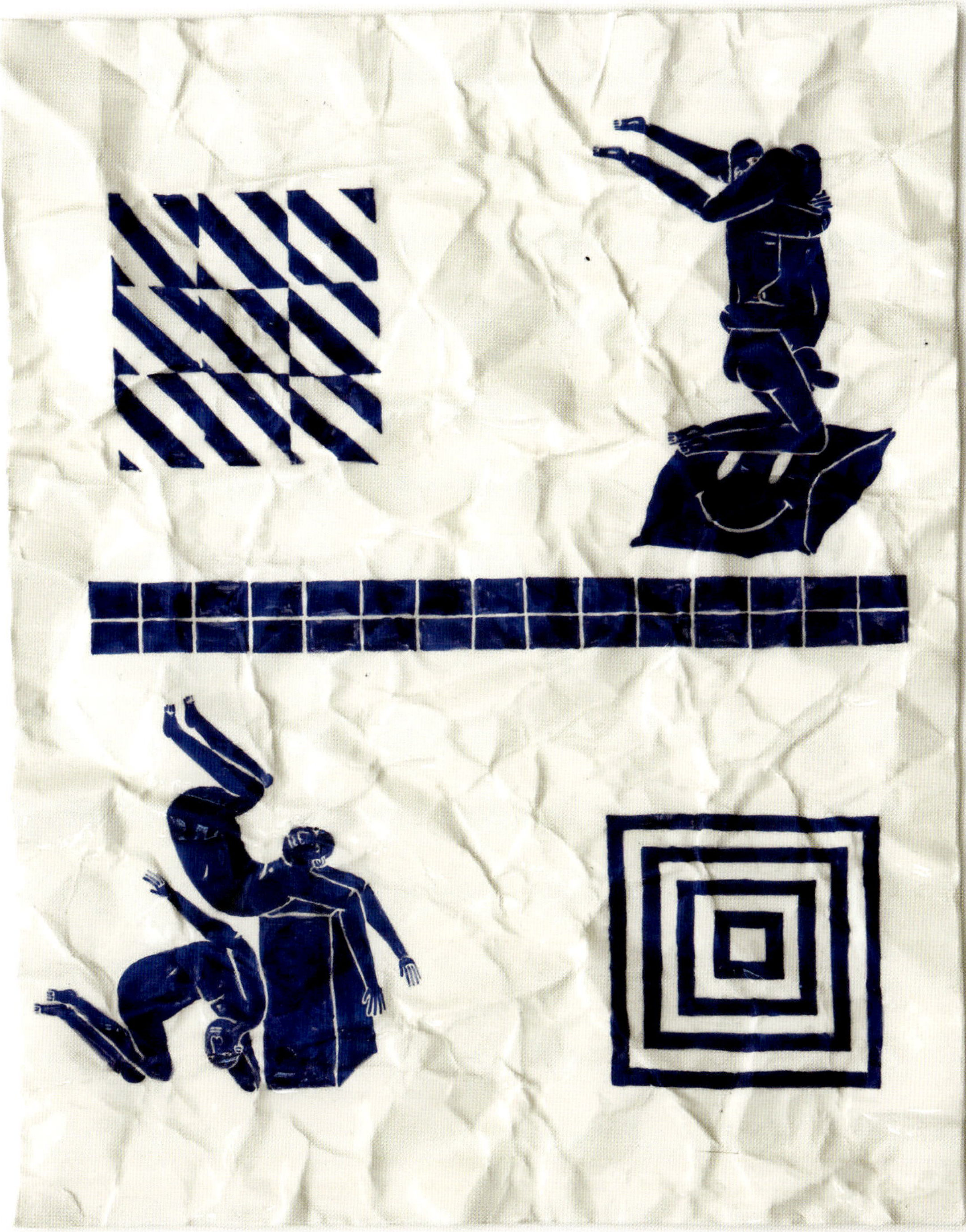

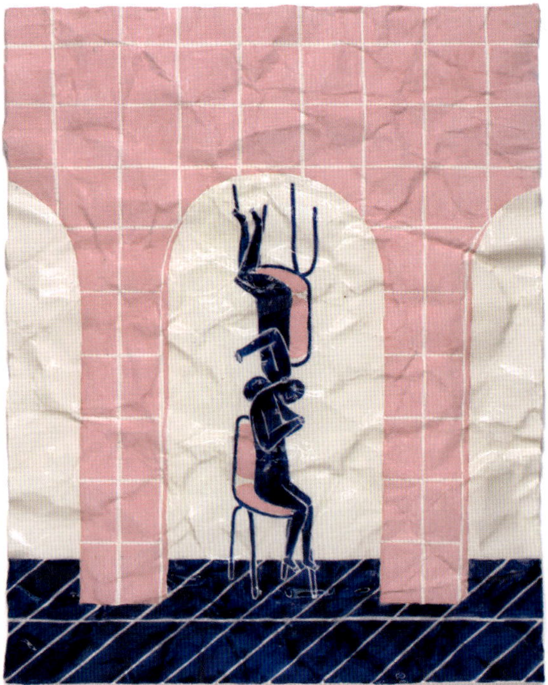
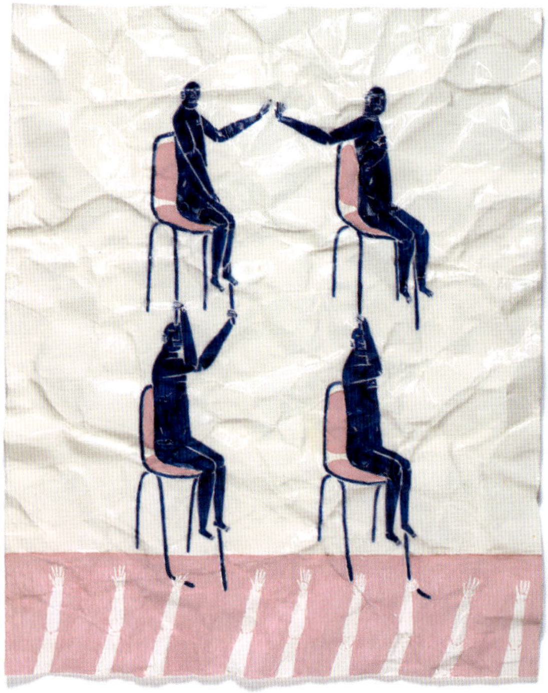
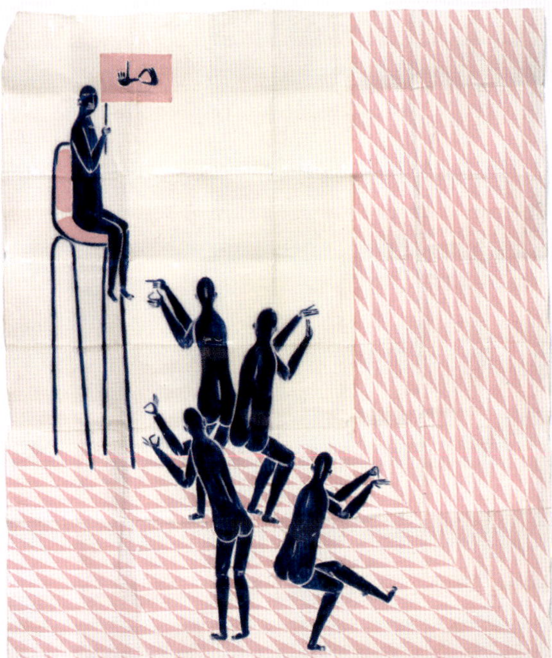
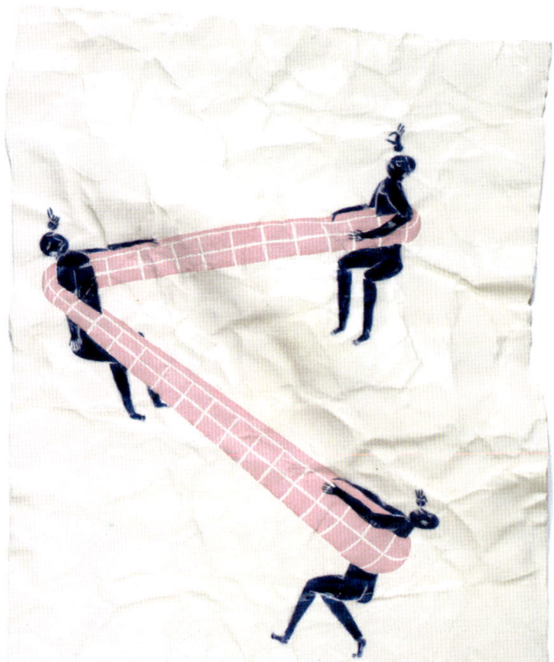

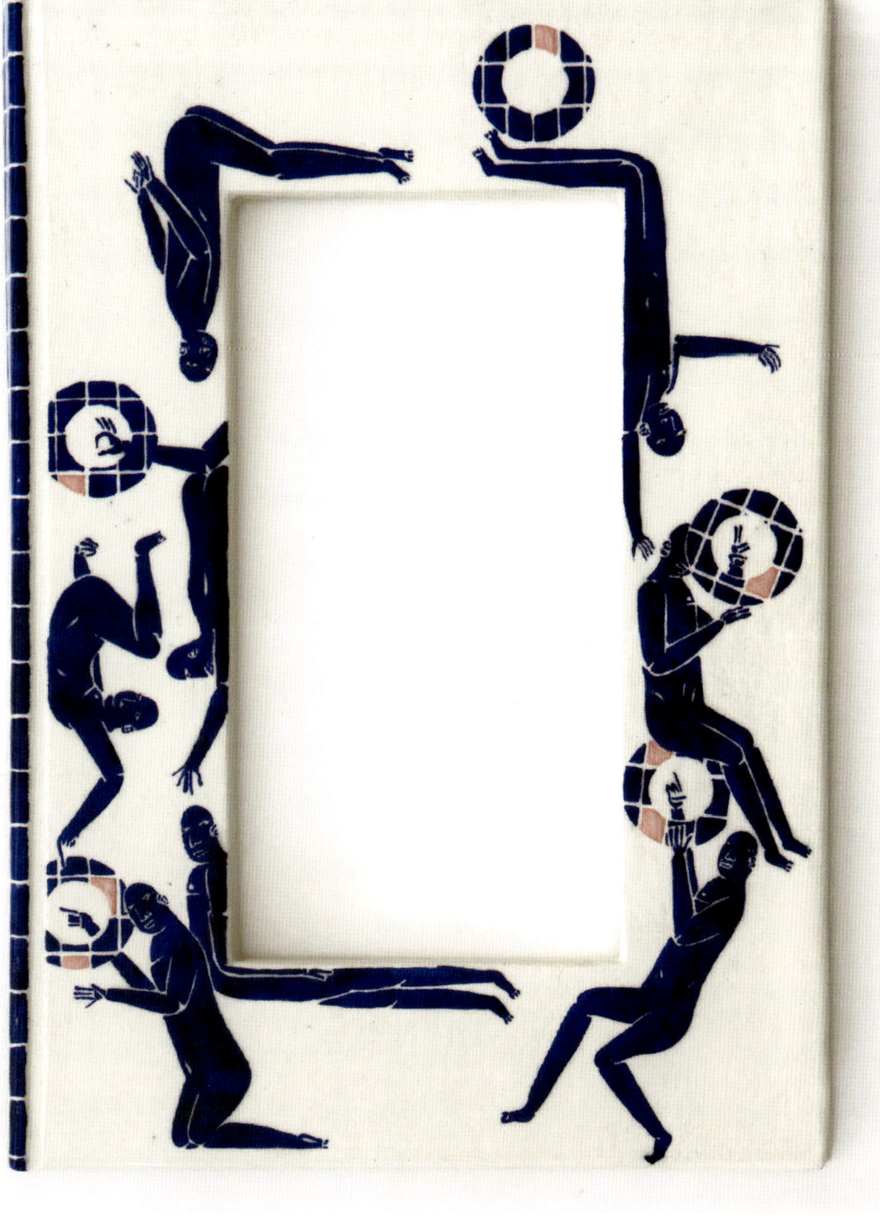

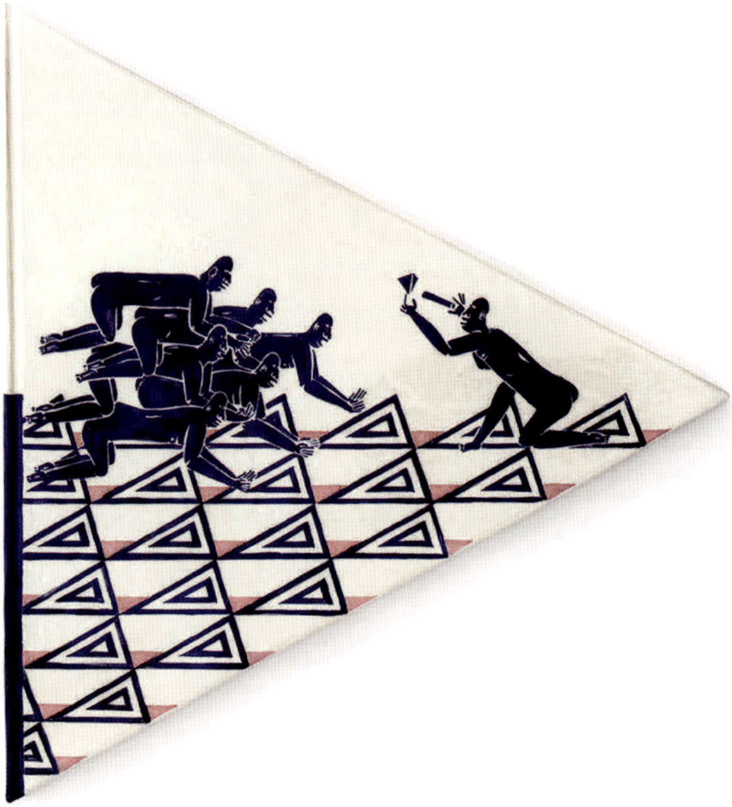

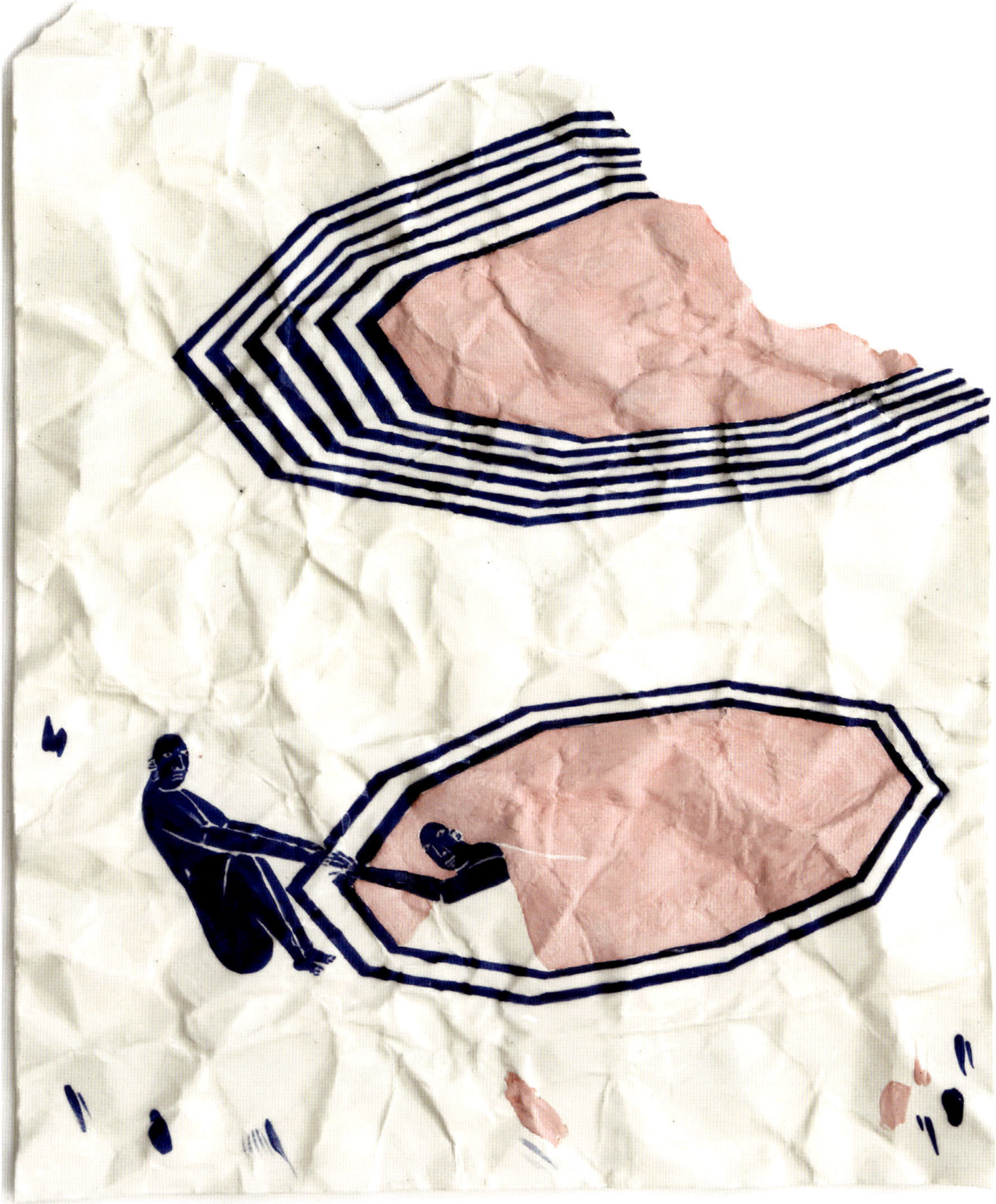

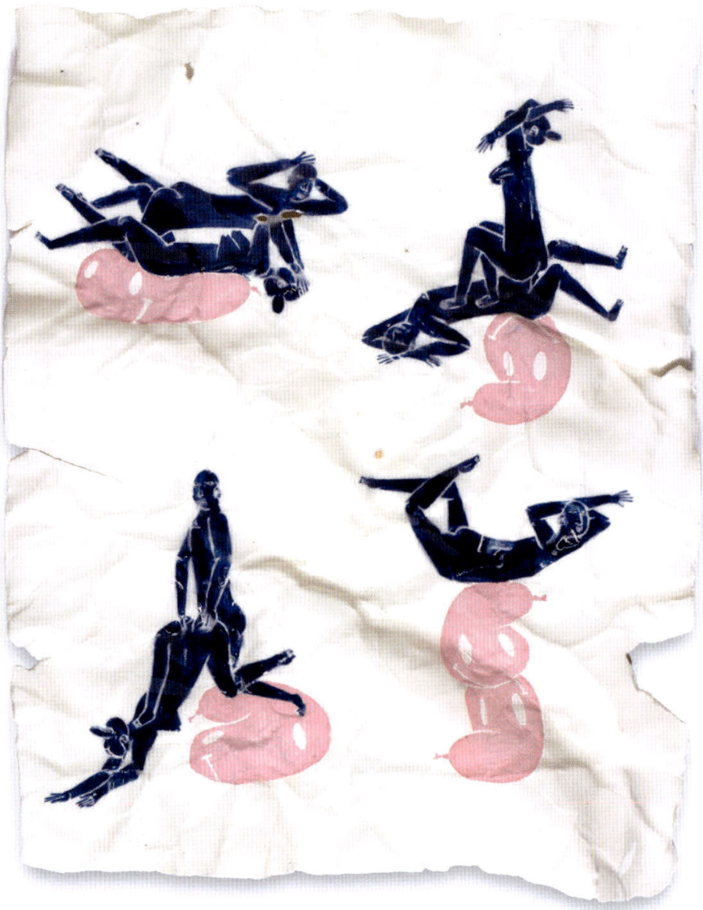

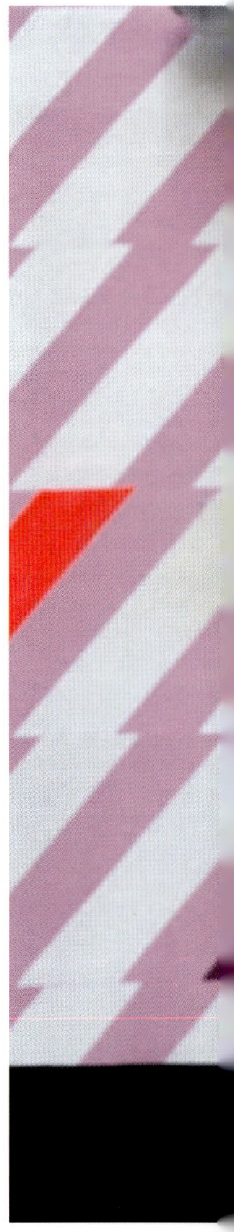

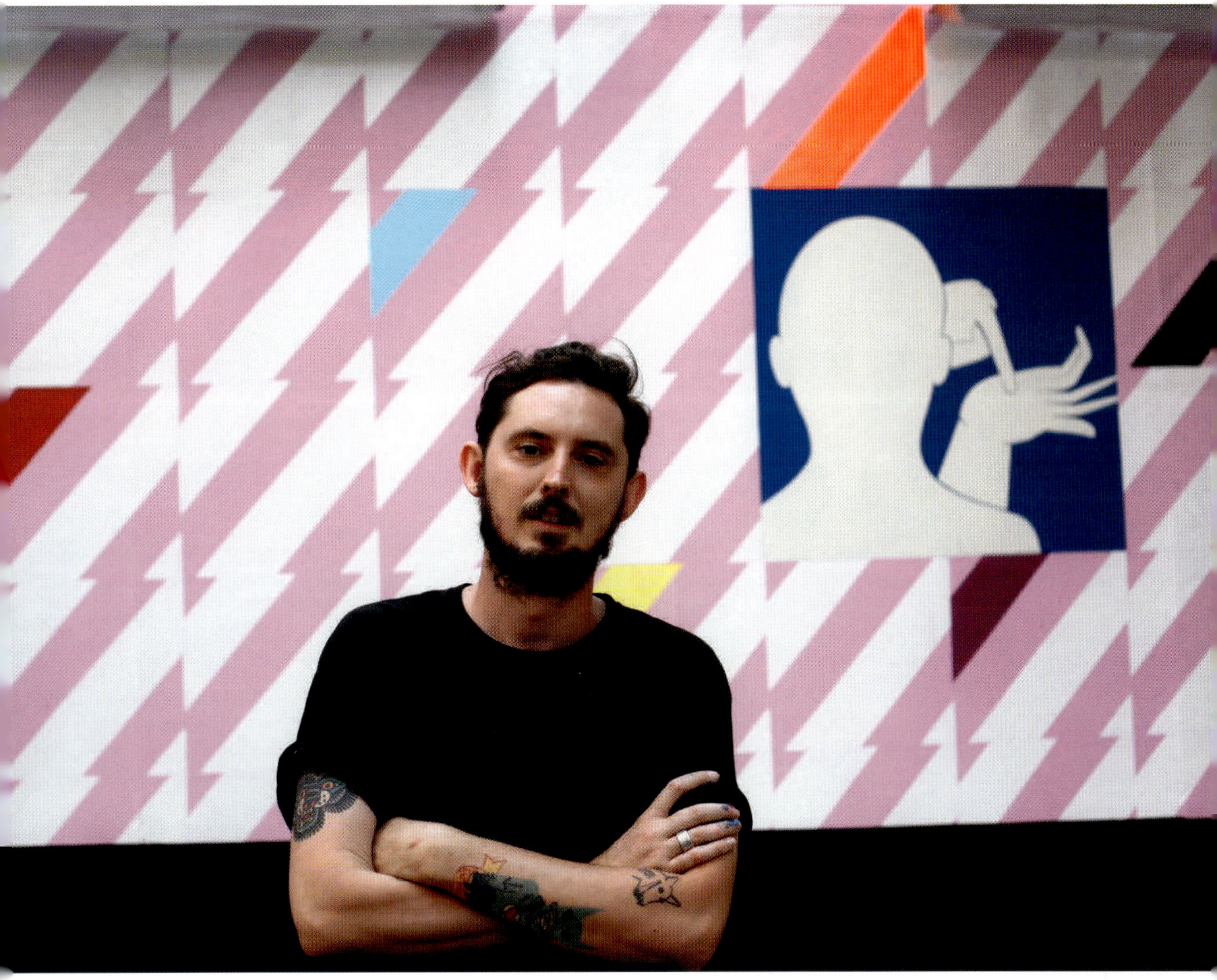

CASE STUDYO

LOCATION Gent, Belgium
YEAR OF BIRTH 1971
MATERIALS Ceramic Porcelain
EXPERIENCE 8 Years

Case Studyo is inspired by the changing boundaries of the art world, different forms and places of creative expression, and the abolishment of predefined rules. We do not limit ourselves to a certain style or visual language, we let our eyes and feelings go towards artists we consider to be important within the interesting and changing times we live in. Our collaborations result in fine art works, furniture, or other functional objects. We strive to push the new boundaries of art a bit further with our projects, to grab the art of today and share it through our limited series.

The CASE STUDYO artworks are the result of a unique collaboration between our studio and the artist. We choose not simply to repeat or reproduce an existing artwork. We opt for a new creation, where interaction between CASE STUDYO and the artist creates its own story — a case study — resulting in a unique object, edited in several pieces.

Photos courtesy of Case Studyo Opposite Page by Parra

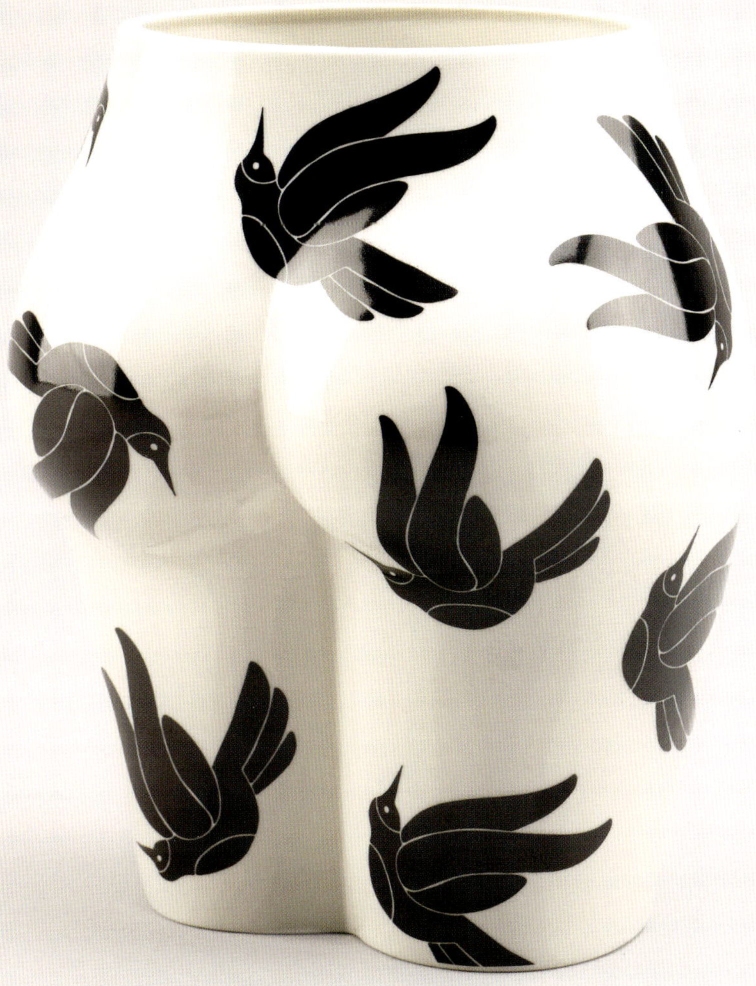

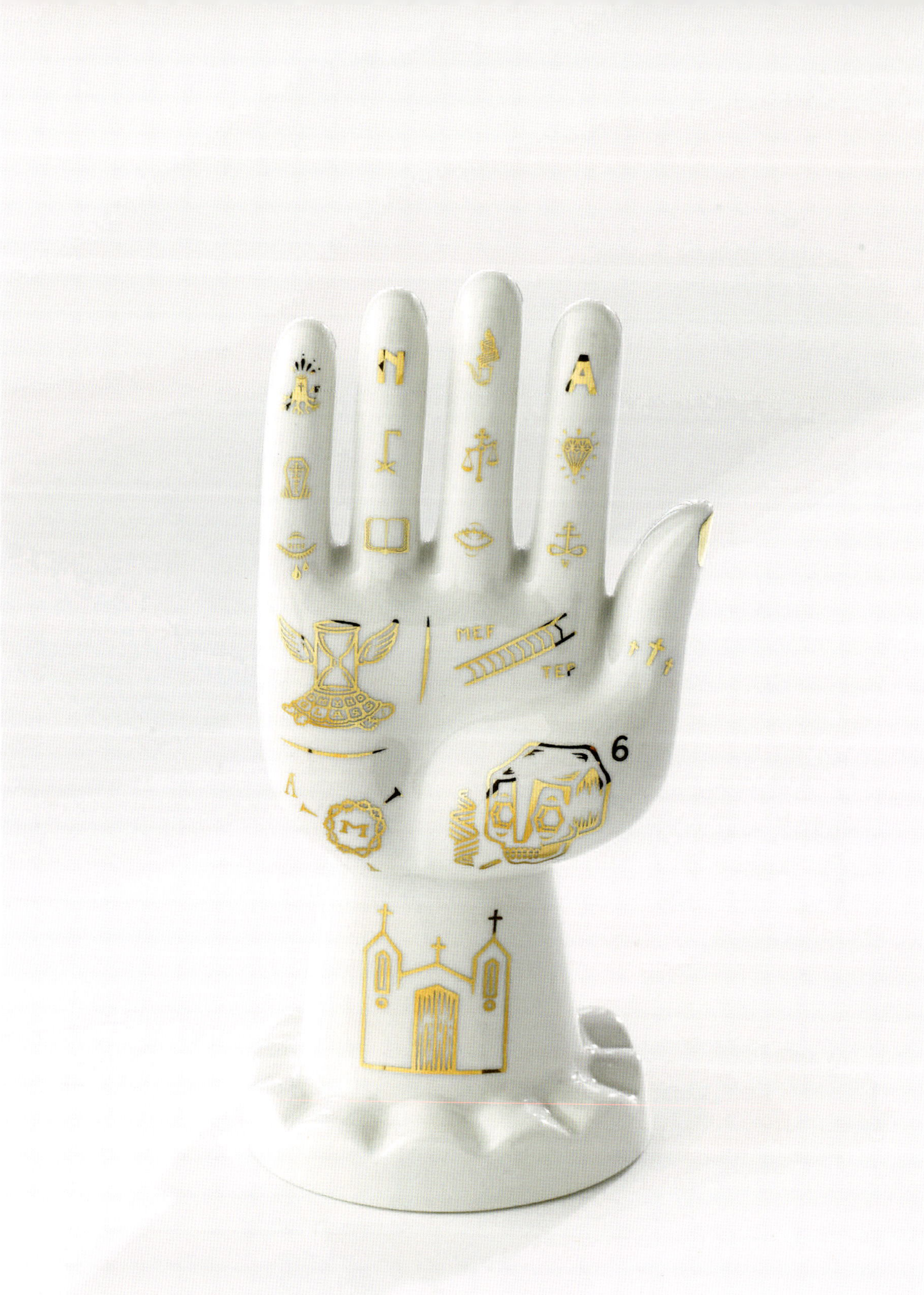

This page Stephan Doitschinoff   Opposite page Cody Hudson

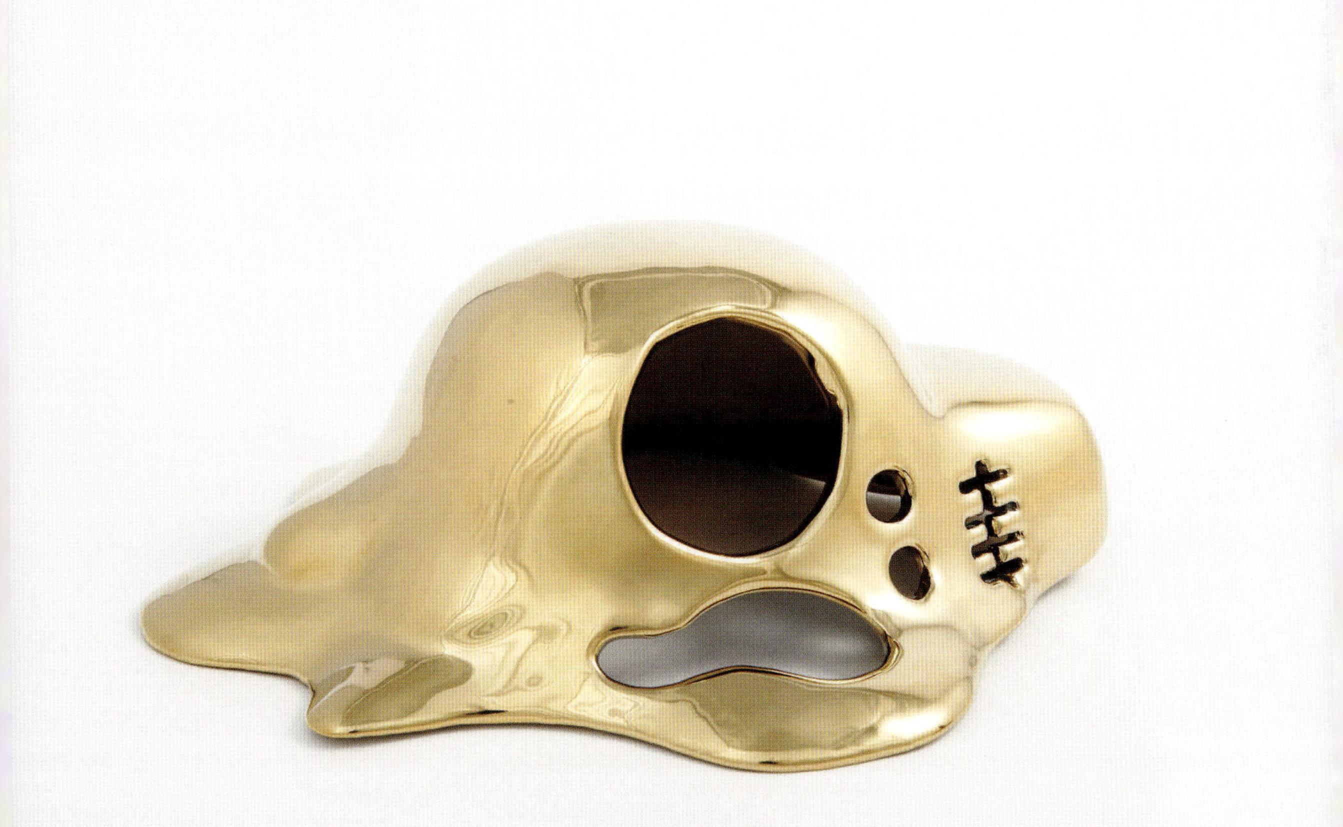

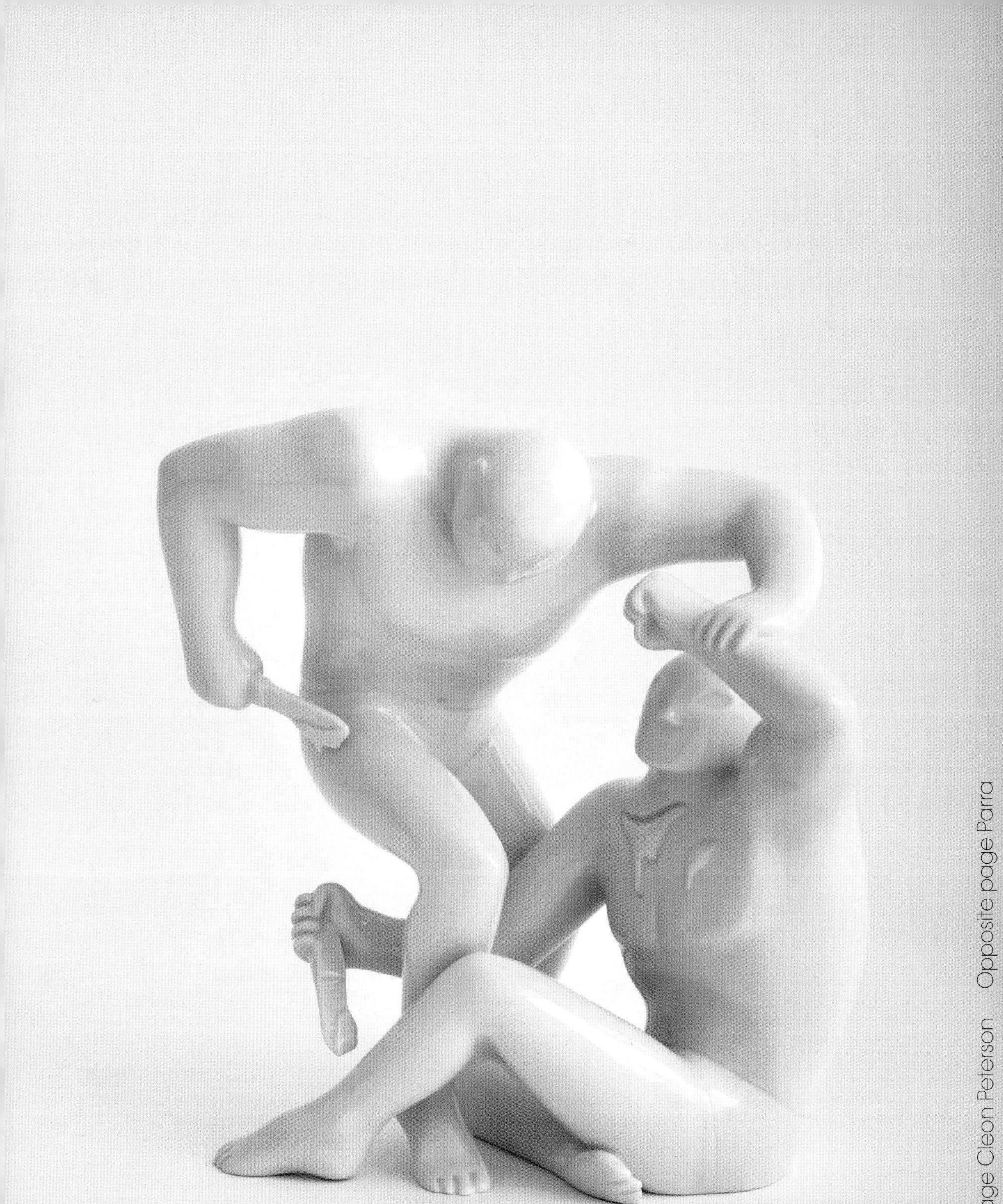

This page Cleon Peterson Opposite page Parra

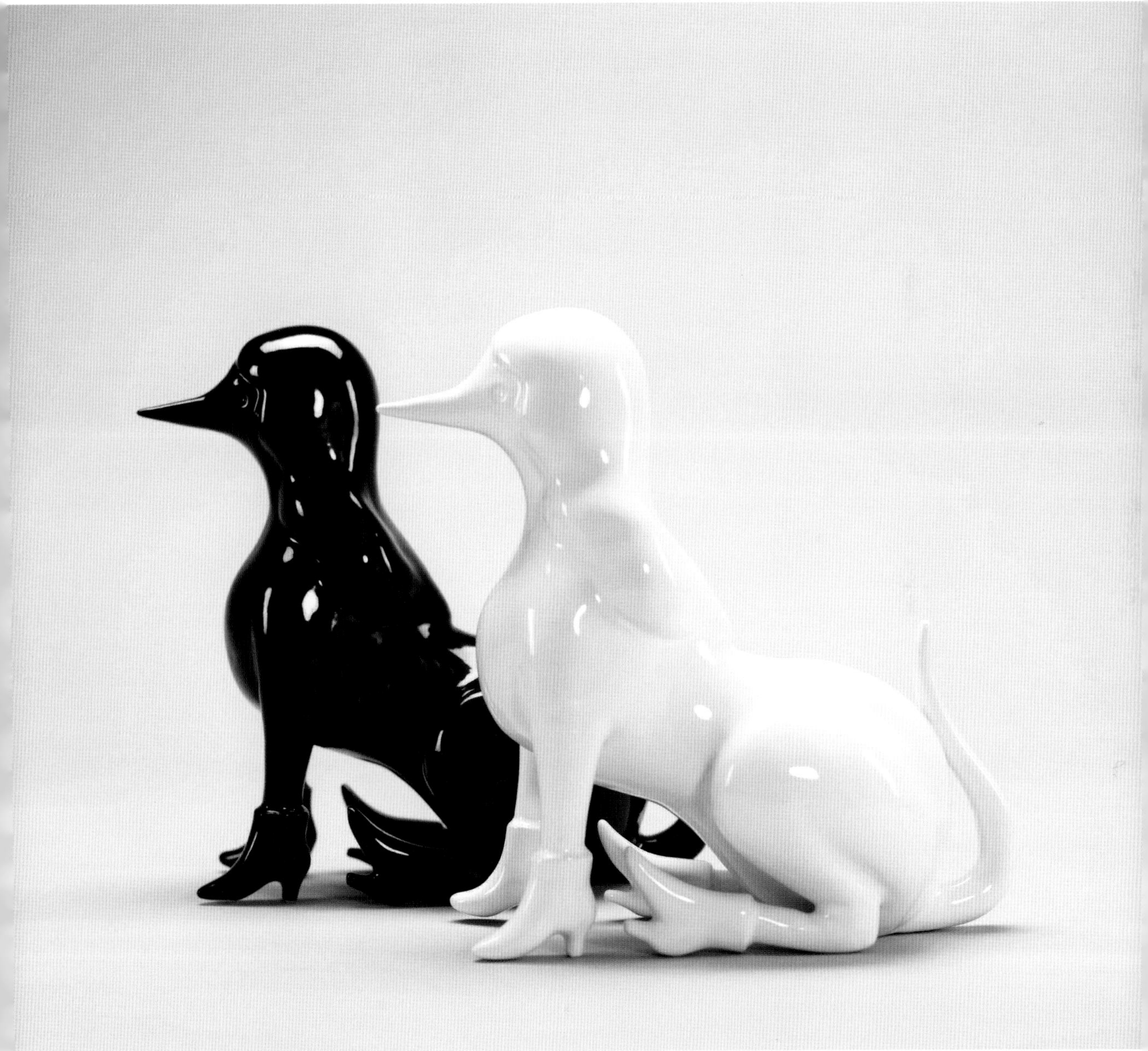

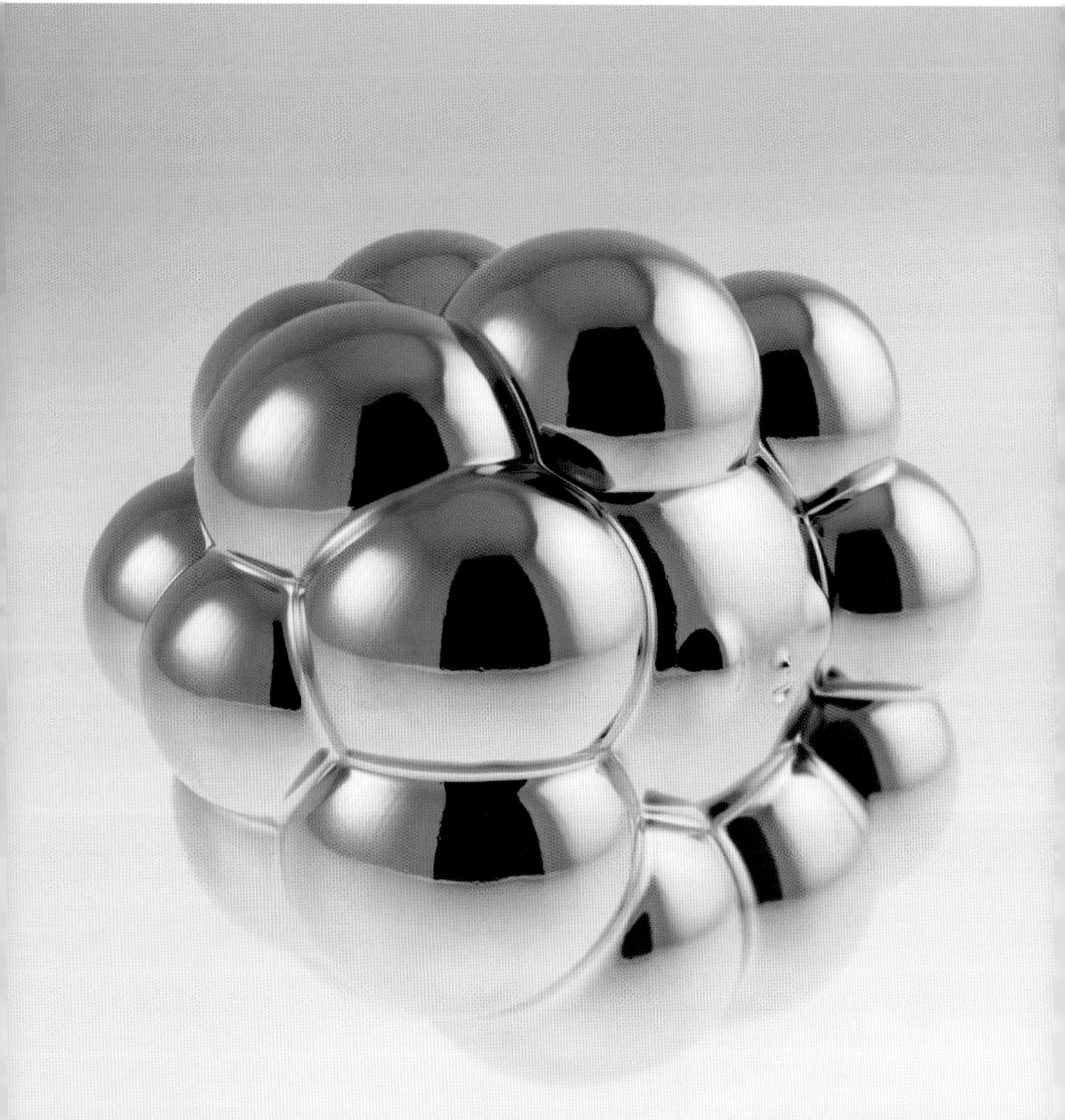

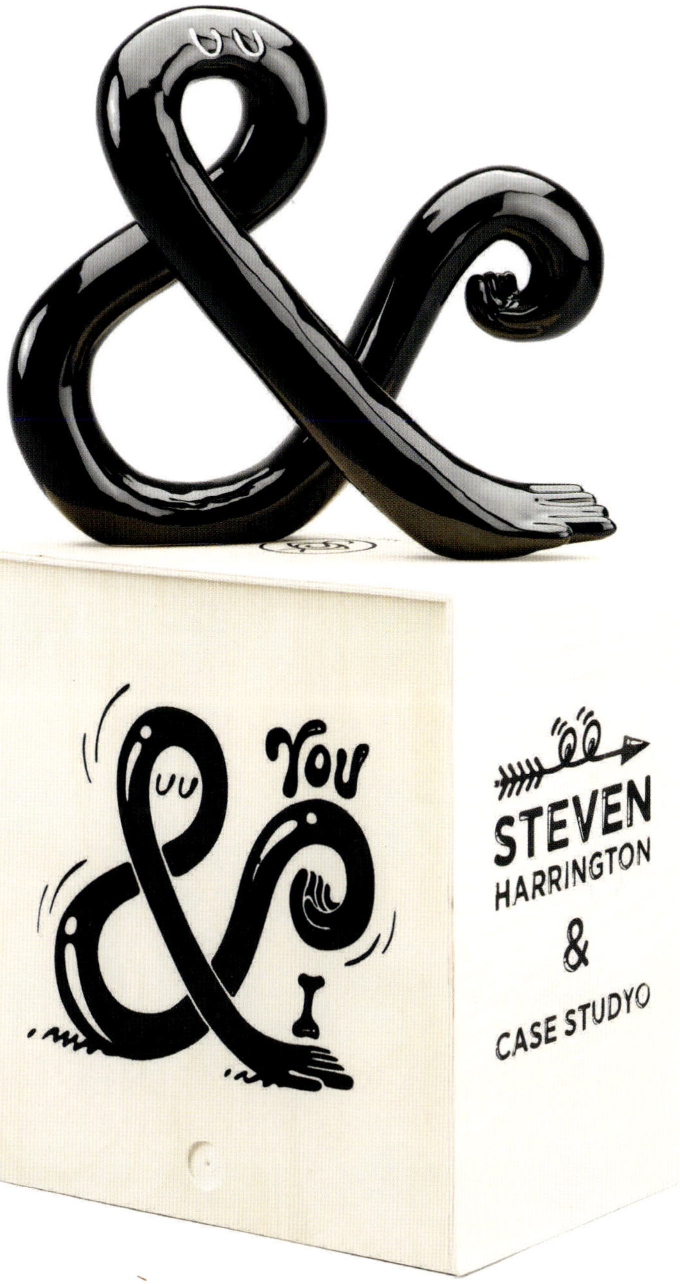

This page Steven Harrington   Opposite page Friends With You

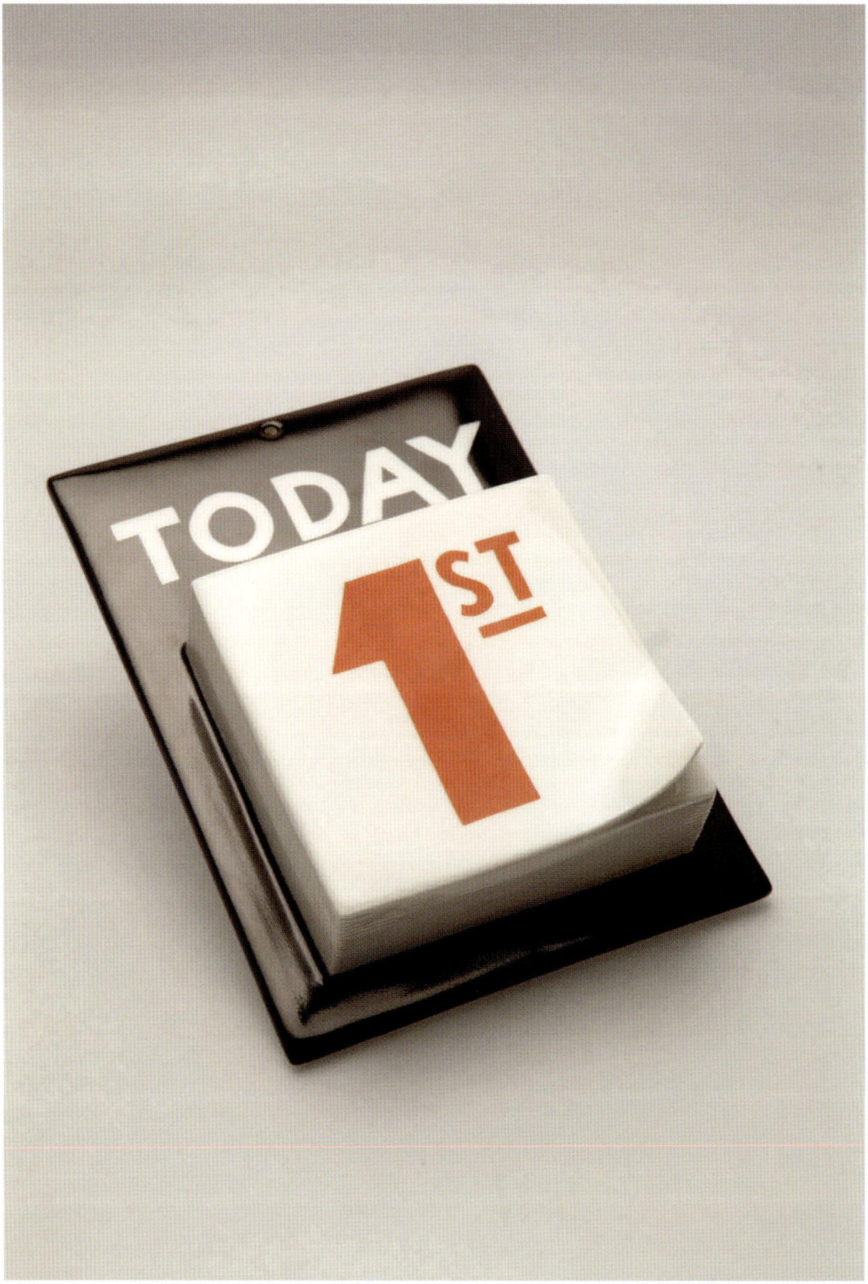

This page Steven Powers Opposite page Todd James

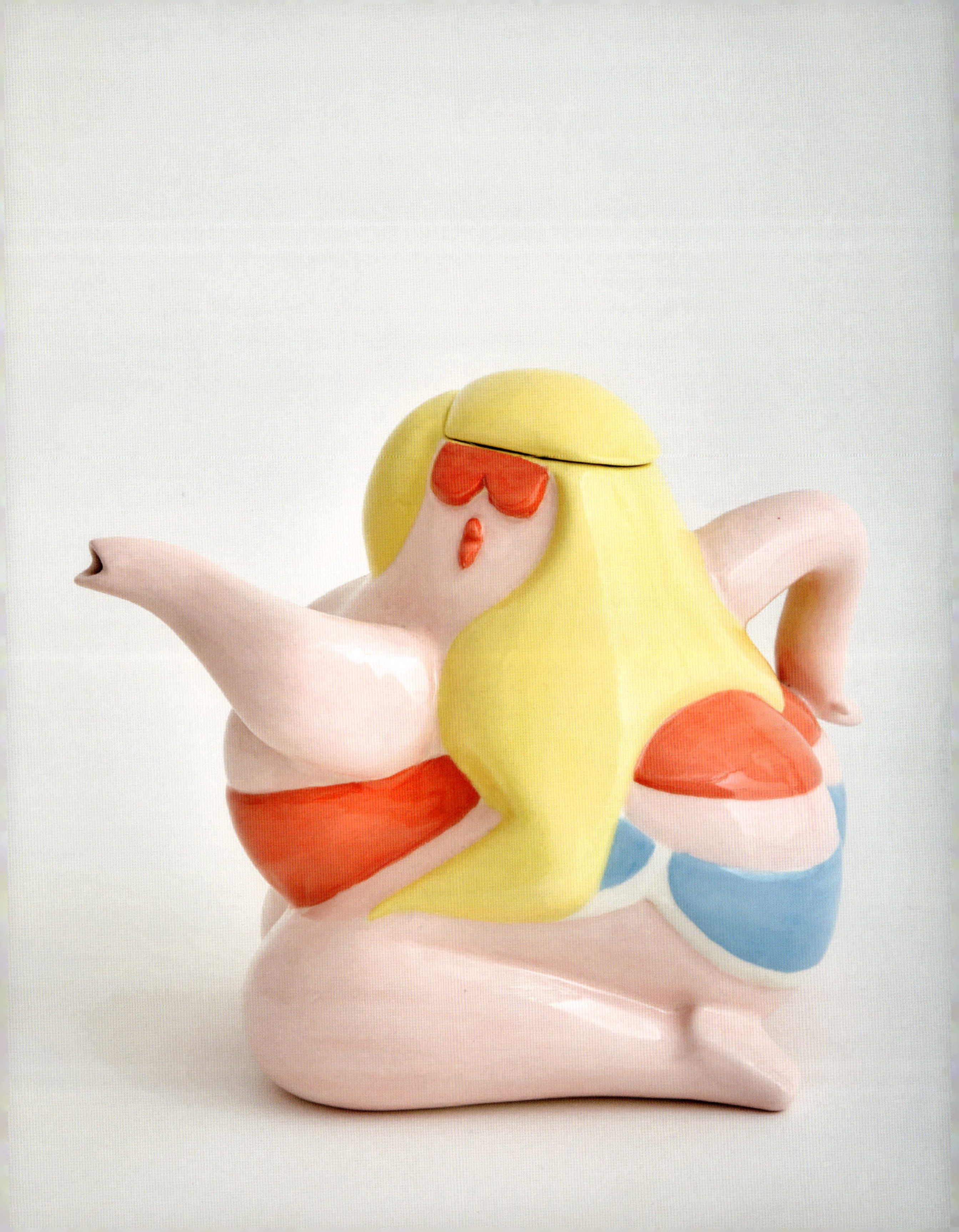

JEN DWYER

LOCATION Brooklyn, NY
YEAR OF BIRTH 1988
MATERIALS Hand built and slipcast ceramics
EXPERIENCE 8 Years

'I fell in love with clay my sophomore year of college when I took a class from my professor, Akio Takamori. After college I spent a few years exploring a variety of mediums including photography, printmaking, and mixed media, but in the end found that clay is the best medium for me to express my voice. Although I use ceramics as the main material in my method of expression, I also incorporate found objects and video into my work. My recent work explores time, anxiety, and preservation. I use various decorative styles, patterns, and symbolic forms to represent my subject matter. I've always been very sensitive to my surroundings and an avid observer, so my work manifests itself as a reaction to my environment. I created Frankenstein-like fusings to investigate human necrosis, and people's phobia of disease and mortality. My work also analyzes how people respond to the fear of their own demise. Our habitual tendency to collect inanimate objects could be seen as our way to fight off the fear of our mortality, human decay, and memory loss, allowing us to leave a trace of ourselves behind.'

Jen Dwyer was born and raised in the San Francisco Bay Area, CA, attended school in Seattle, WA and is currently living and working in Brooklyn, NY.

Photos courtesy of the artist

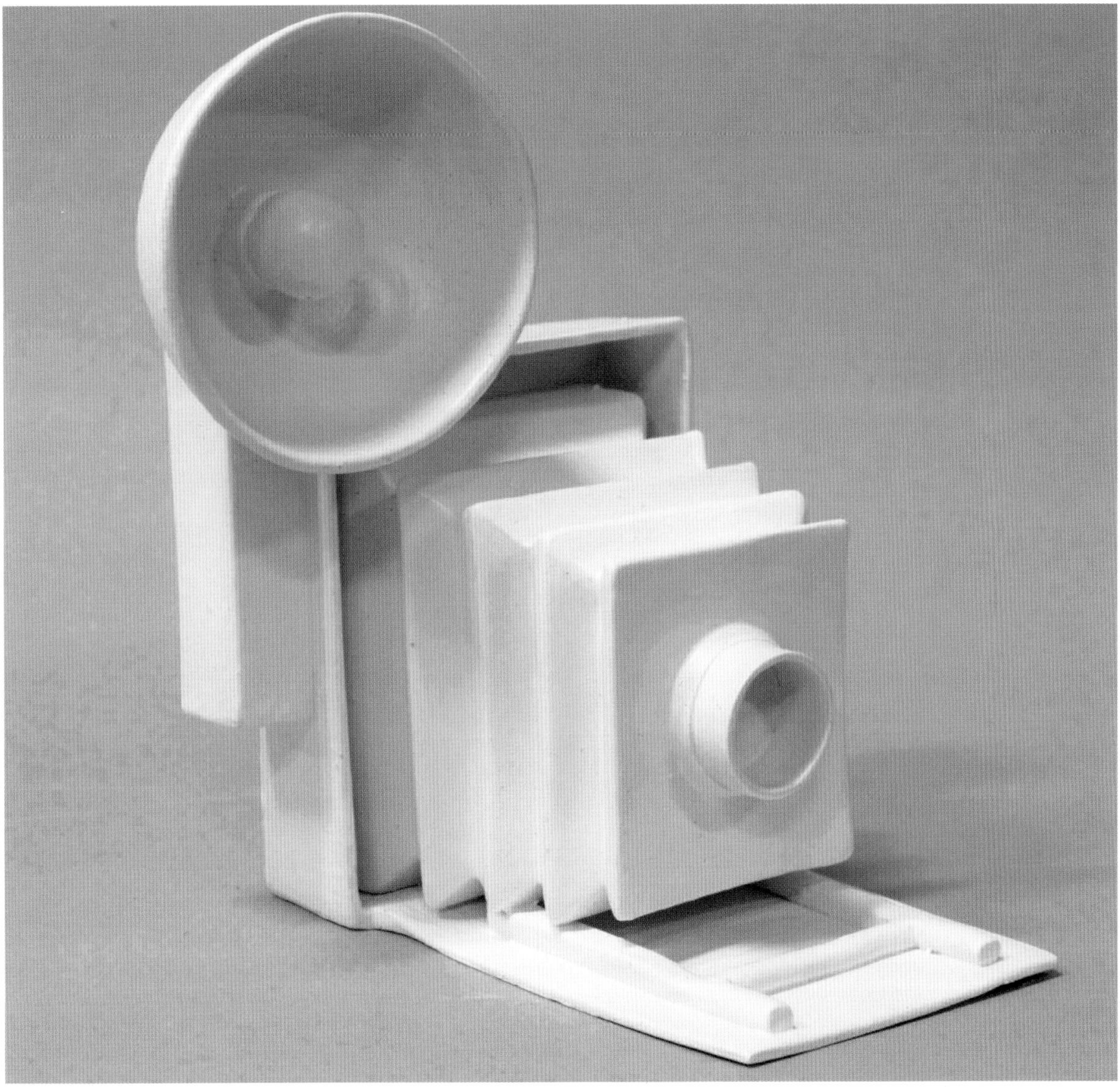

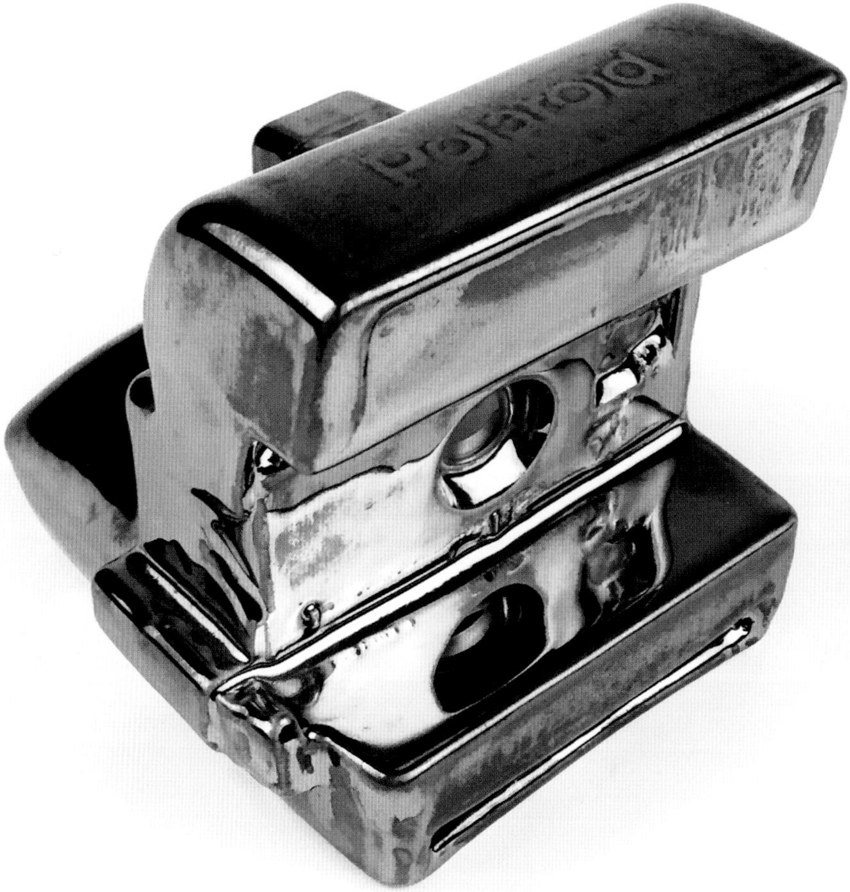
Polaroid

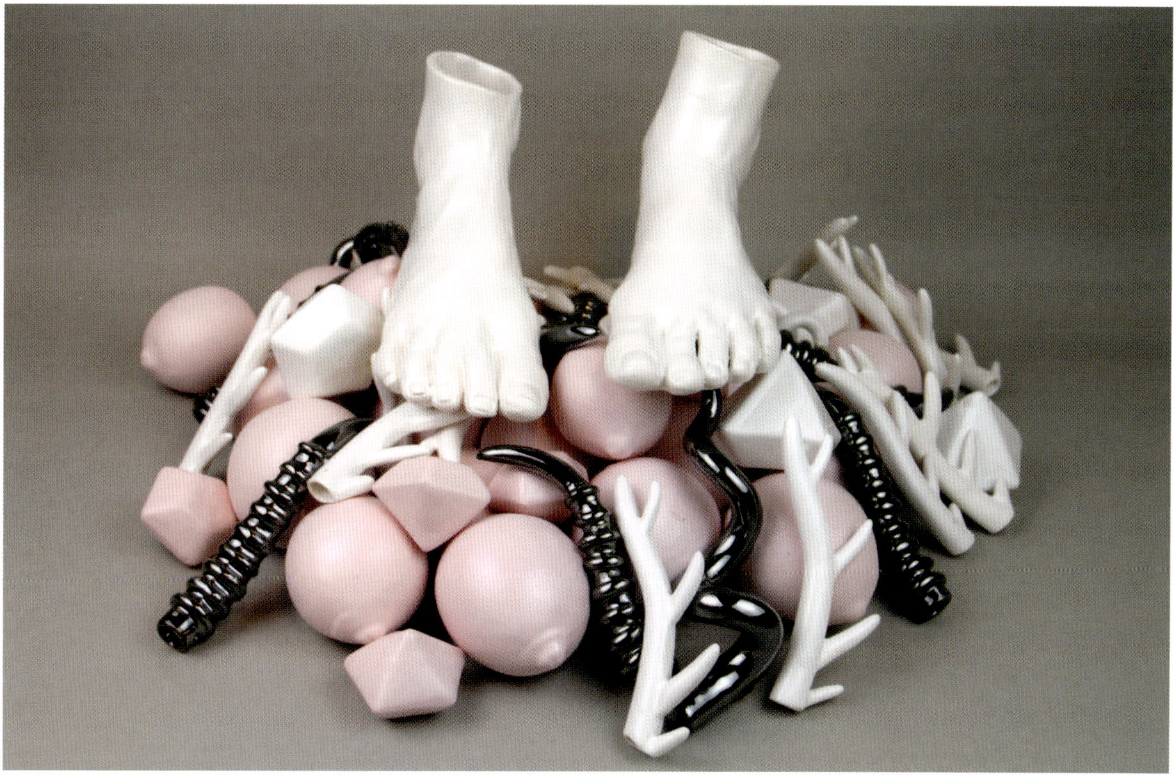

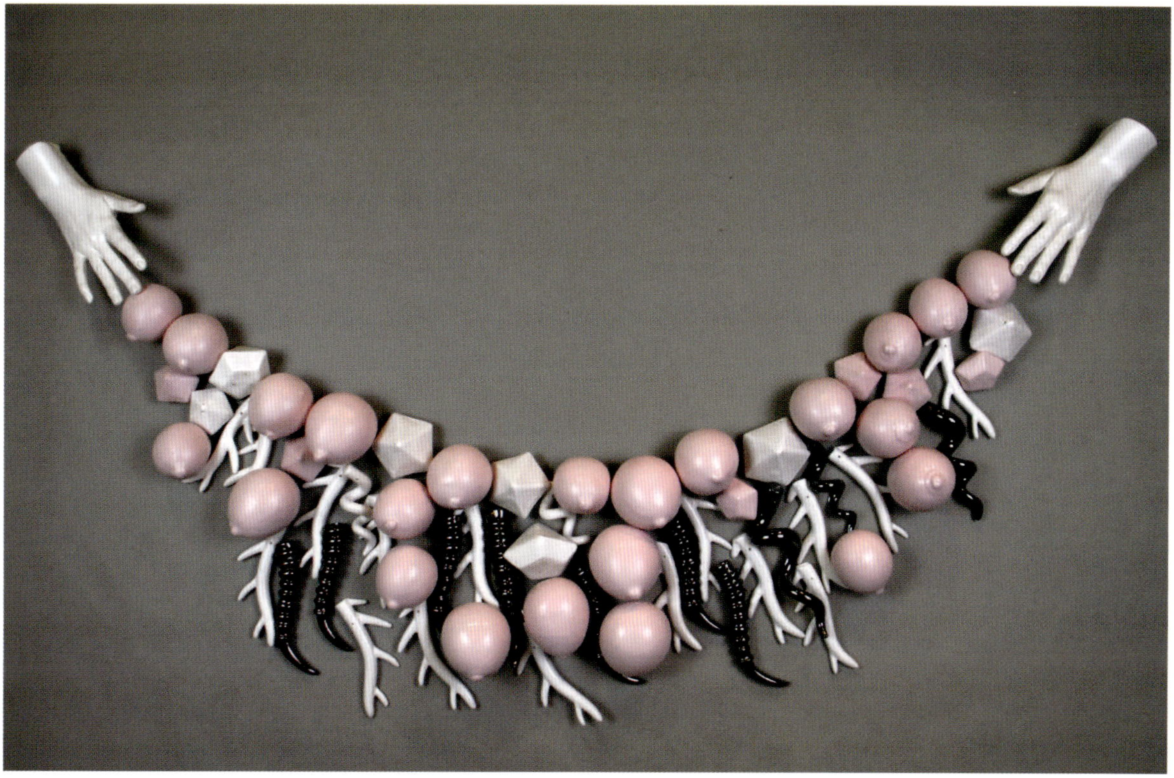

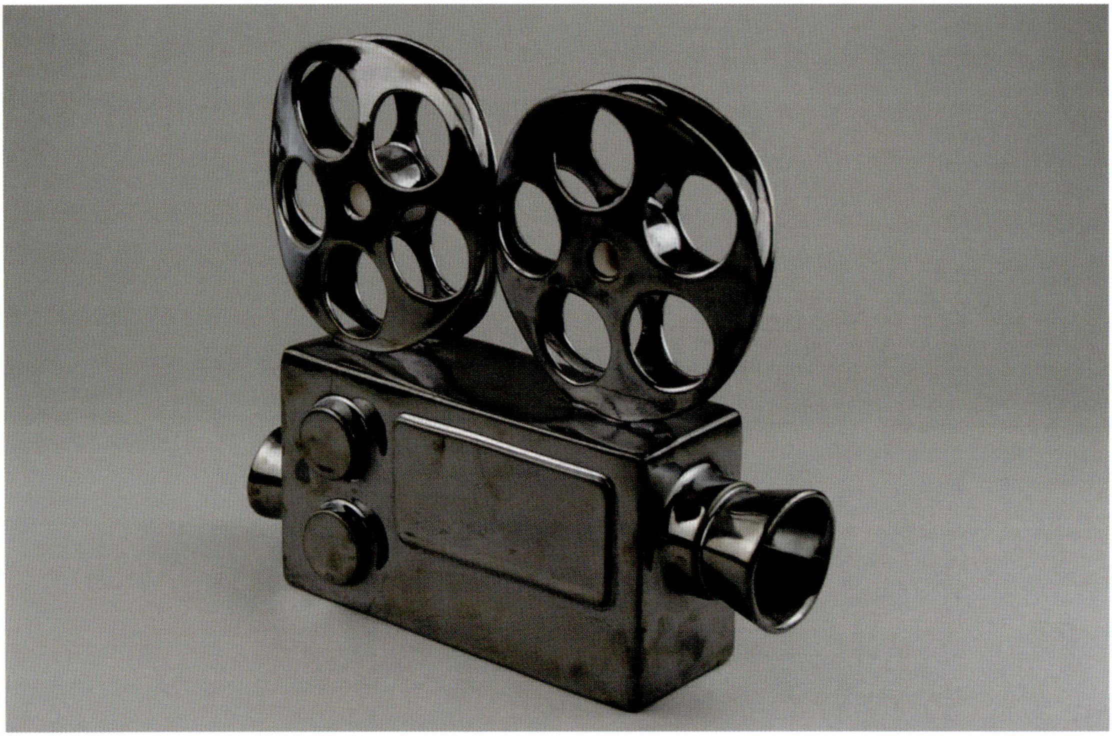

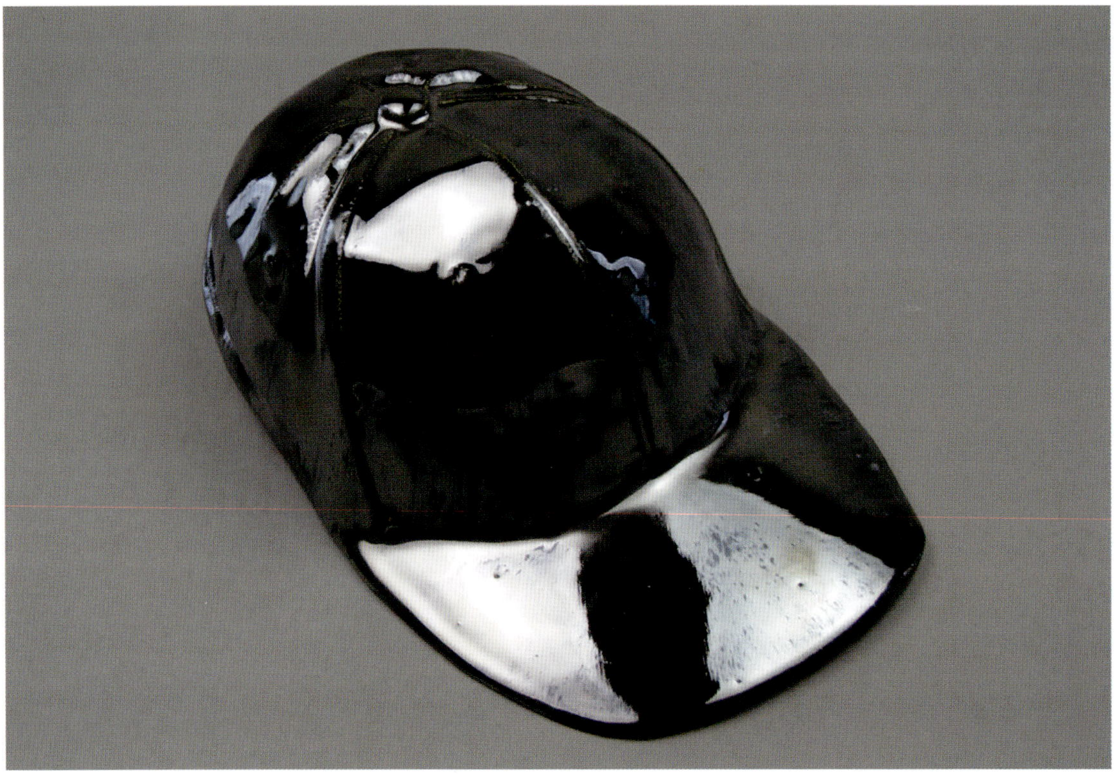

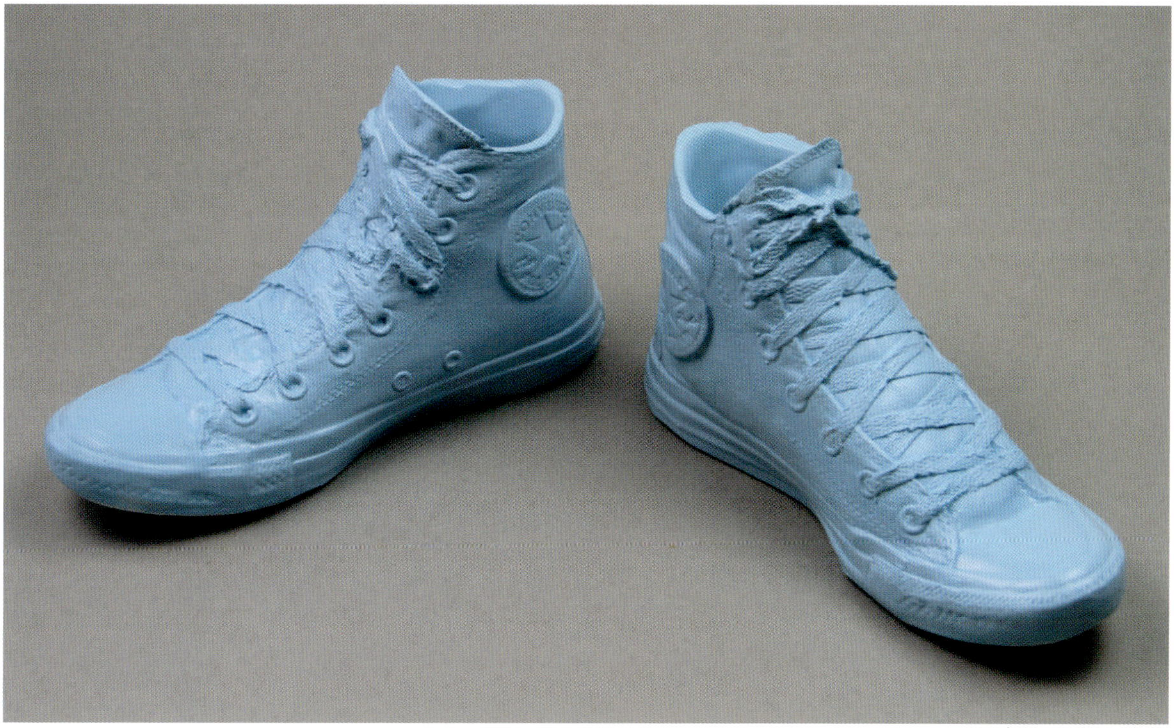

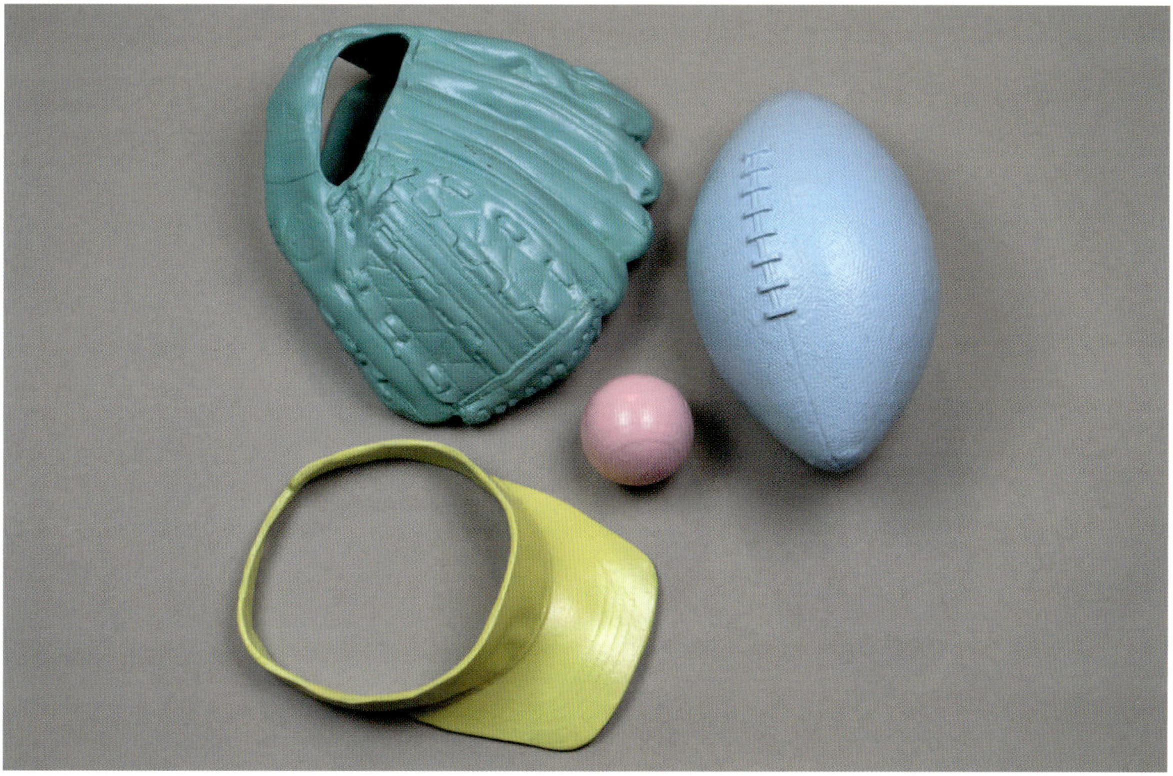

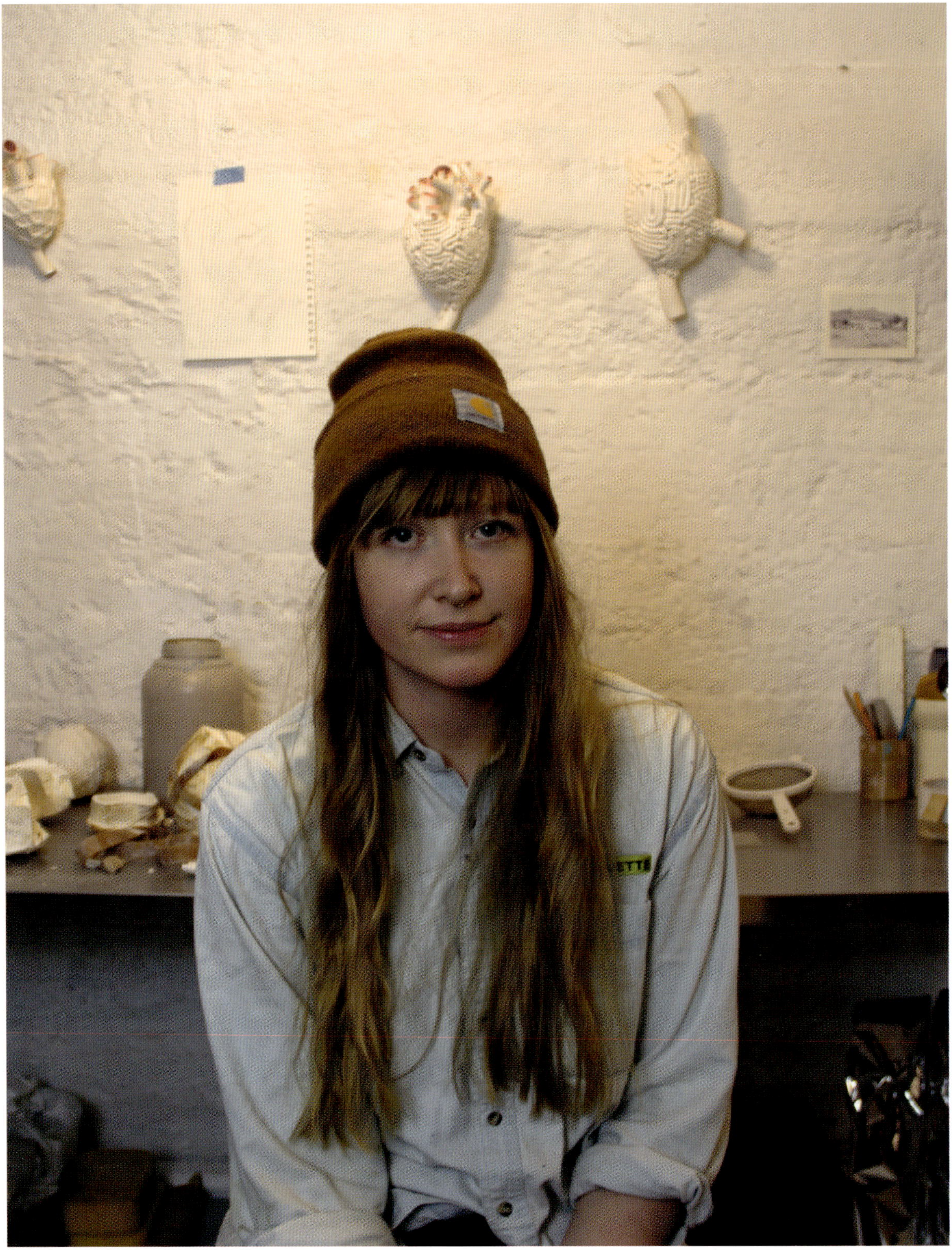

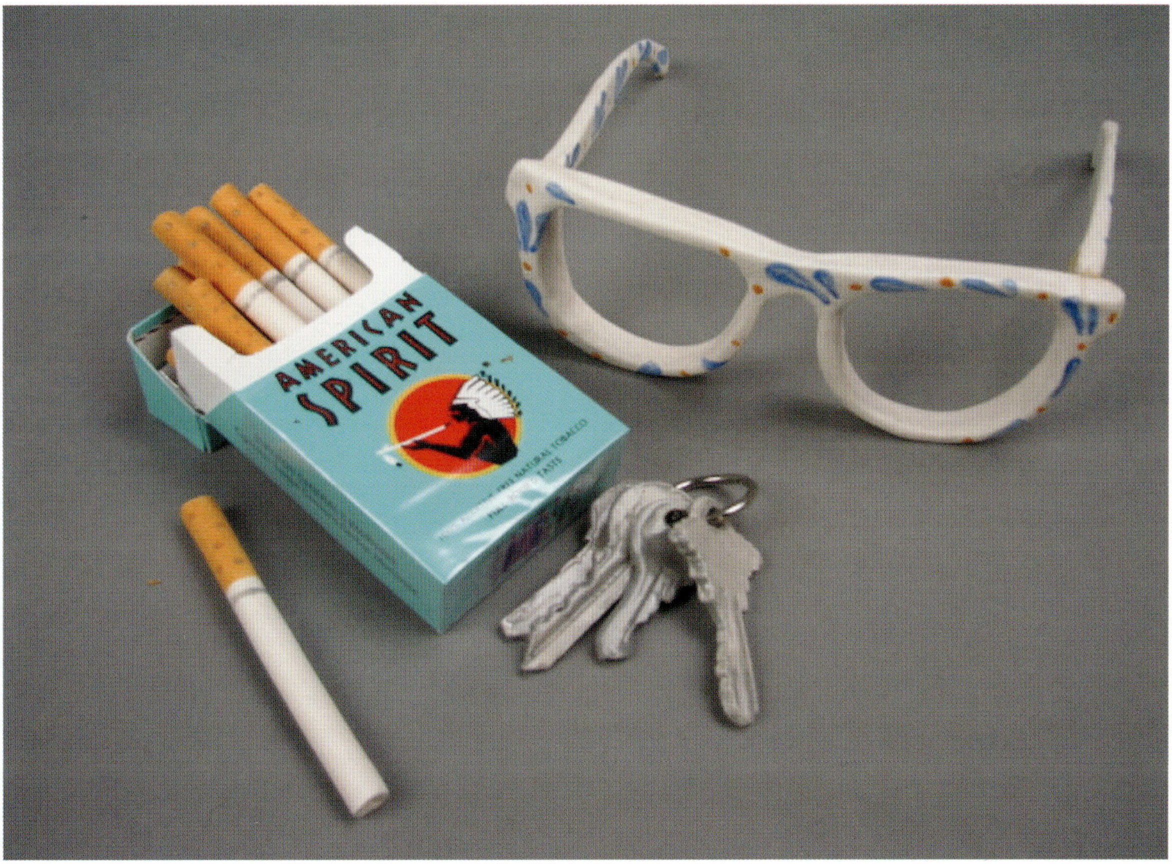
AMERICAN
SPIRIT

RICHARD SHAW

LOCATION London, UK
YEAR OF BIRTH 1941
MATERIALS Slip cast ceramics
EXPERIENCE 50+ years

In the world of contemporary ceramics, Richard Shaw is the master of trompe-l'oeil sculpture. He has developed an astonishing array of techniques, including perfectly-cast porcelain objects and overglaze transfer decals. By combining the commonplace with the whimsical, the humorous with the mundane, Shaw captures the poetic and the surreal with the sensibility of a comedian.

Shaw is one of the most respected and collected artists in contemporary ceramics. He came out of the San Francisco Bay Area art scene in the late 1960s, and he continues to add to his skills and appropriate from mass culture. He has developed a vocabulary of found objects that form intimate still life sculptures, complex figures, and personally referential assemblages. He brings life to the detritus of the studio as a cartoonist animates the page.

Photos courtesy of the artist

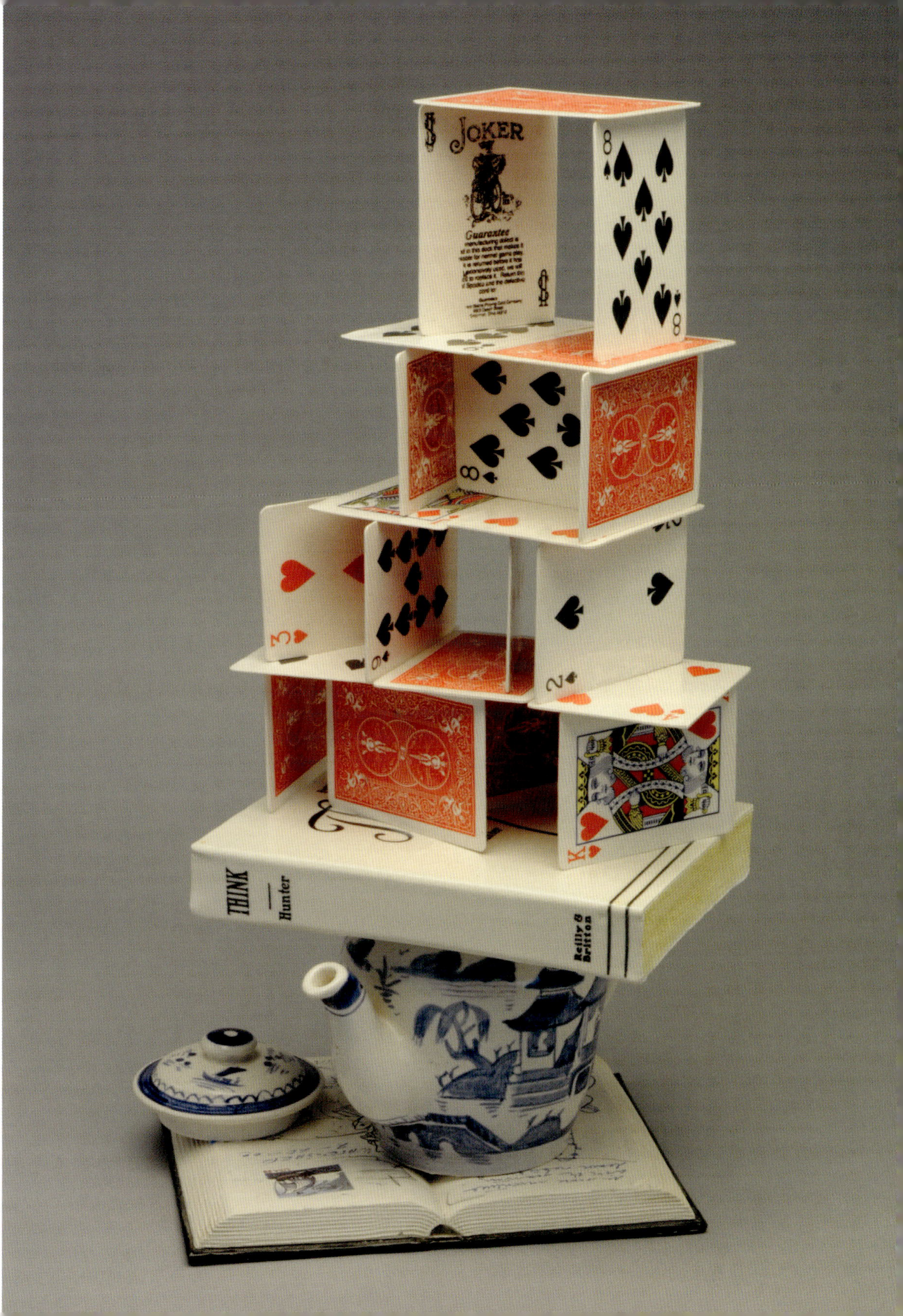
JOKER
Guarantee
THINK
Hunter
Reilly & Britton

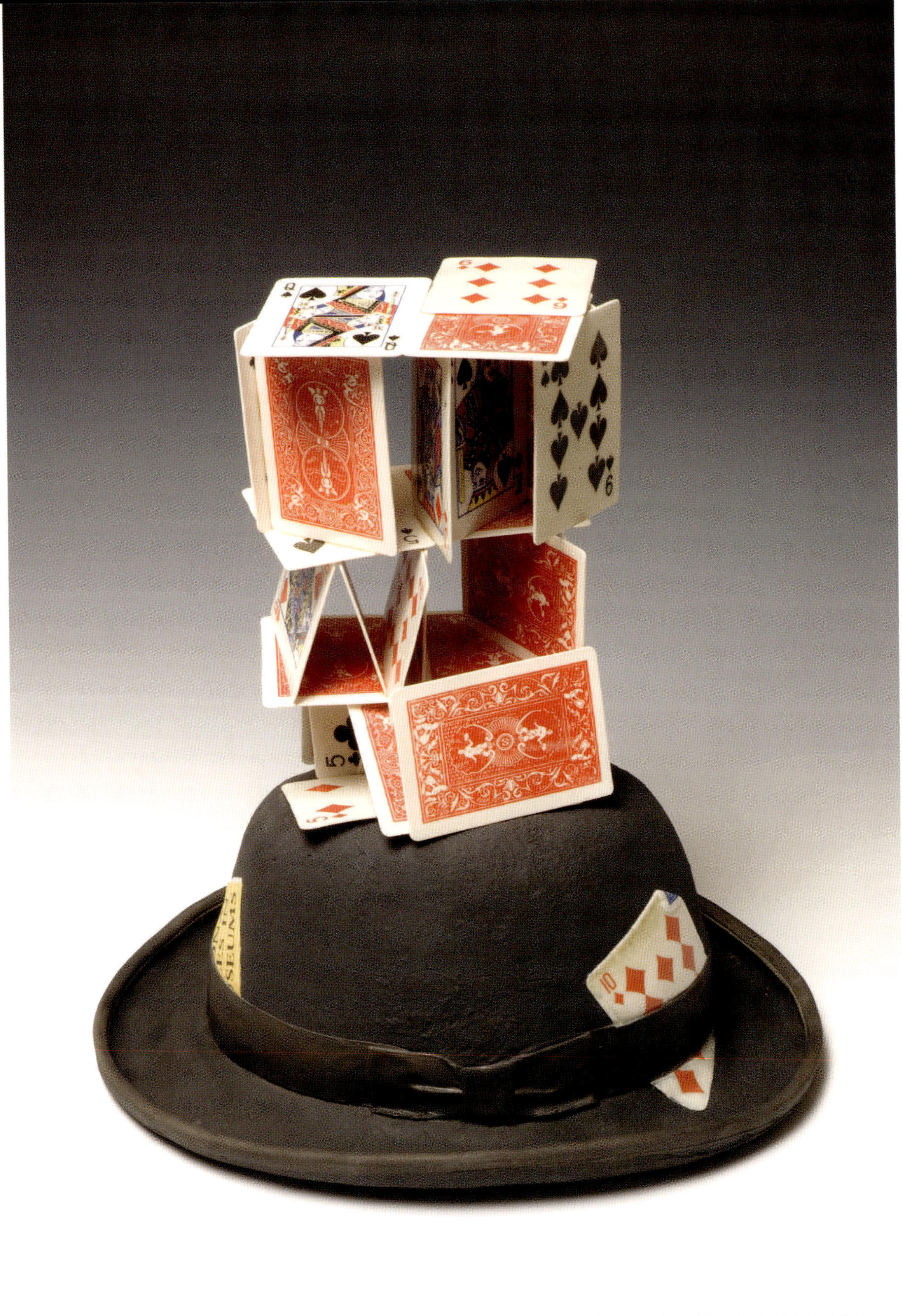

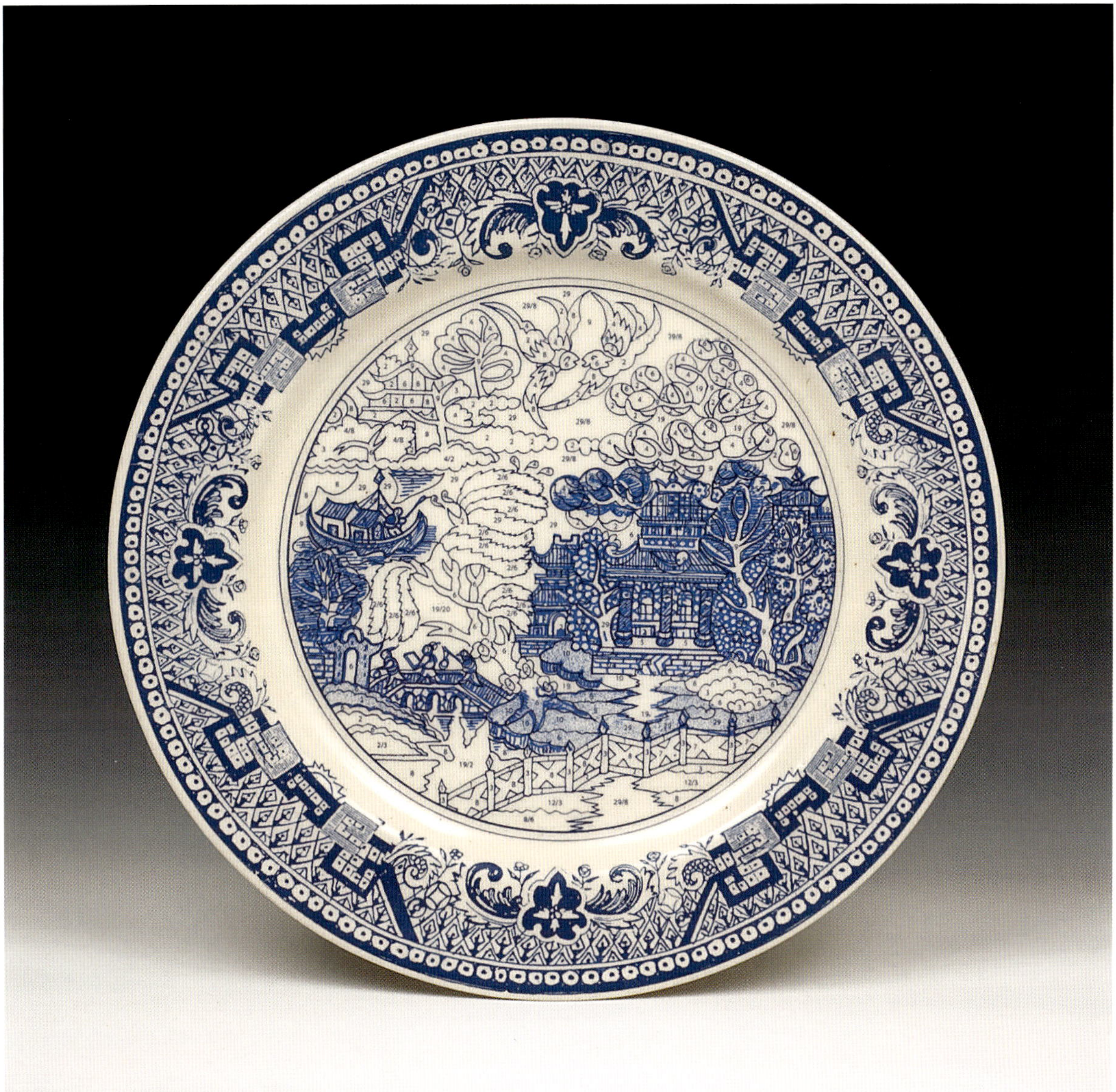

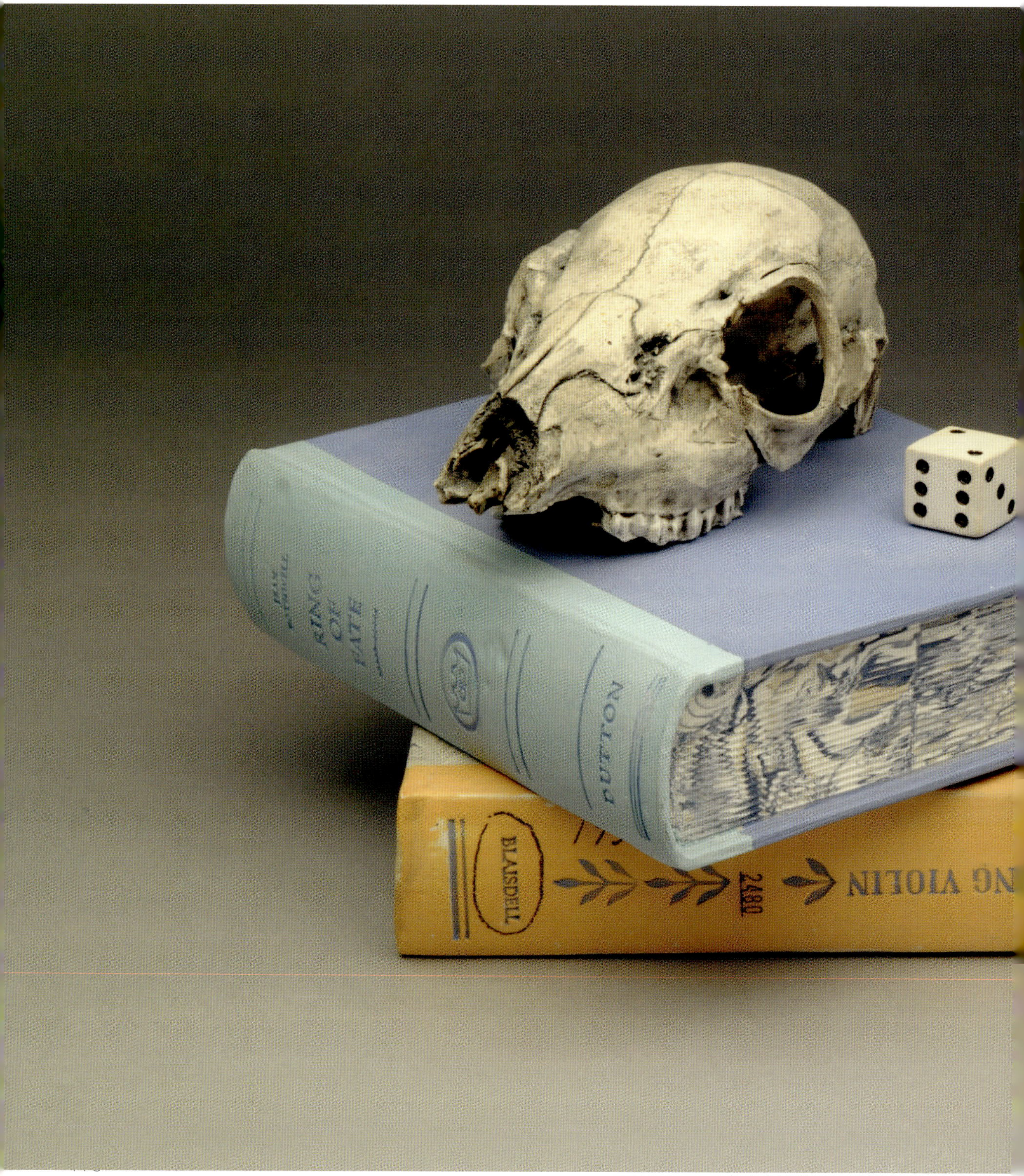
RING OF FATE
DUTTON
BLAISDELL
2480

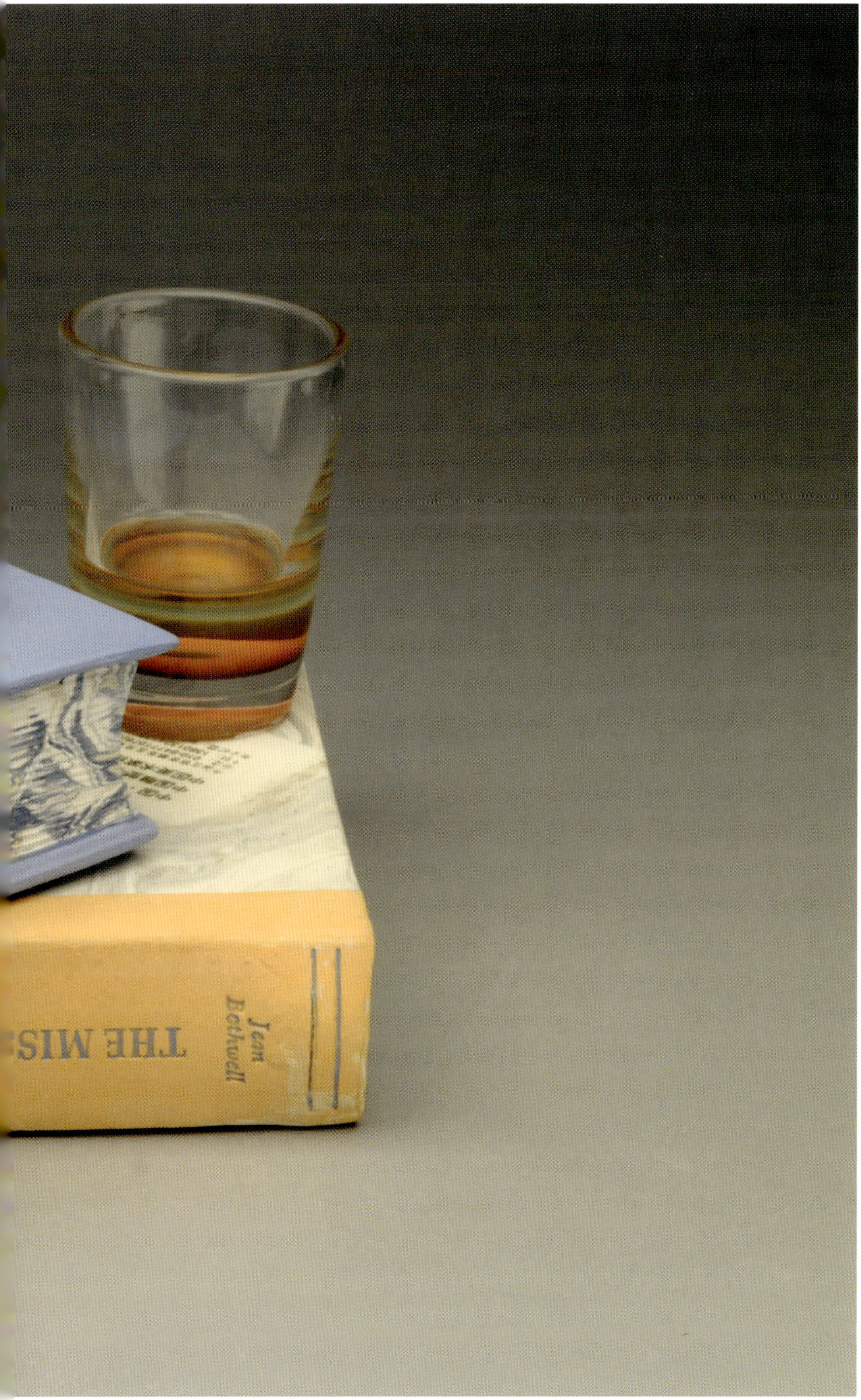
Jean Bethwell

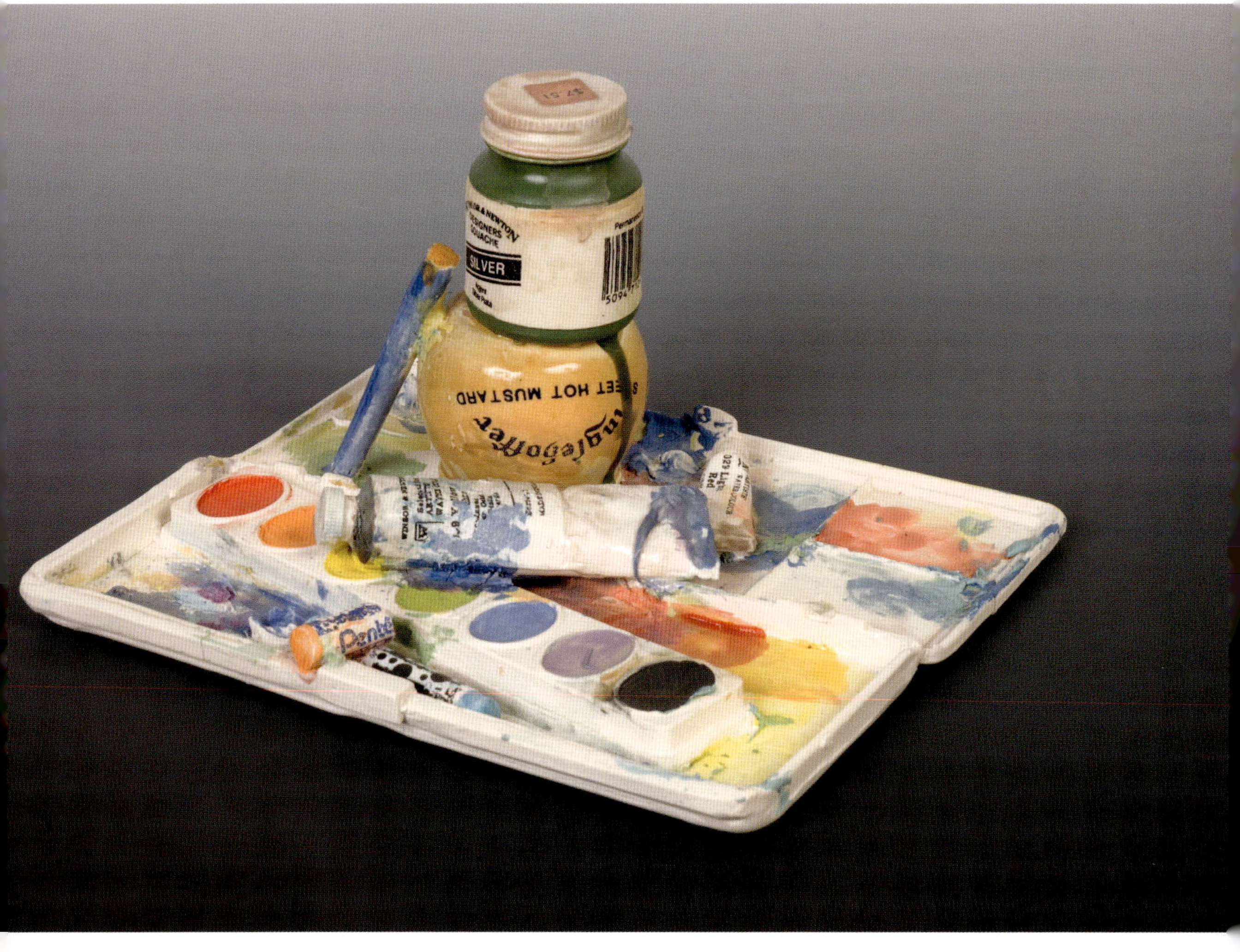
DESIGNERS
GOUACHE
SILVER
HOT MUSTARD

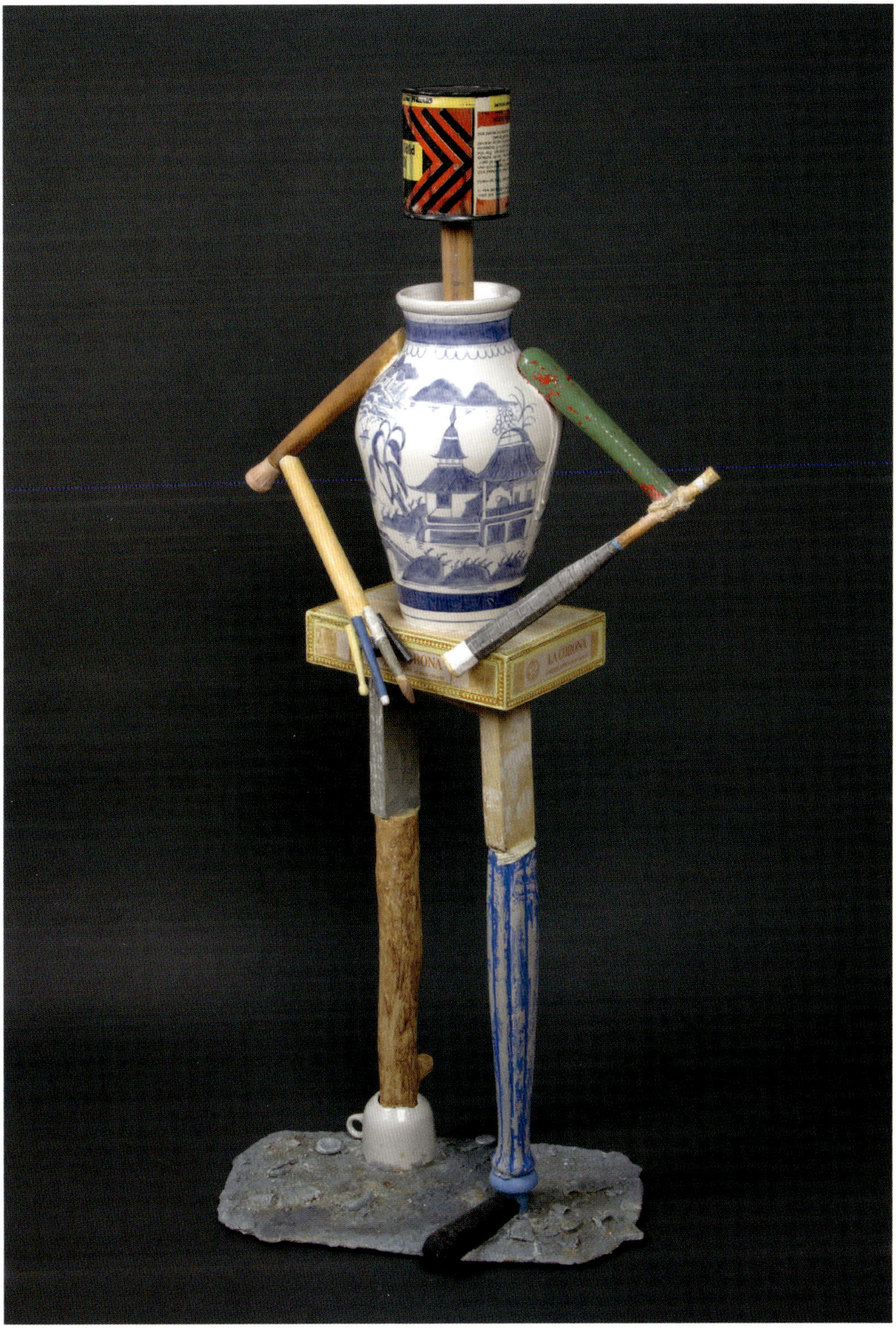

MATTHEW GROVES

LOCATION Chicago, IL
YEAR OF BIRTH 1965
MATERIALS Glazed earthenware
EXPERIENCE Since youth

'Preoccupied with figuration, my practice stems from a keen interest in historical forms, and a desire to ask new questions and raise fresh perspectives on contemporary figurative ceramics practices. Collectively, these sculptures bear witness to my artistic journey. Singularly, they do not represent or illustrate any one particular moment or event; rather they interweave many significant milieus, associations and obsessions. While the work incorporates recognizable components which have universal appeal, it is my intention that the meaning can never be described or interpreted in any strict or absolute terms, and should be dependent upon the viewer's own poetic imaginings.

I aim to achieve a balance of opposing static and fluid elements in each work, which reflects the key aspects of my process, control and spontaneity. I usually begin with some drawing and mixed-media model making, followed by a phase of mold-making. After this comes a period of clay construction, which includes a variety of pressing and handbuilding techniques. After bisque-firing, I end with spontaneous applications of colorful, translucent glazes, which lend fluidity and emotional impact to what might otherwise be considered stiff, wooden portraits.'

Matthew Groves received his undergraduate degree in Ceramics from Staffordshire University, and his master's degree in Ceramics and Glass from the Royal College of Art. Before becoming a full-time studio artist and educator, Matthew was a professional fine art and design technician in colleges, universities, private workshops and museums.

Photos courtesy of the artist

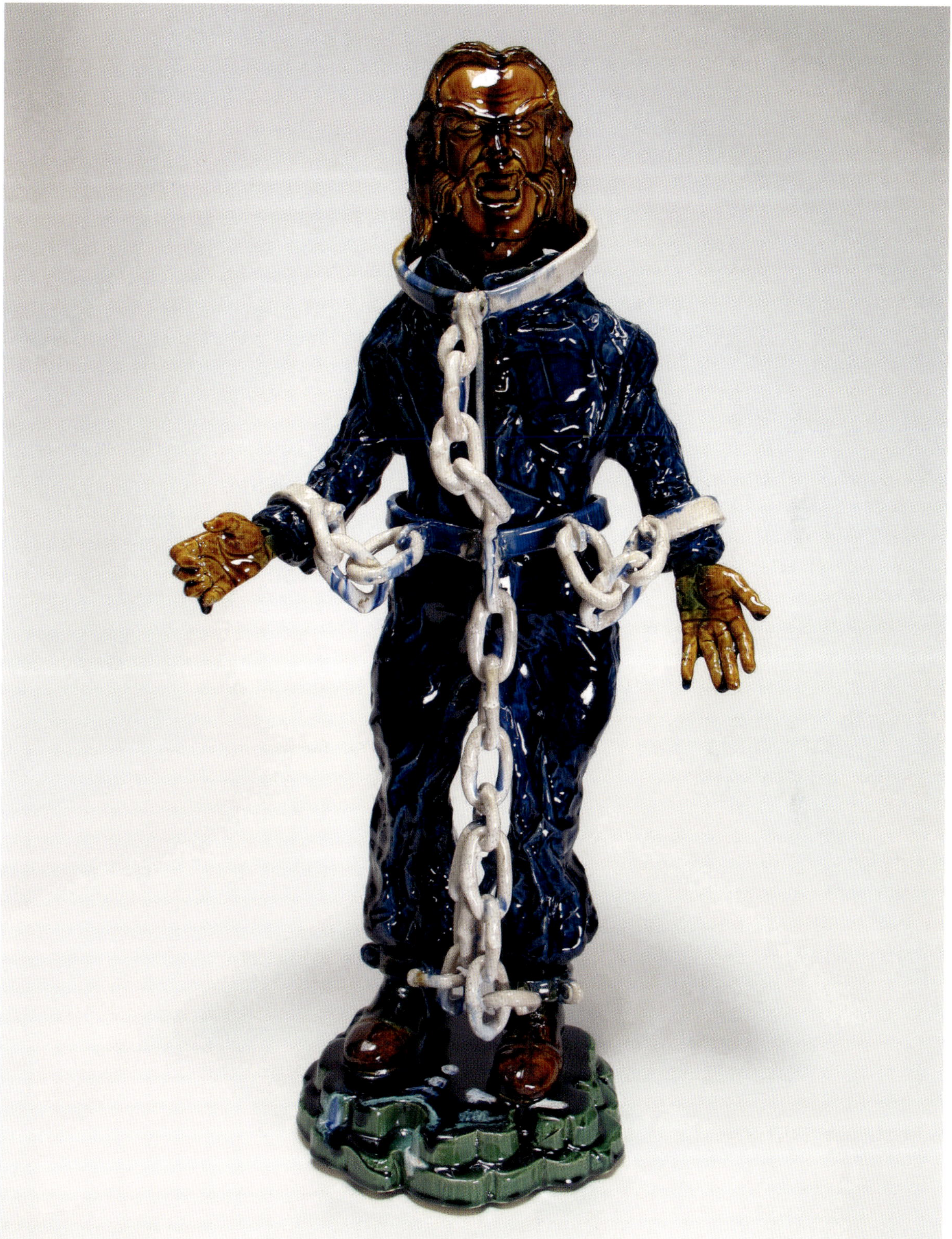

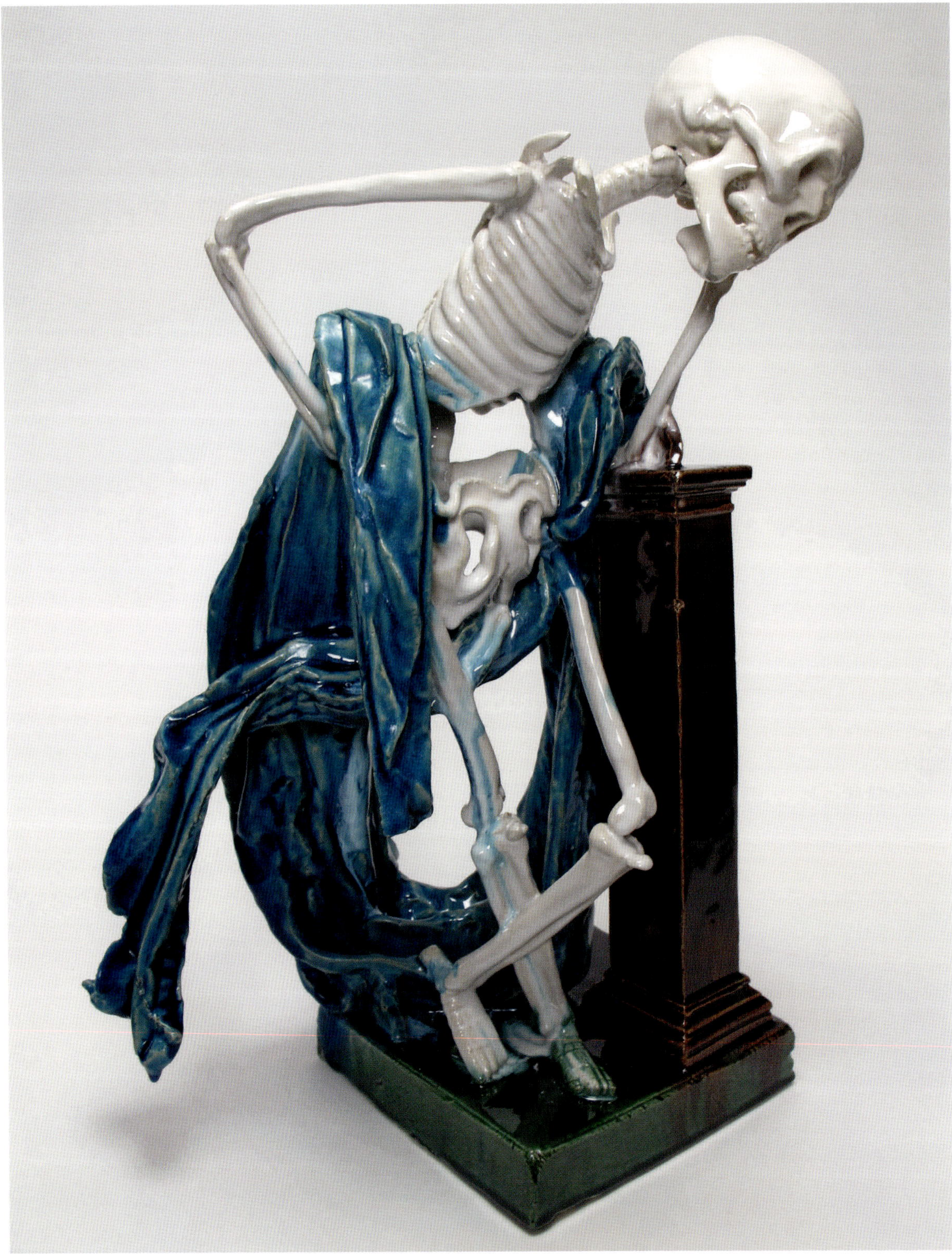

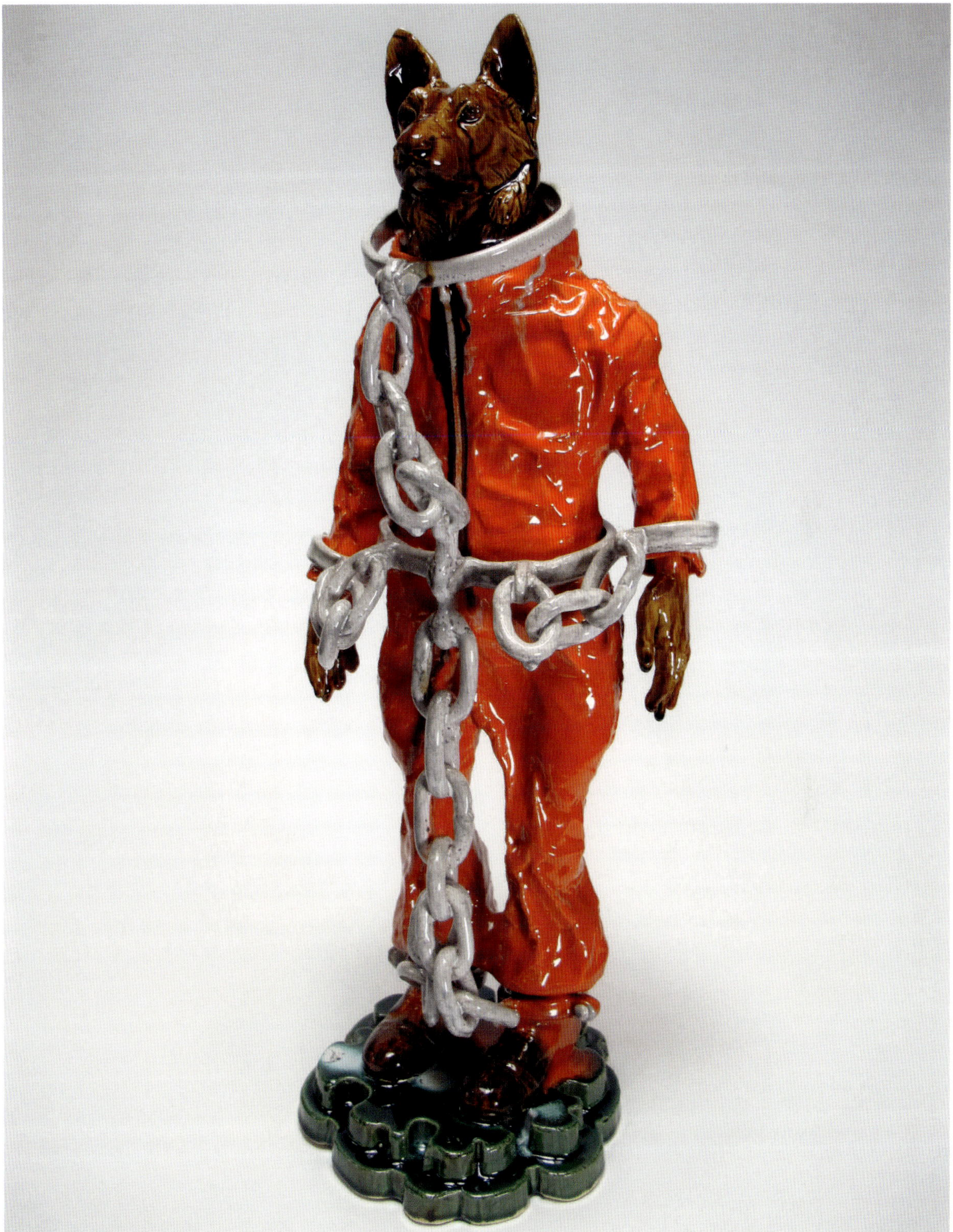

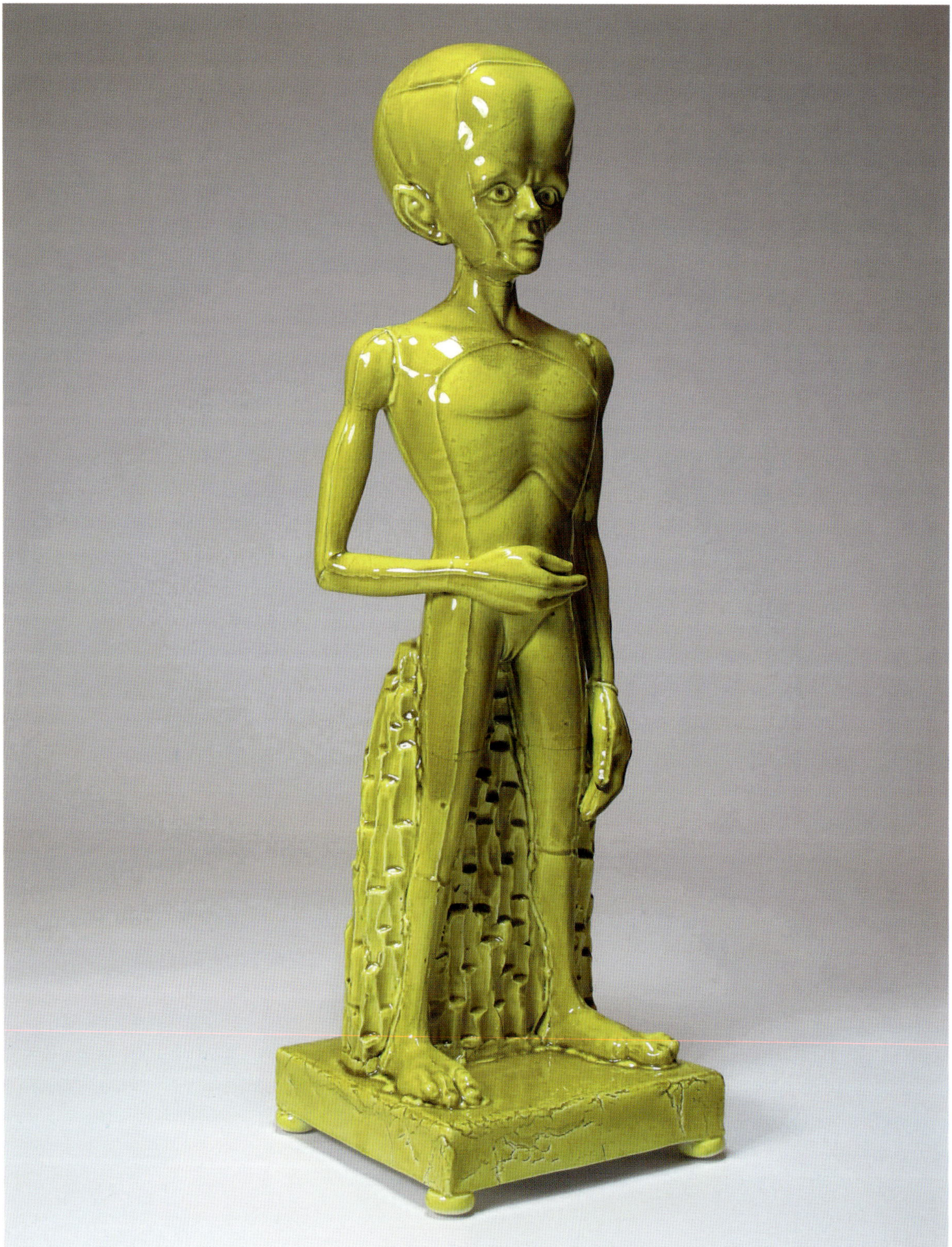

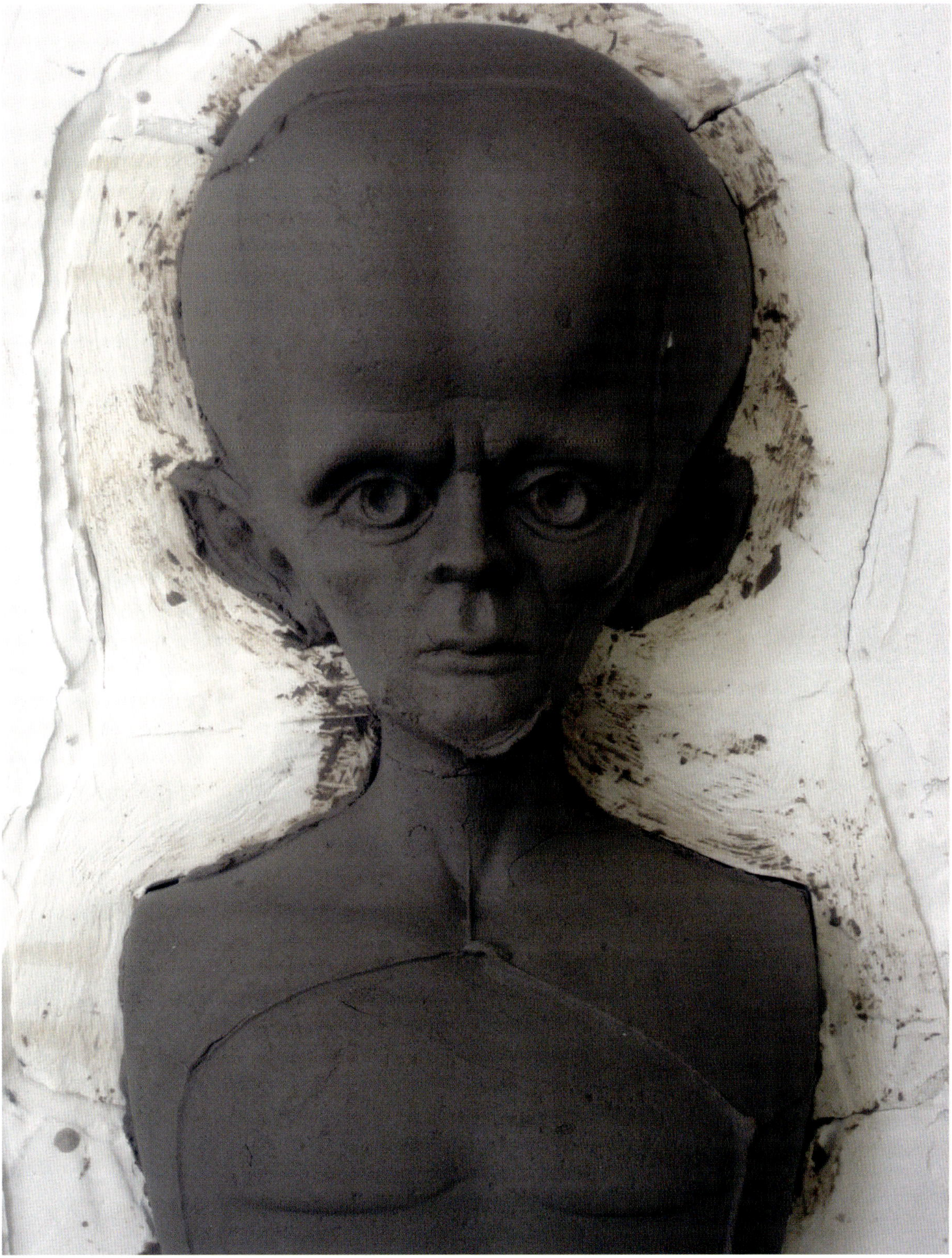

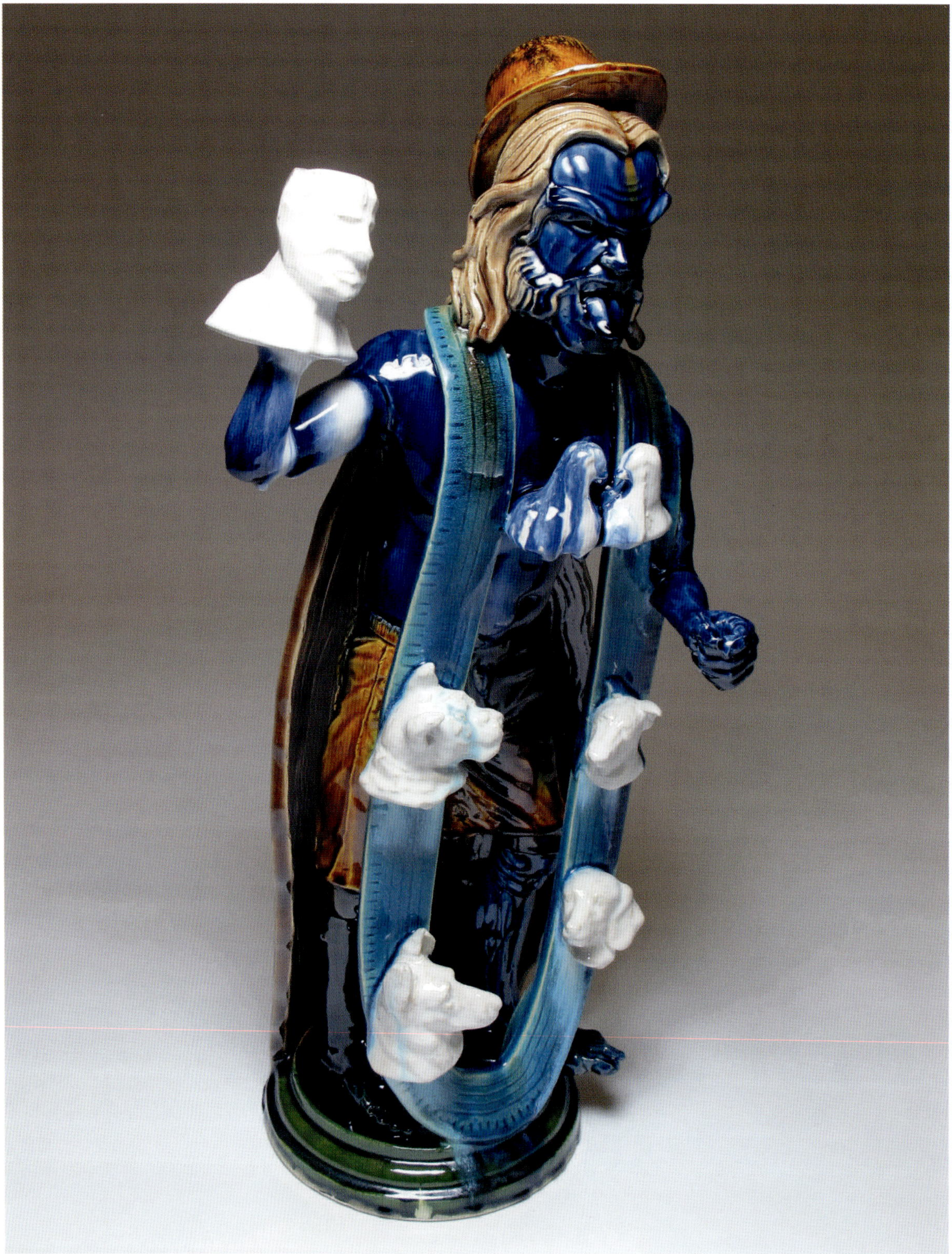

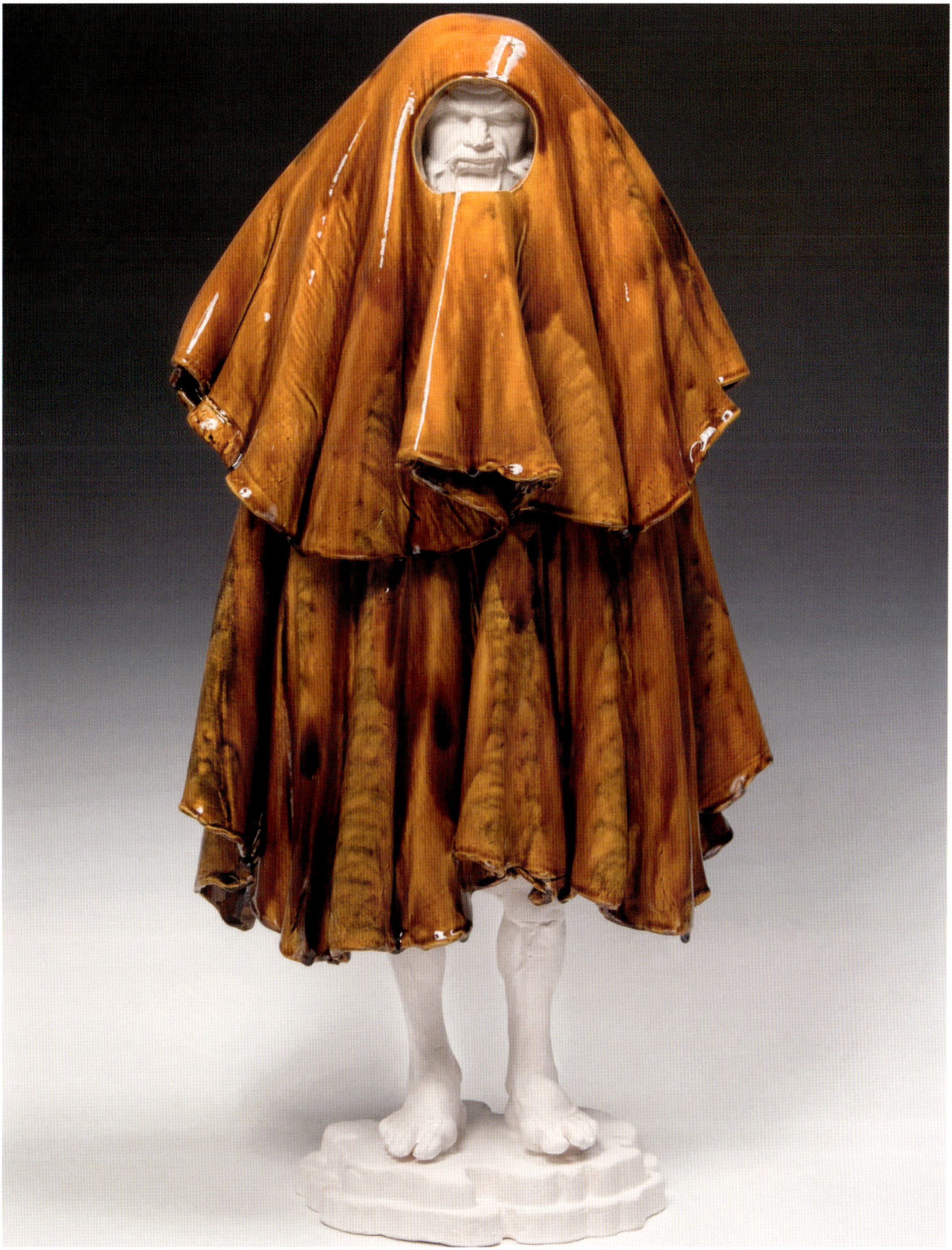

SUMMERLAND CERAMICS

LOCATION San Francisco, CA
YEAR OF BIRTH 1985
MATERIALS Slip-cast ceramics
EXPERIENCE 5 years

'Garage saleing with my father every Saturday when I was young imbedded within me a weird obsession with objects. I could always fish something really cool out of a pile of junk. I think this is where I gained my eye for design. Once I started smoking 'the good stuff,' I took a stroll down Haight Street to find a cool bong to call my own. In my head I imagined this pristine, immaculate piece, almost like a bong from Japan. Unfortunately, everywhere I looked, I could only find the same typical swirly rainbow bullshit. So, I set out to create a new kind of bong for the contemporary stoner. A bong that would look good in a clean room; a bong you wouldn't feel the need to hide when company arrived. I chose ceramic as a medium because of its purity, elegance, and timelessness.'

Liam Kaczmar, the creator of Summerland Ceramics, is an artist and filmmaker with a background in graphic design and art direction.

Photos courtesy of StianRasmussen and the artist

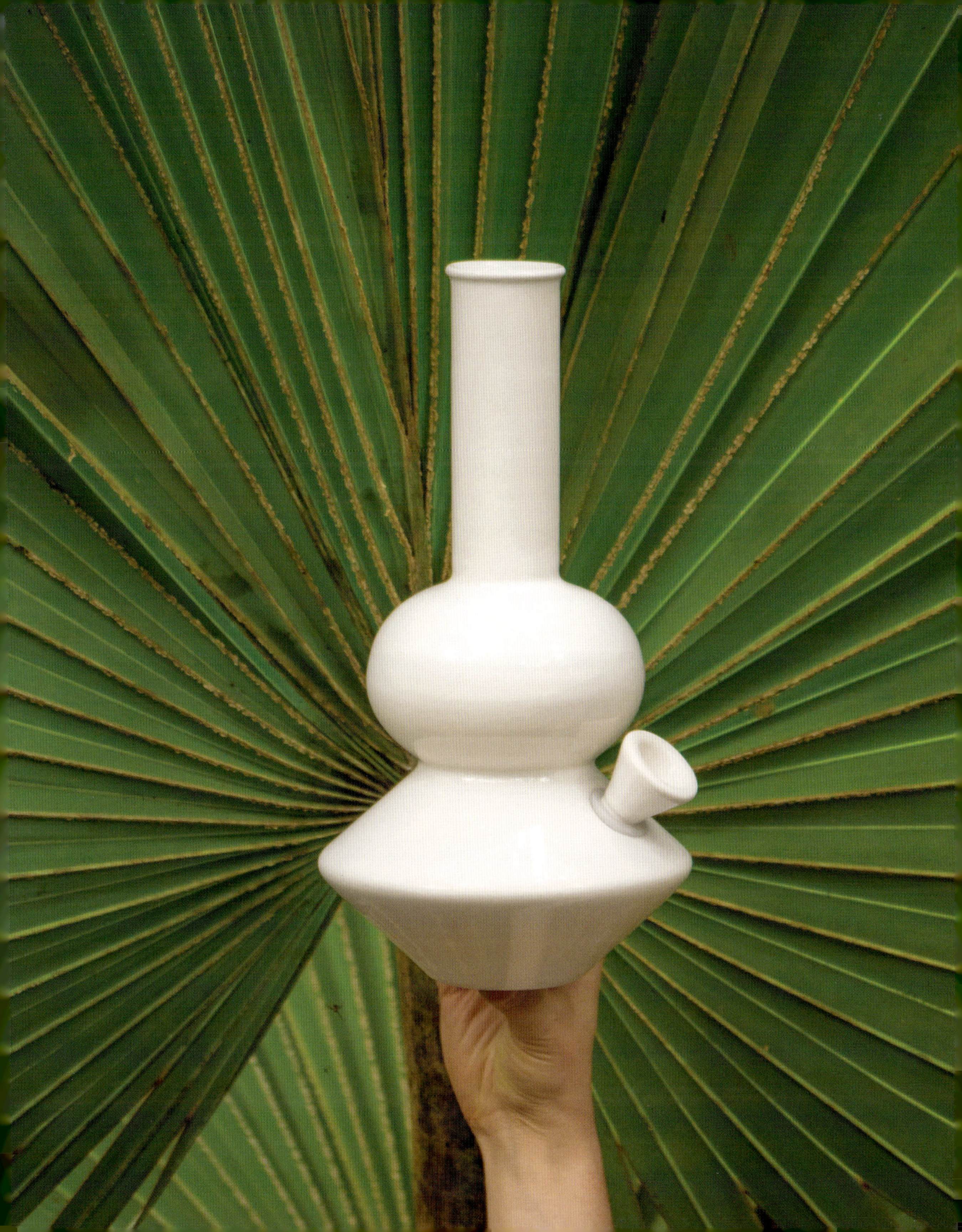

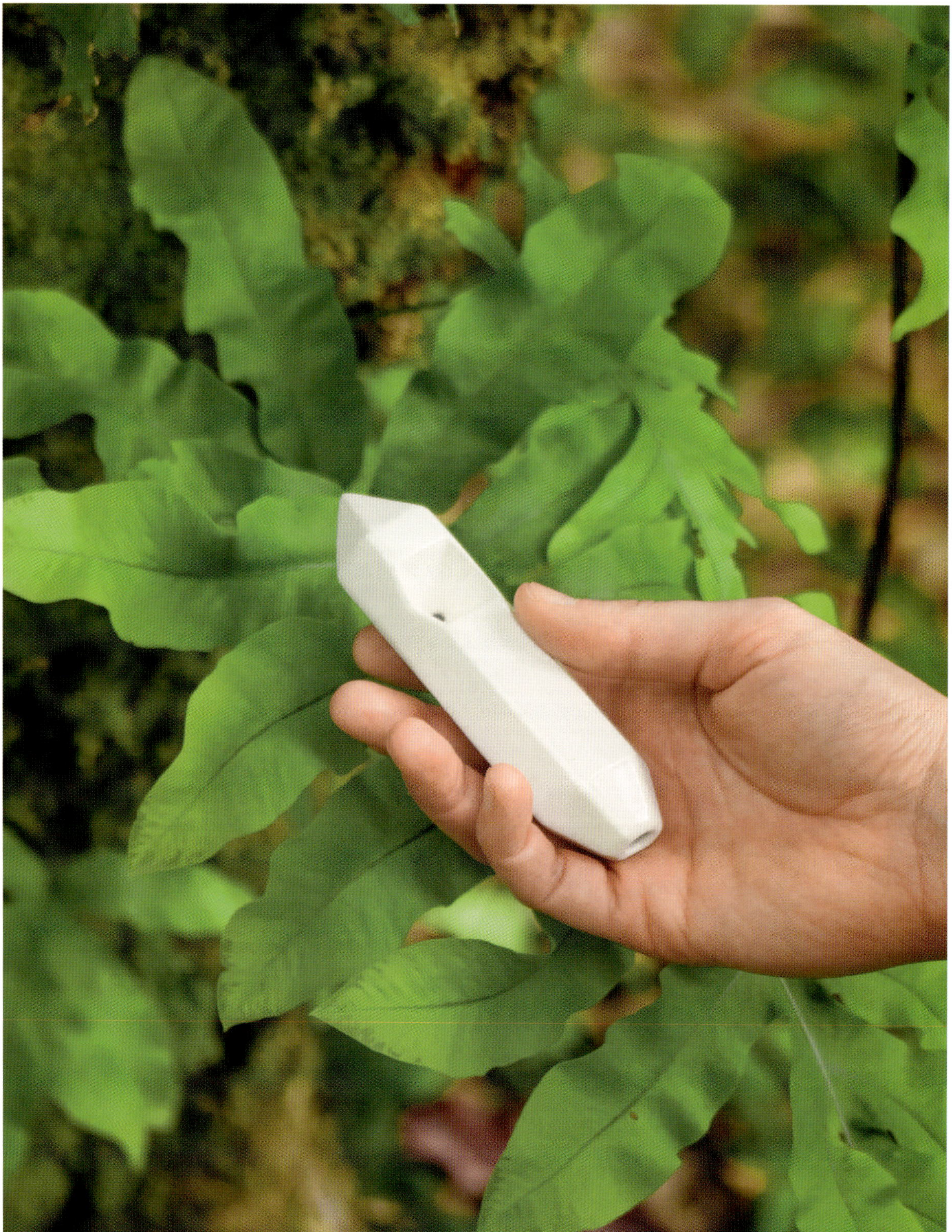

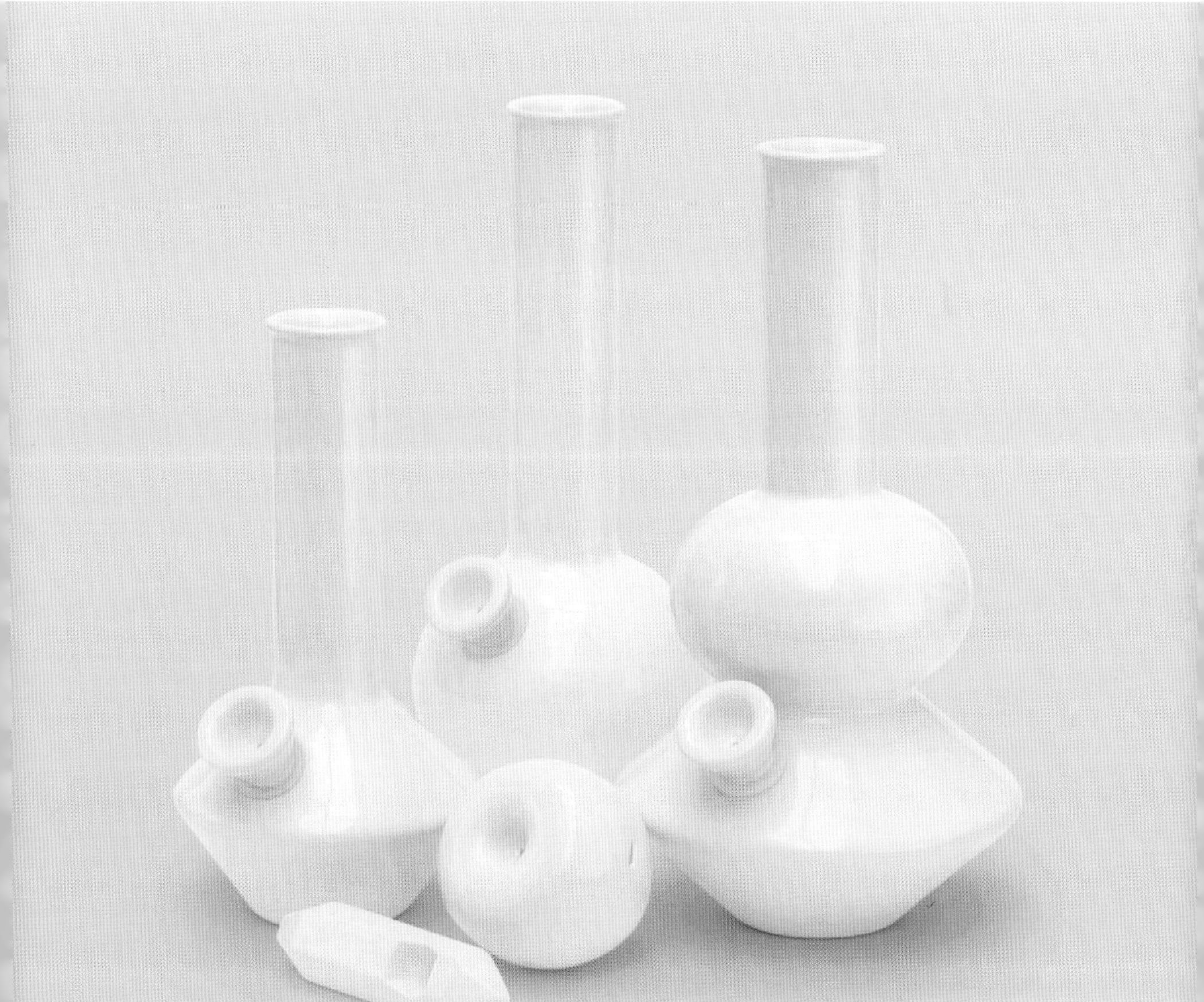

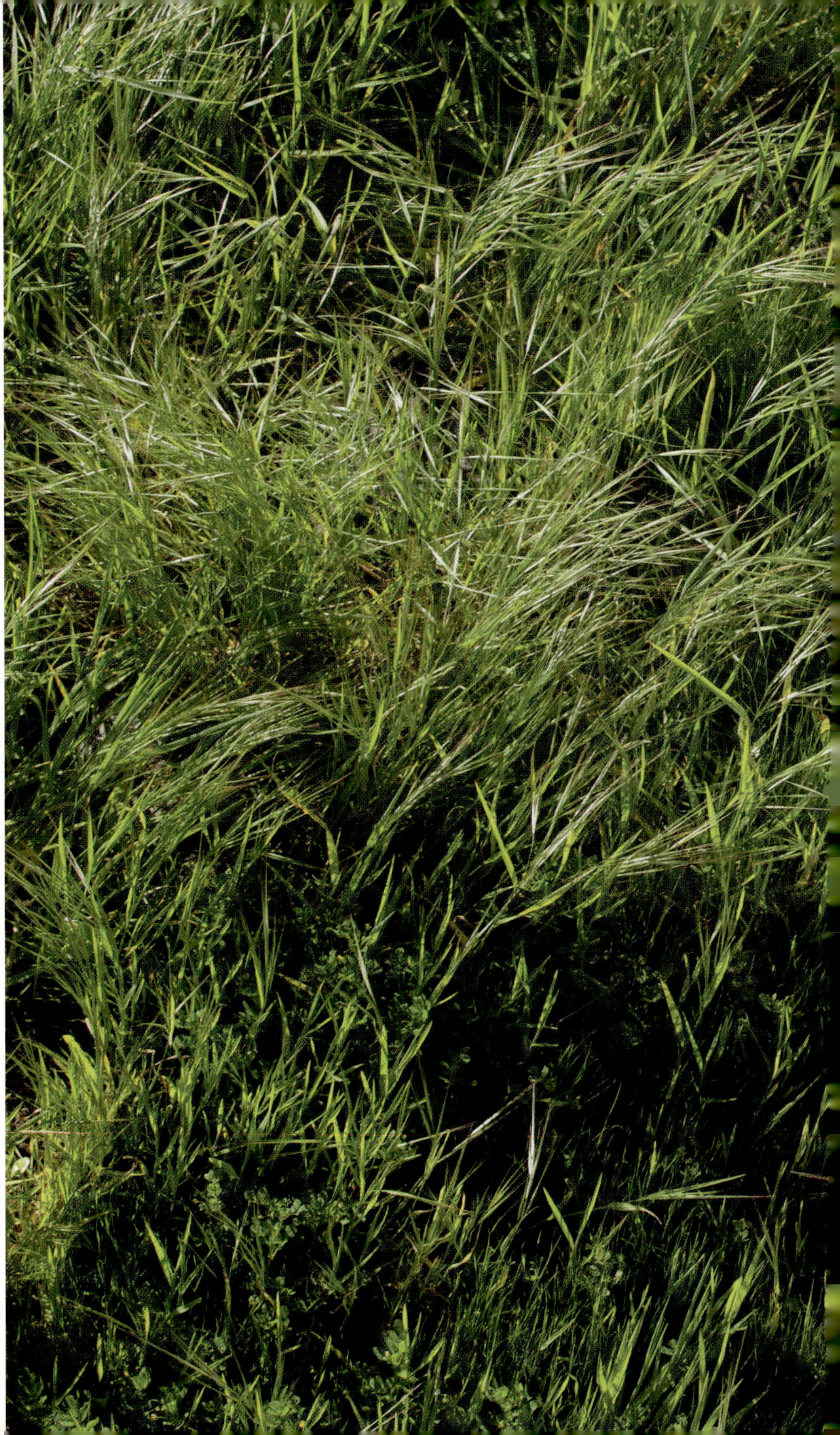

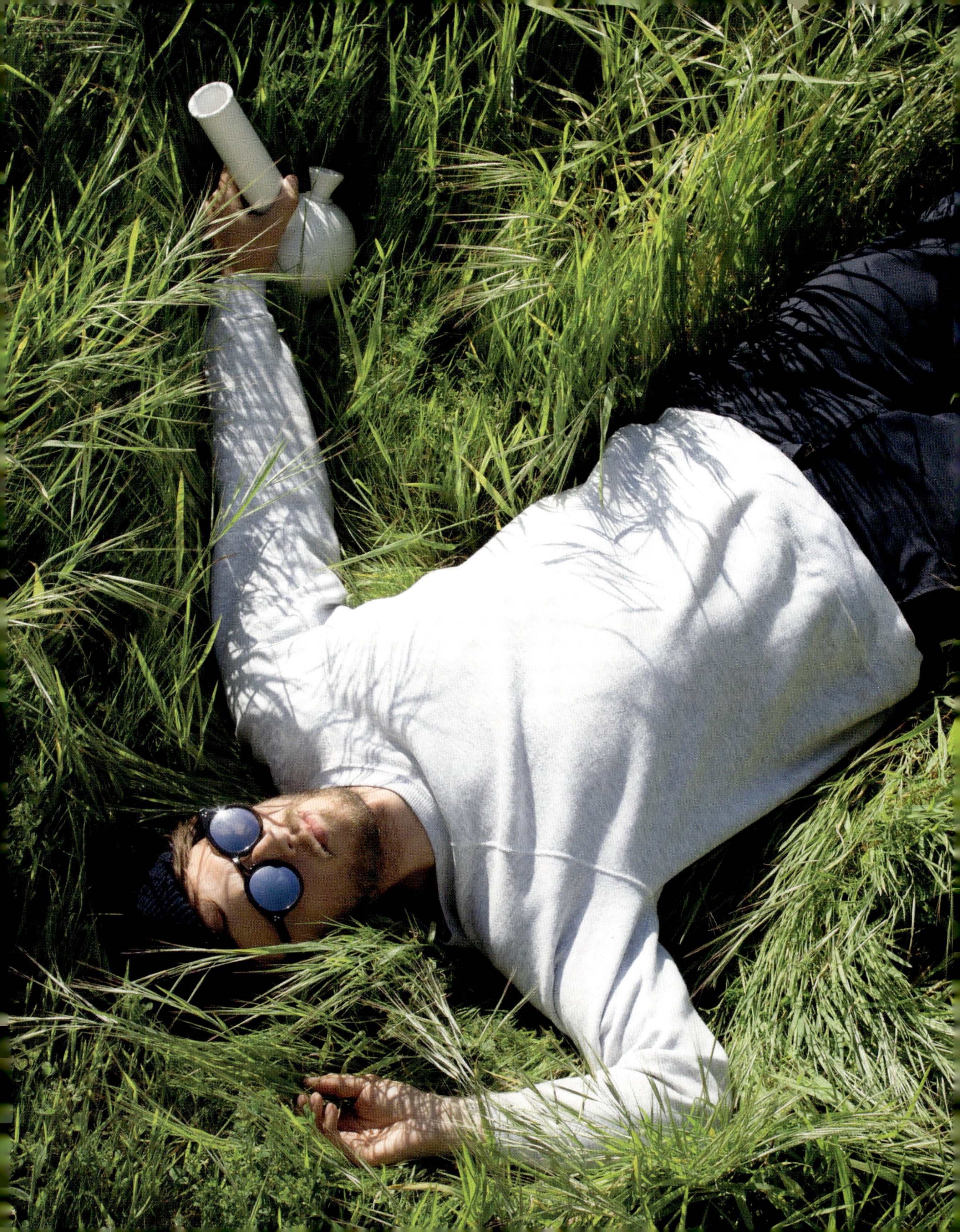

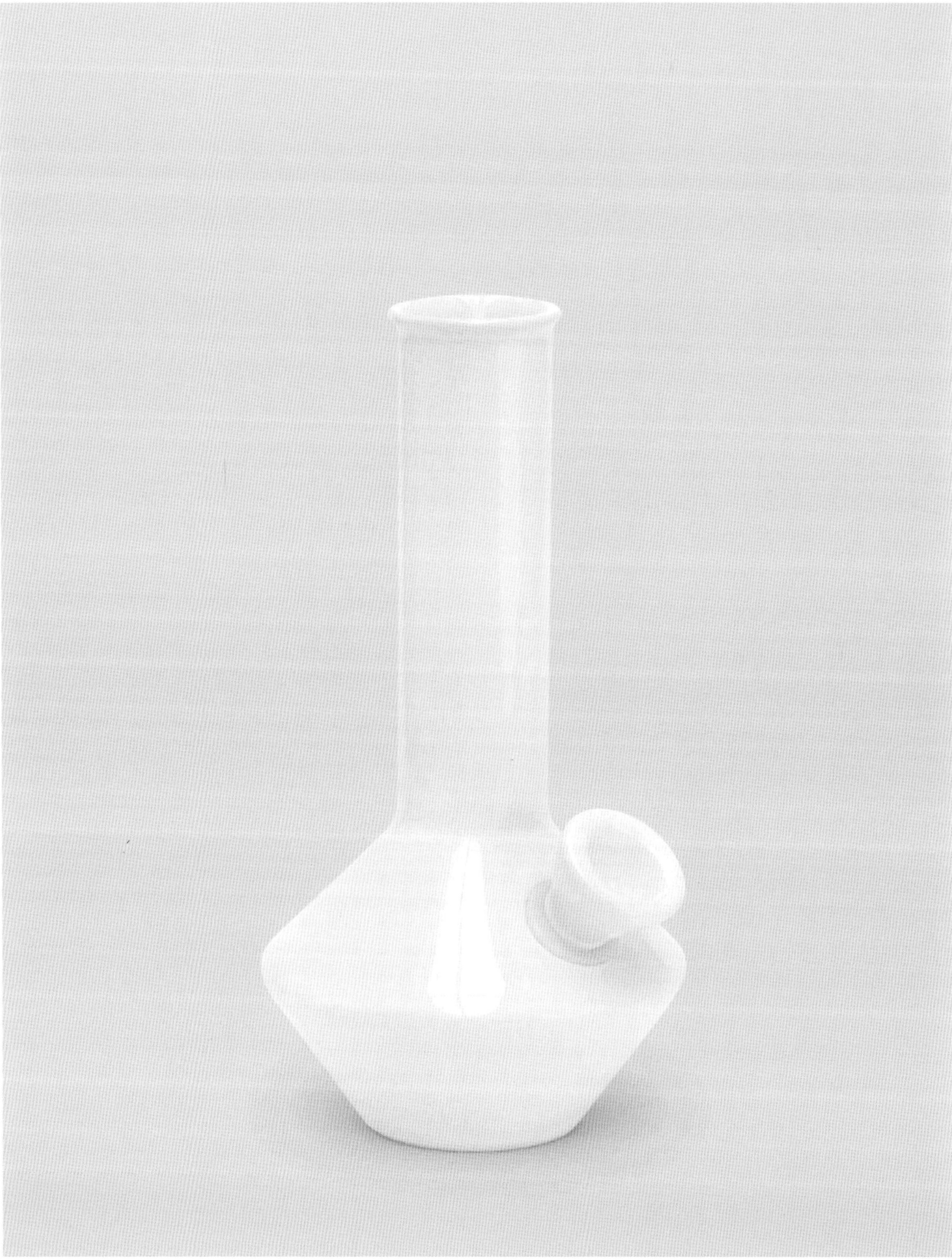

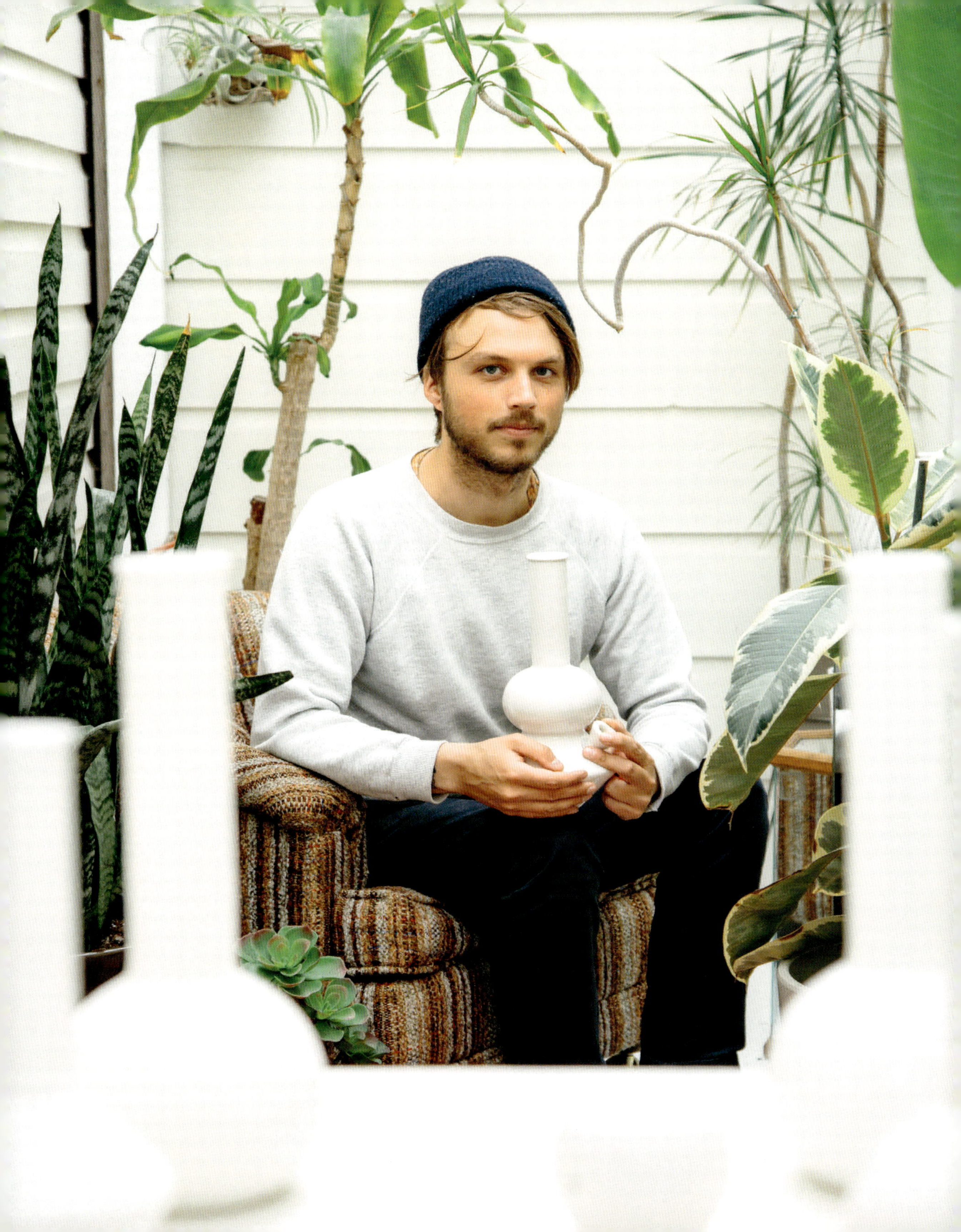

EXPLORING SHAPES

KATSUYO AOKI

MATT WEDEL

ZEMER PELED

BRENDAN MONROE

LAUREN SHAPIRO

NICK MAKANNA

KATSUYO AOKI

LOCATION Tokyo, Japan
YEAR OF BIRTH 1972
MATERIALS Porcelain
EXPERIENCE 18 years

Katsuyo Aoki uses porcelain as her medium of expression, incorporating various decorative styles, patterns and symbolic forms as the principal axis in creating her works.

'The decorative styles and forms I allude to and incorporate into my pieces each contain a story based on historical backgrounds, ideas, myths and allegories. Their existence in the present day makes us feel many things; adoration, romantic emotions and finally a sense of fruitlessness and languor from their excessiveness and vulgarity. On the other hand, they make us feel tranquility and awe that can almost be described as spiritual, as if in the presence of an object of worship. By referencing such images, I feel I am able to create an atmosphere representing the complexity of our world. Several of the decorative forms I employ, simultaneously portray both divine and vulgar meanings which well represent the irrational and contradictory aspects of our contemporary era. Further, traditional ceramics techniques play an important role within the long history of the decorative arts. I feel that the delicateness and fragile tension of these techniques and materials express my concepts very well.'

Photos courtesy of Mareo Suemasa and the artist

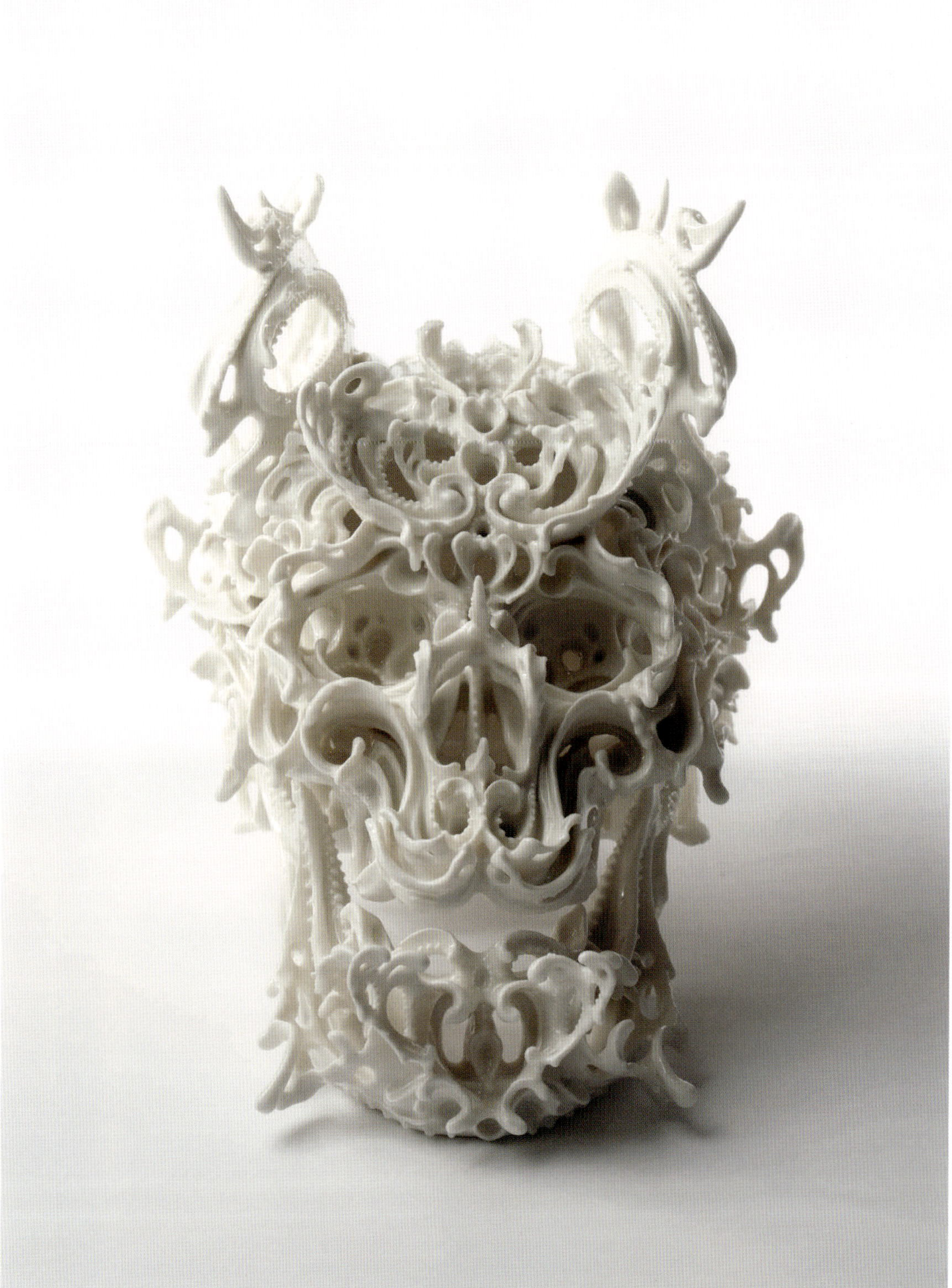

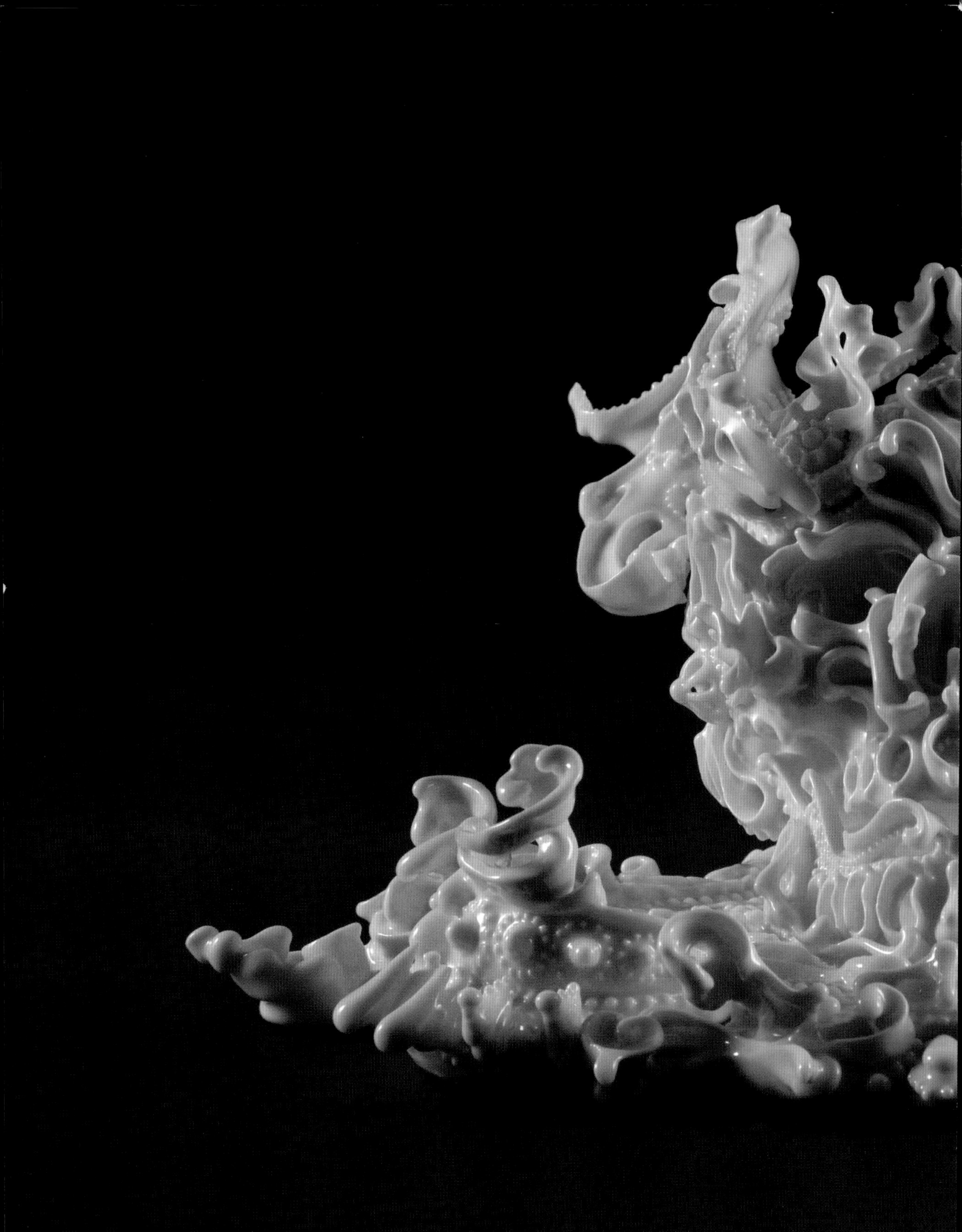

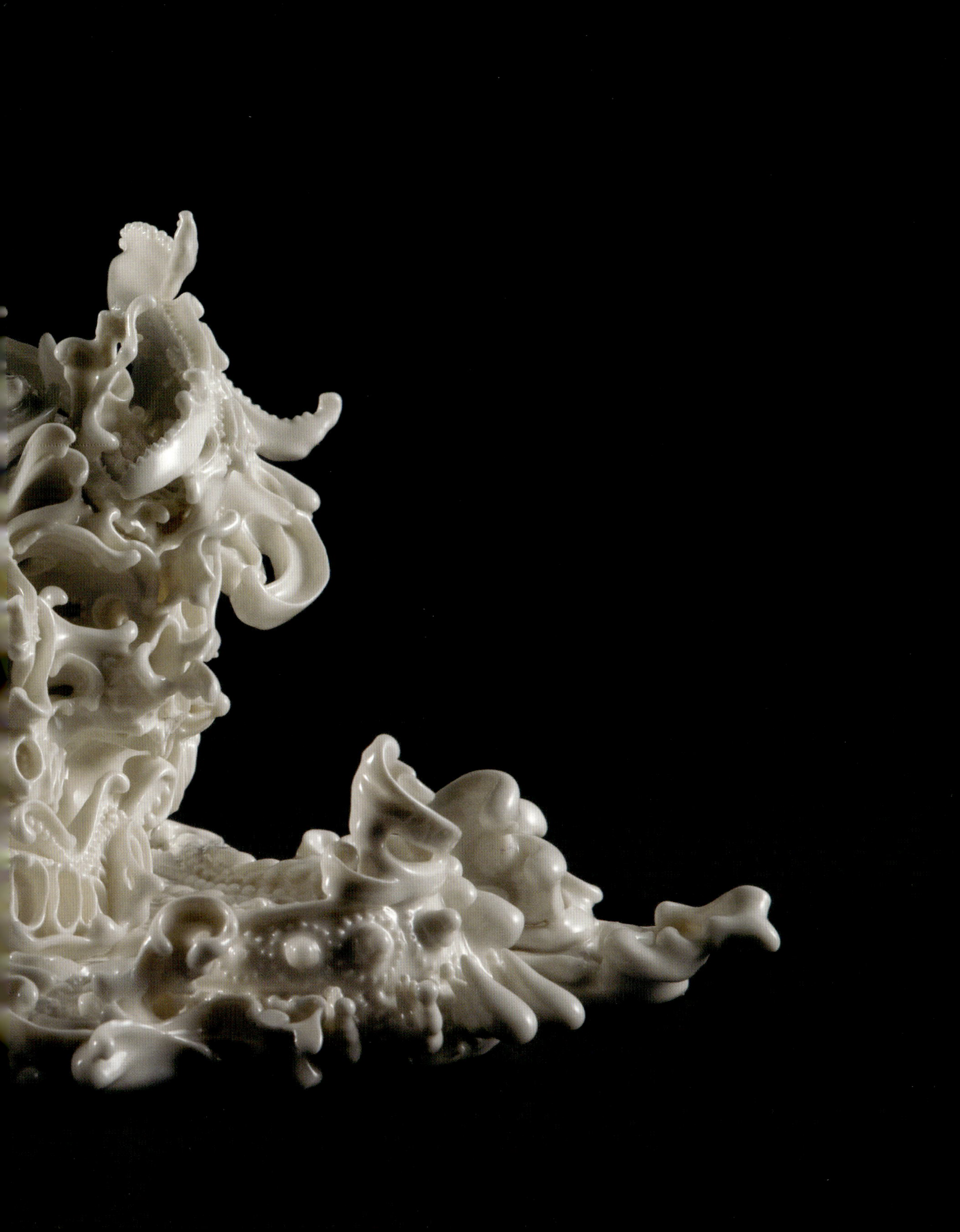

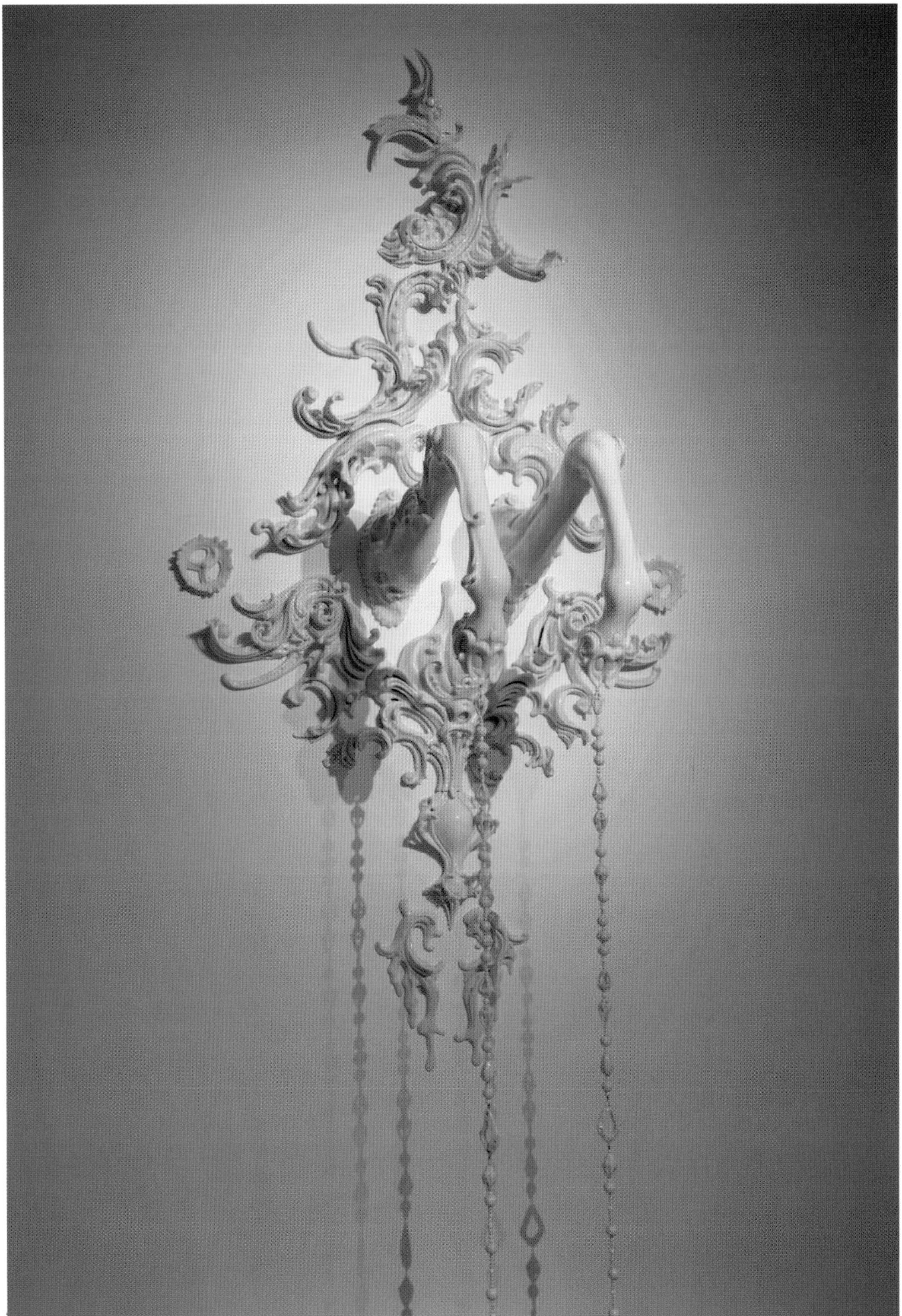

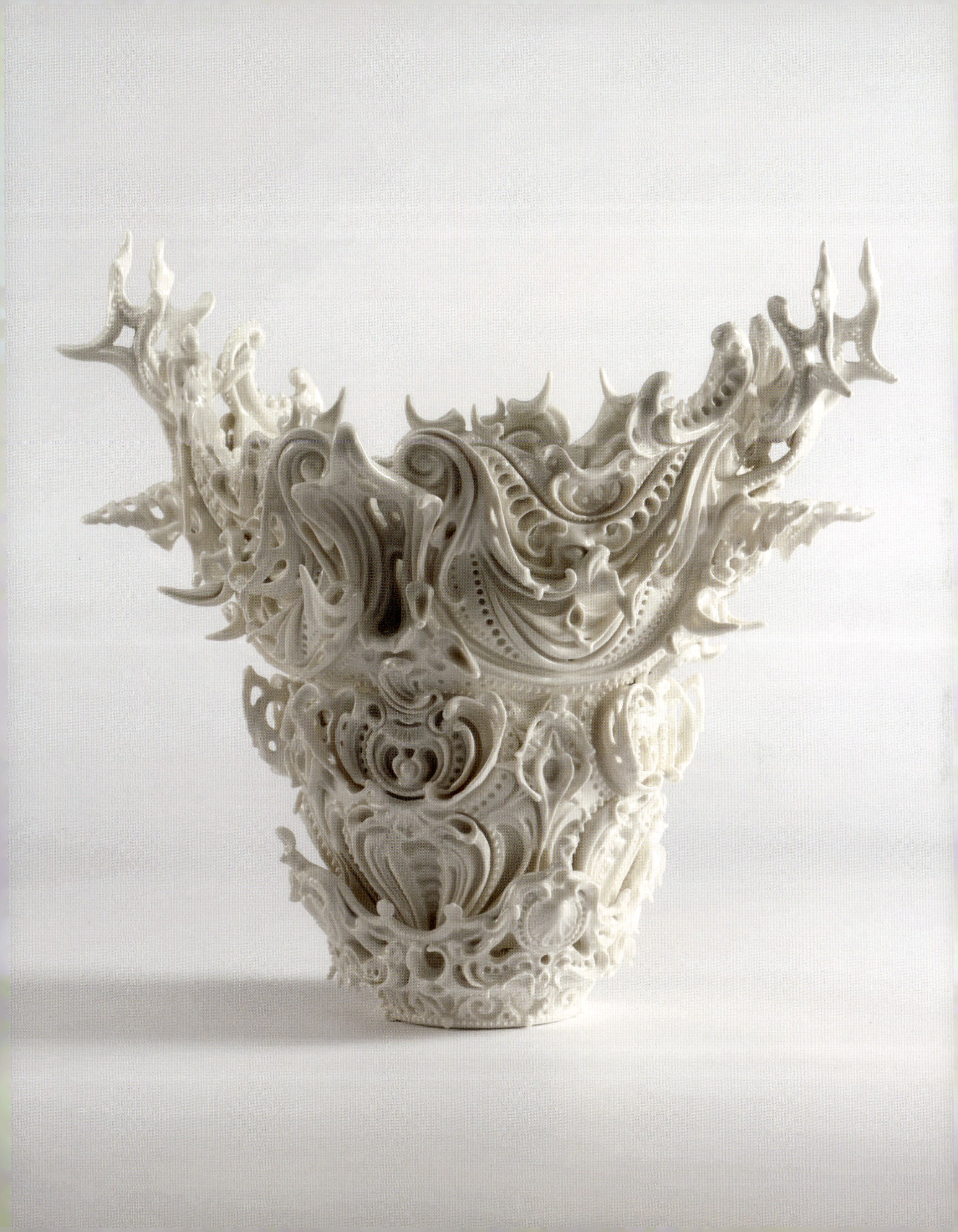

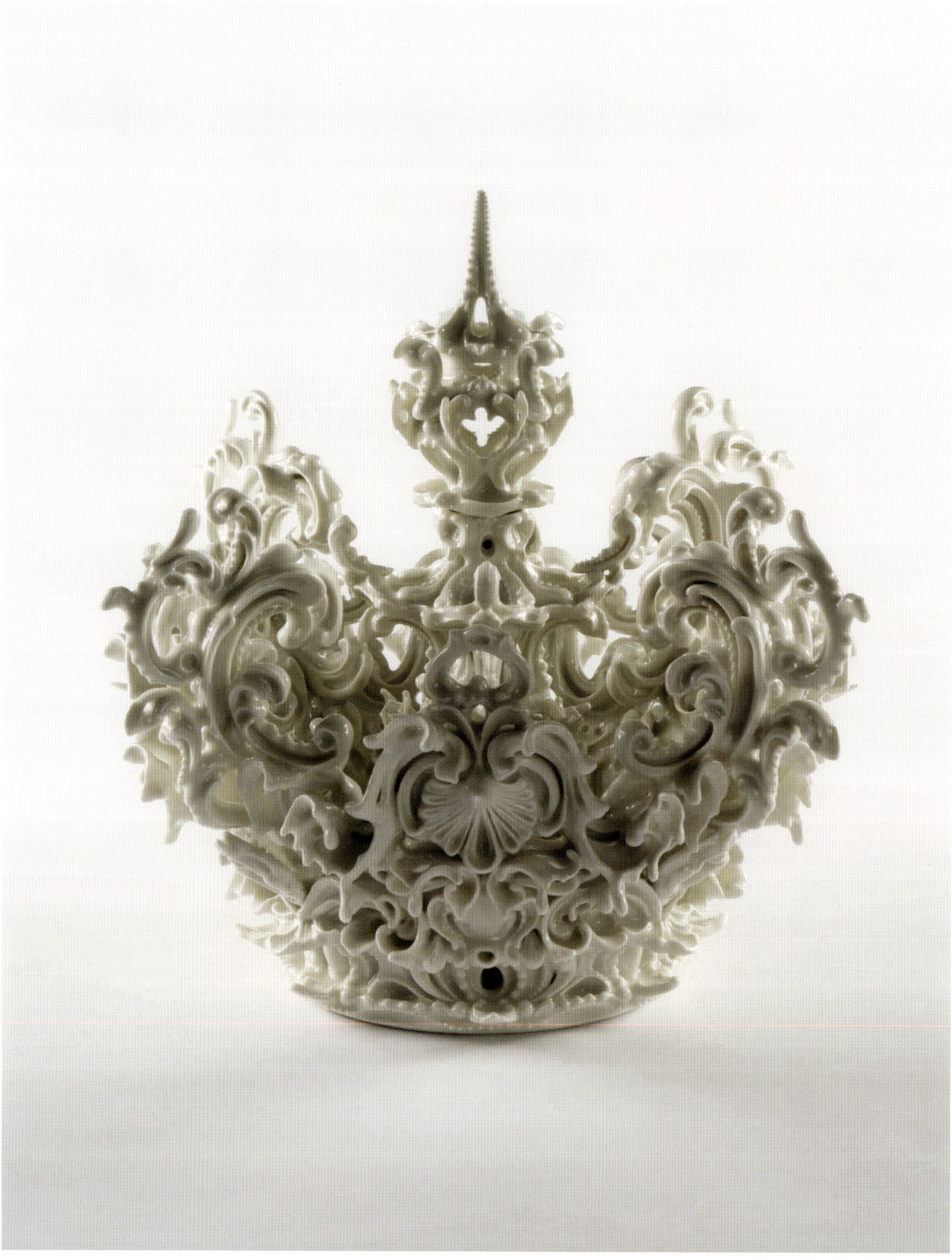

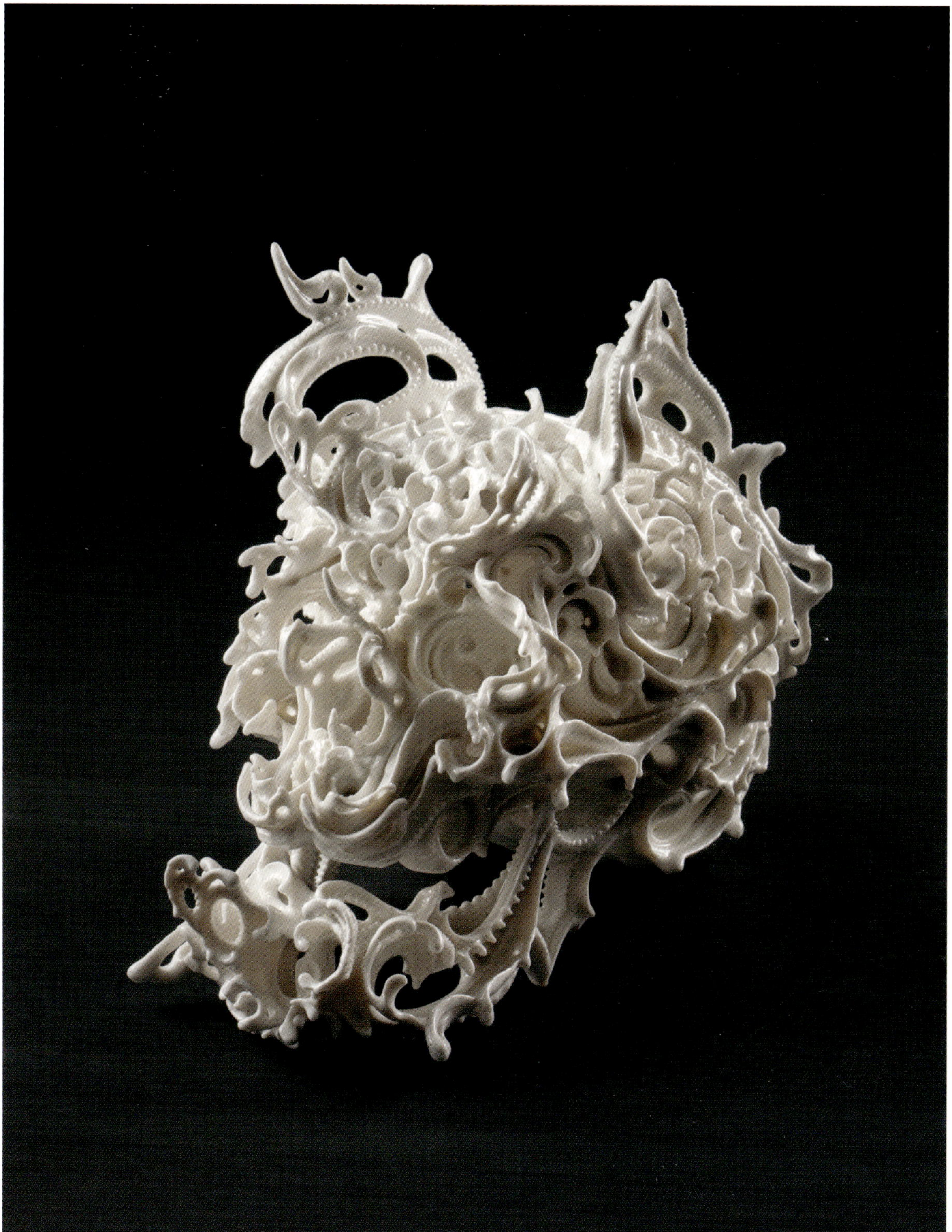

MATT WEDEL

LOCATION Athens, OH
YEAR OF BIRTH 1983
MATERIALS Ceramics, coil construction
EXPERIENCE 20 years

Matt Wedel was born in Palisade, CO and currently lives and works in Athens, OH. He earned his Bachelor of Fine Arts at The School of the Art Institute of Chicago, and then obtained a master's degree in ceramics from California State University, Long Beach. He is currently represented by LA Louver in Venice, CA.

Photos courtesy of the artist

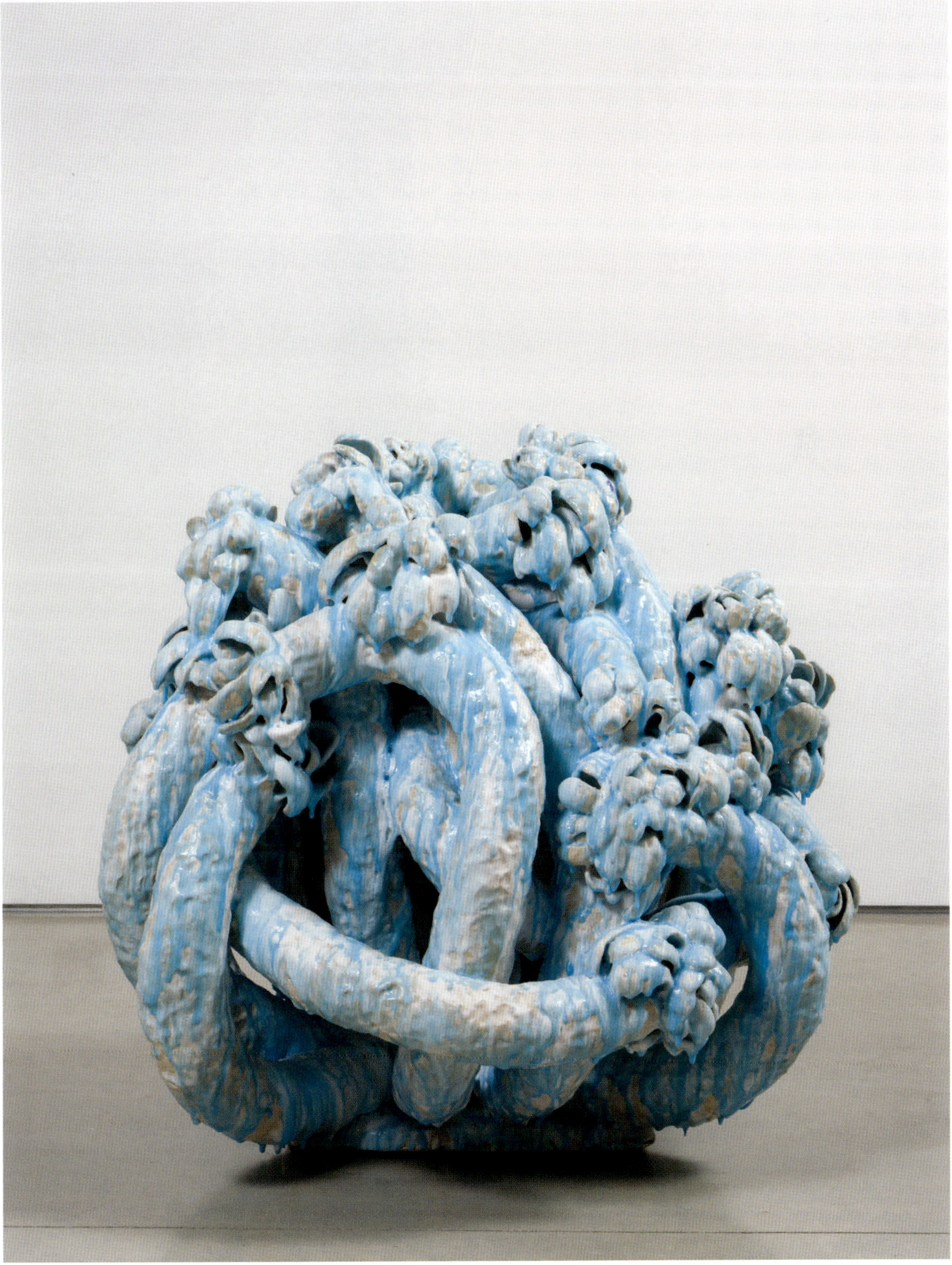

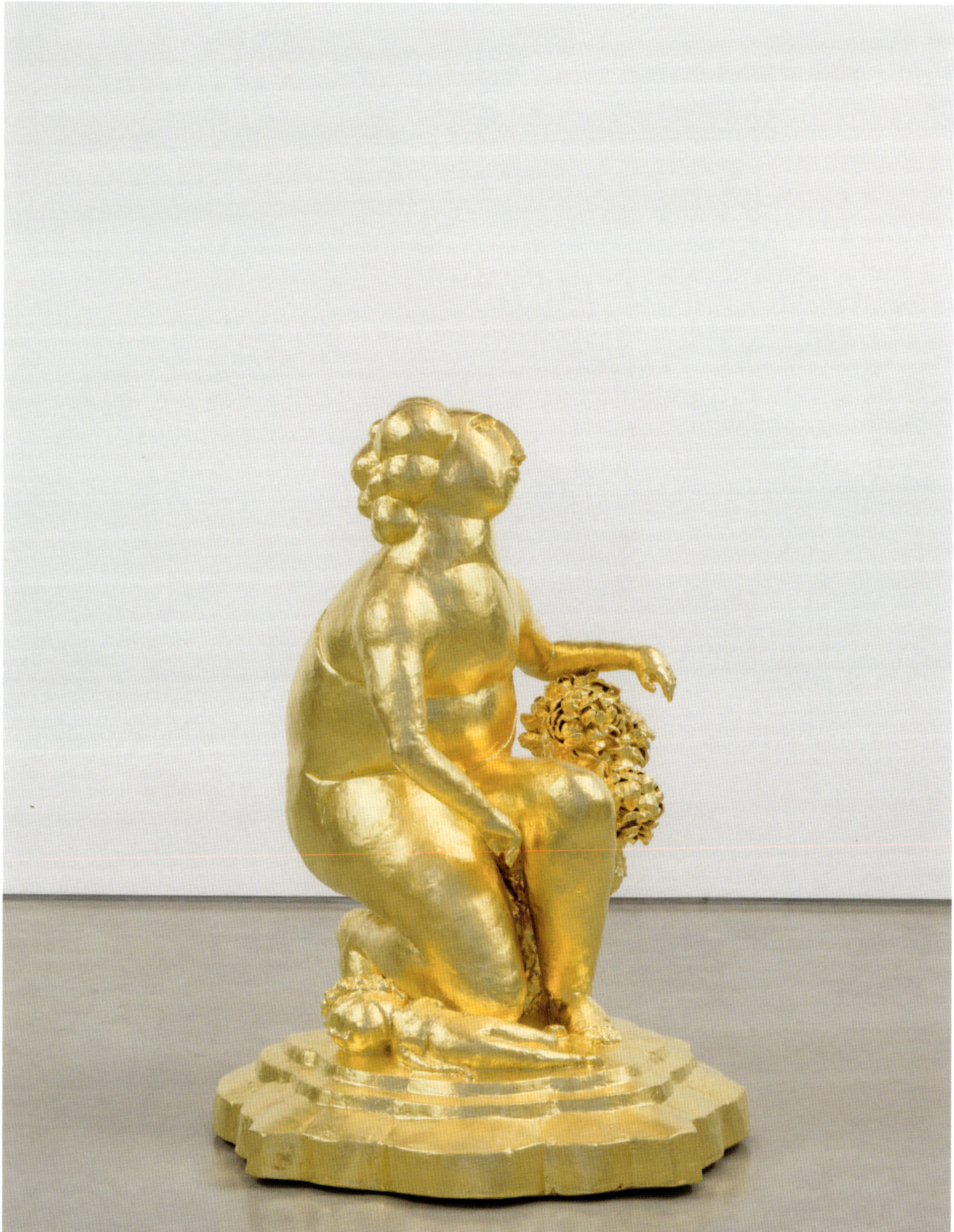

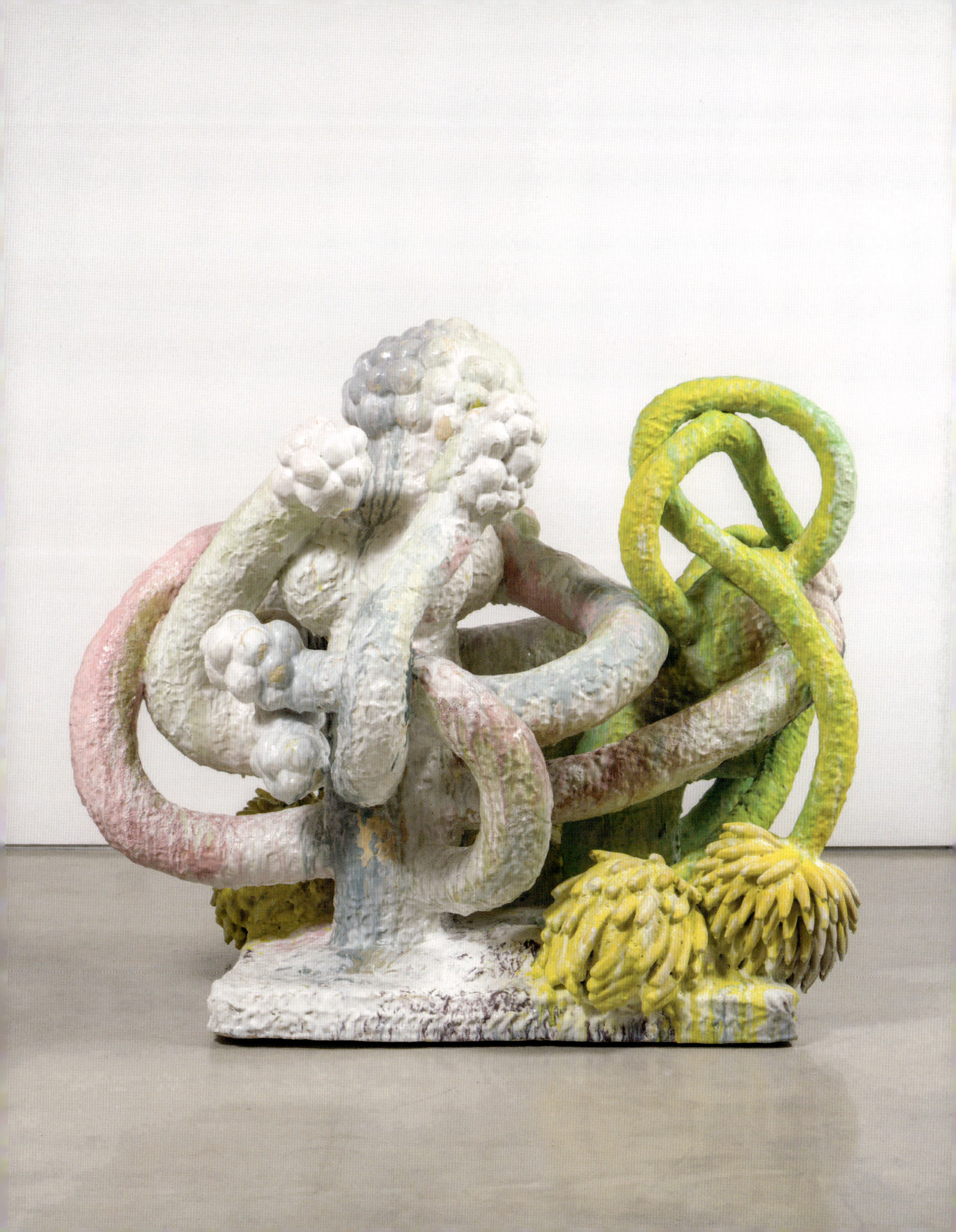

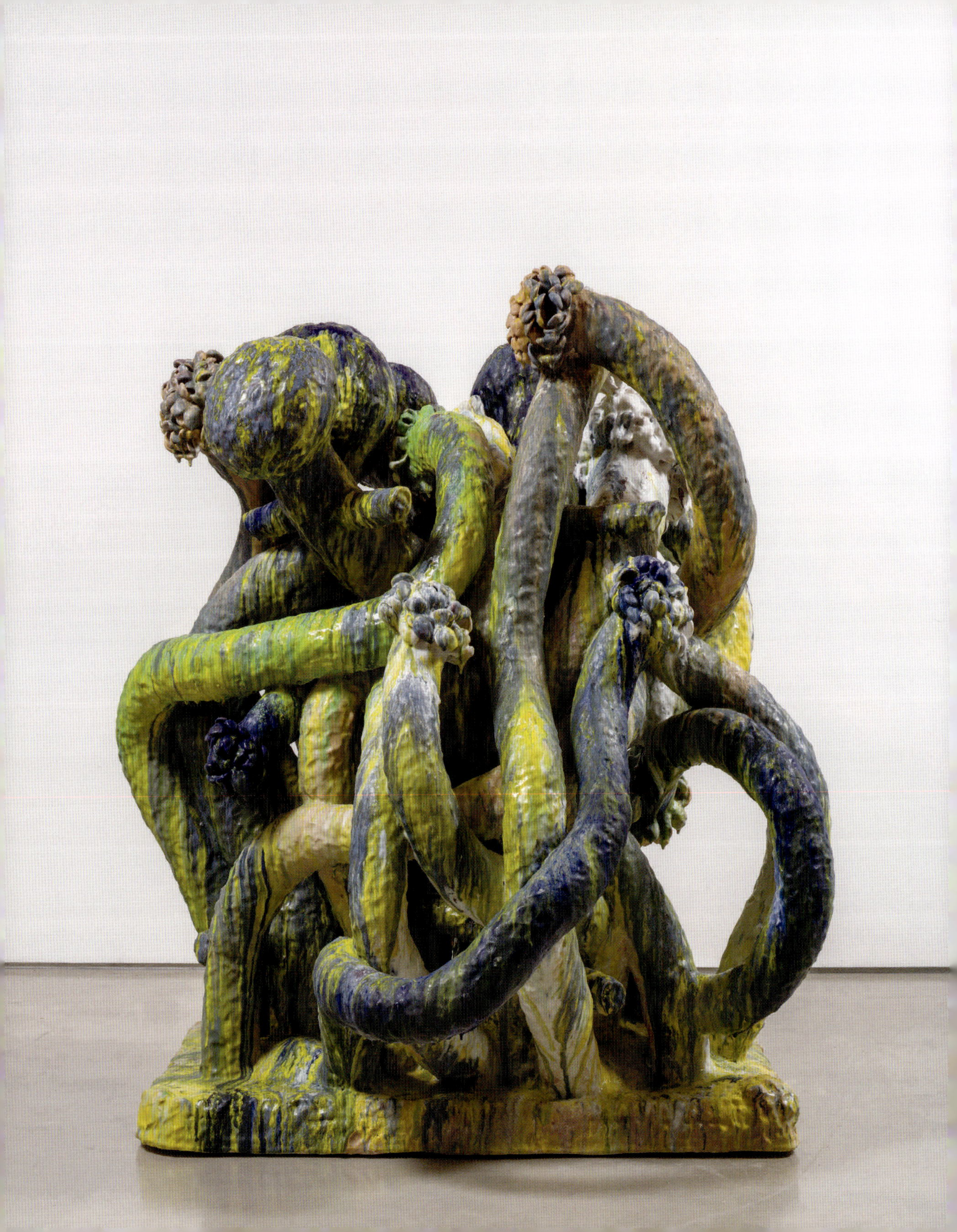

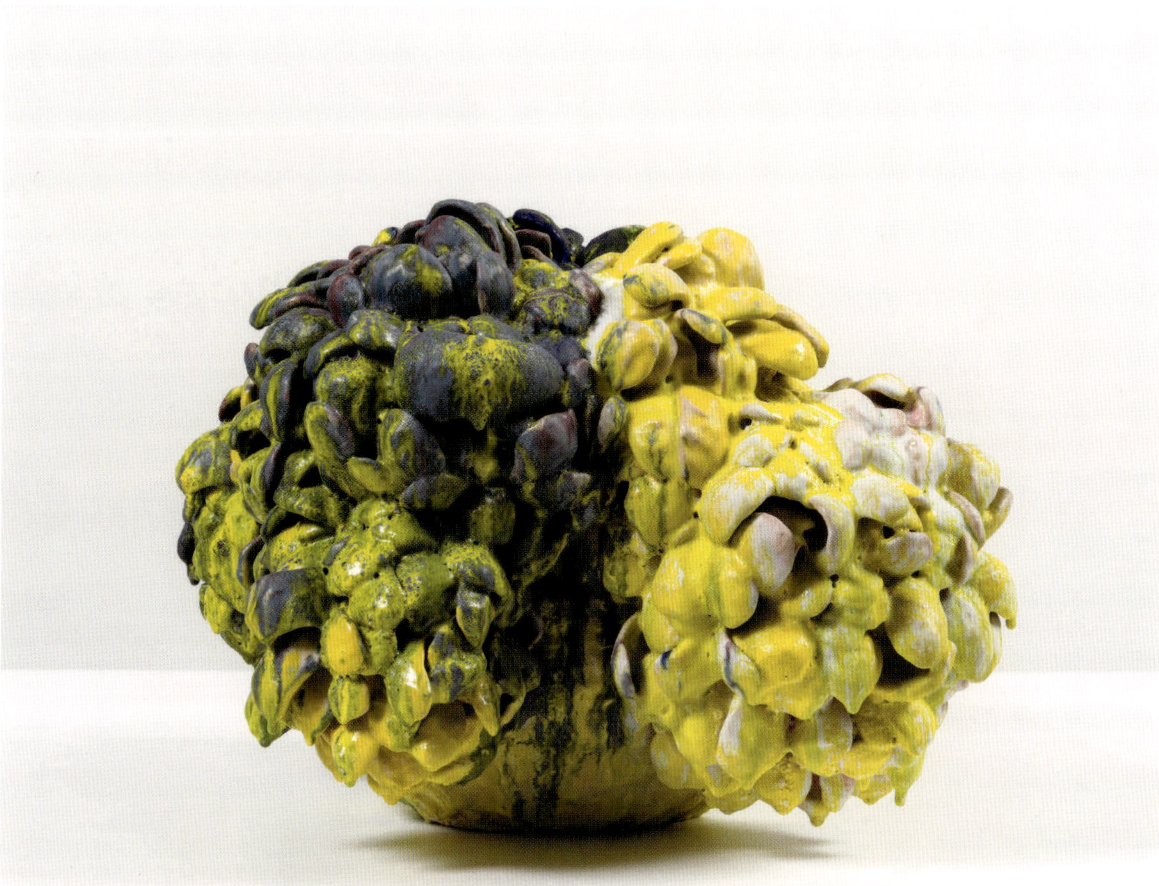

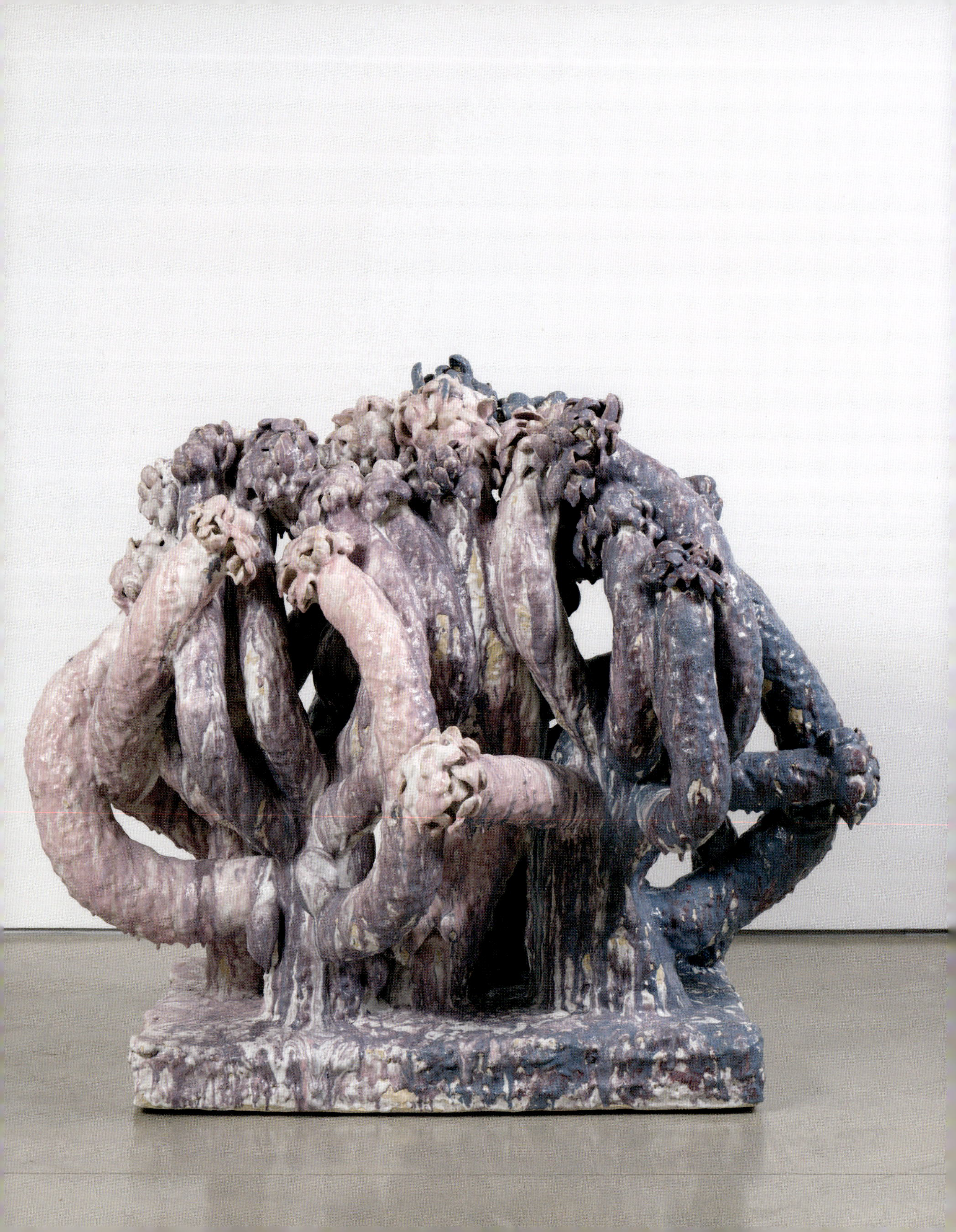

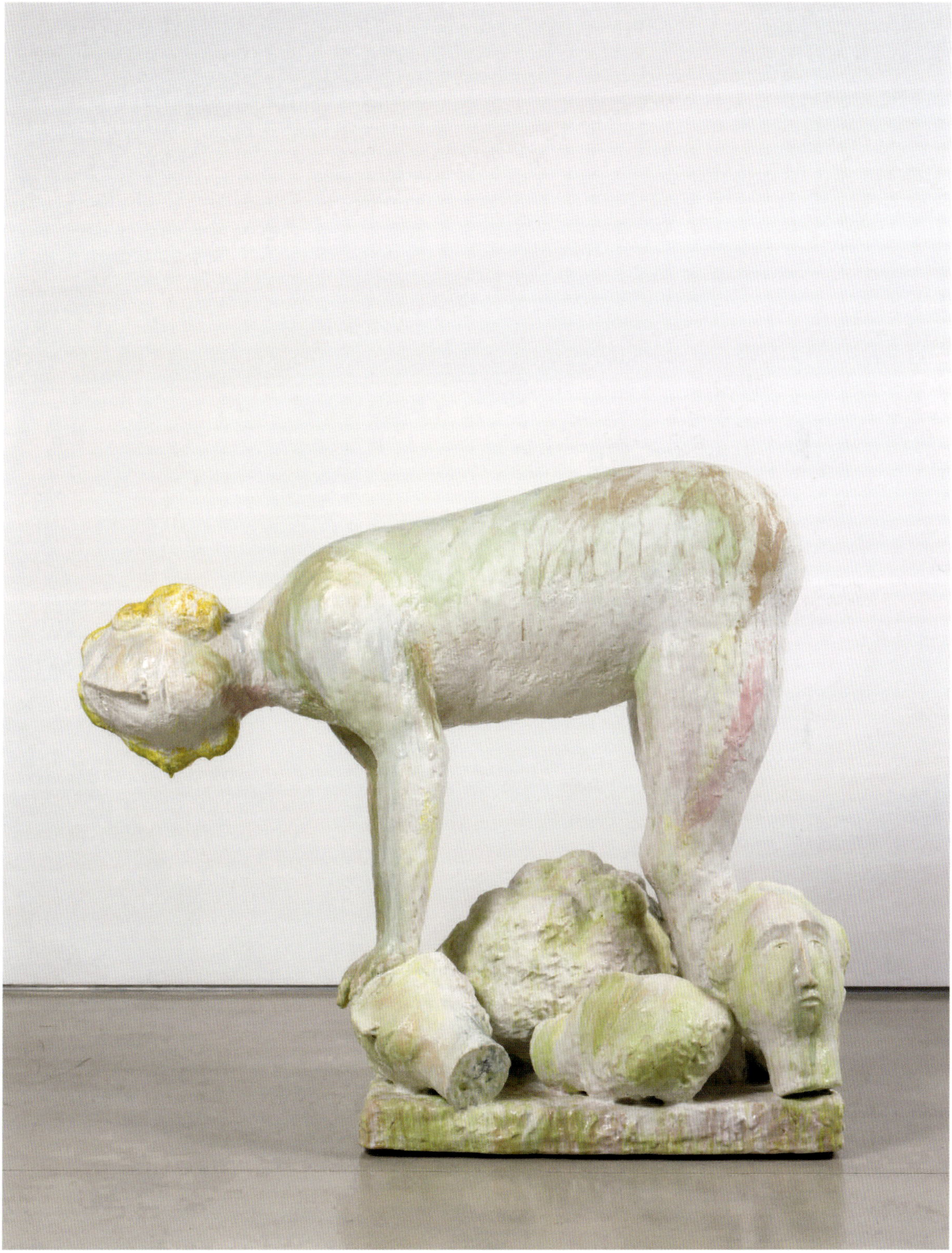

ZEMER PELED

LOCATION Long Beach, CA
YEAR OF BIRTH 1983
MATERIALS Ceramics
EXPERIENCE 12 years

Zemer Peled's work examines the beauty and brutality of the natural world. Her sculptural language is formed by her surrounding landscapes and nature, and engages with themes of memories, identity, and place. The association of porcelain with grace, refinement and civilization gets turned on itself when we are confronted with this material in another state. When a porcelain form is broken down into shards, the brutality of its jagged edges is juxtaposed with its insistent fragility. The material becomes violent and pretty, hard but breakable. When seen in the organic formations of Peled's structures, a whole for the shards is recreated, this time estranged from its original context of neatness, tradition, and cultivation, but nonetheless unified by an overall cohesiveness of movement and composition.

'The sculptures are formed of thousands of porcelain shards constructed into large-scale/small-scale sculptures and installations. All shards are made of porcelain and fired to high temperature. I use the slab roller and the extruder to create different forms, and after the pieces are fired I break them to pieces with a hammer.'

Photos courtesy of the artist

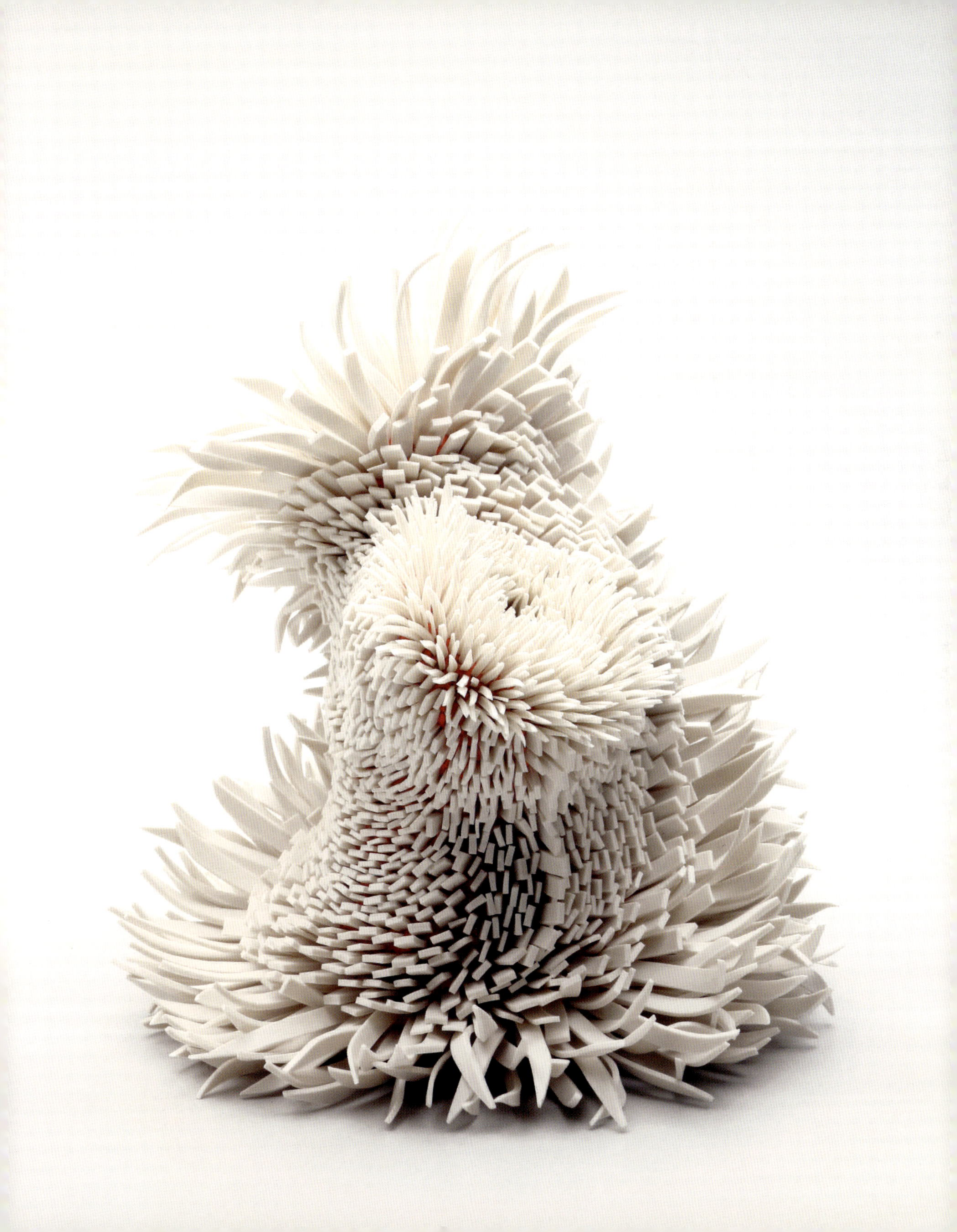

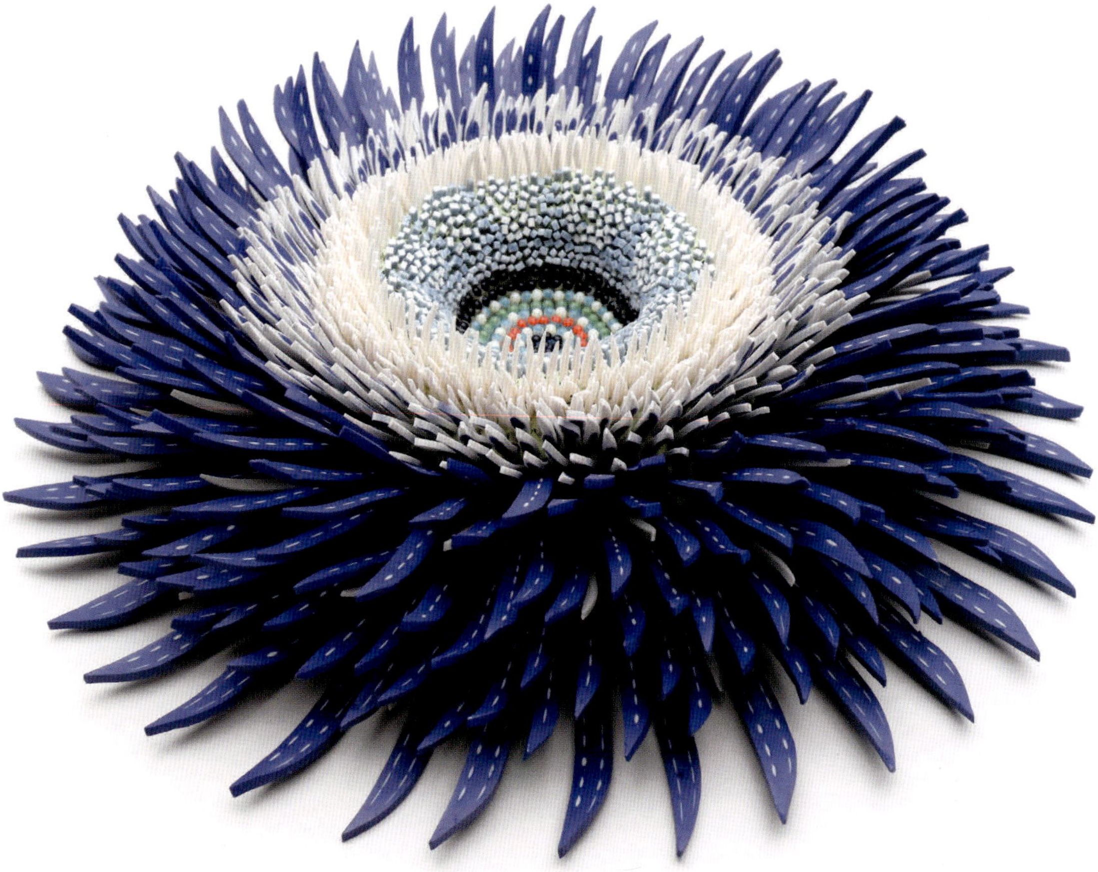

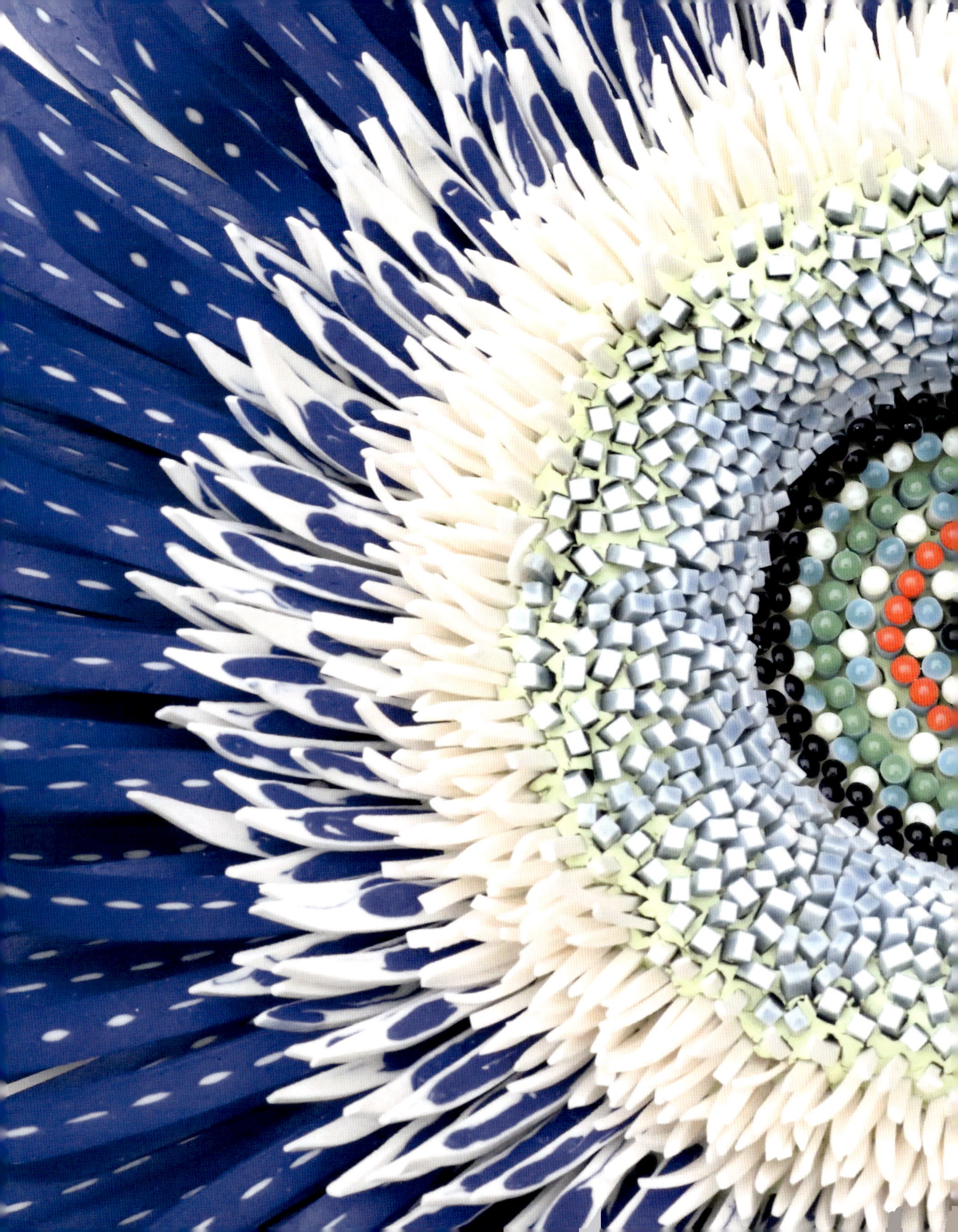

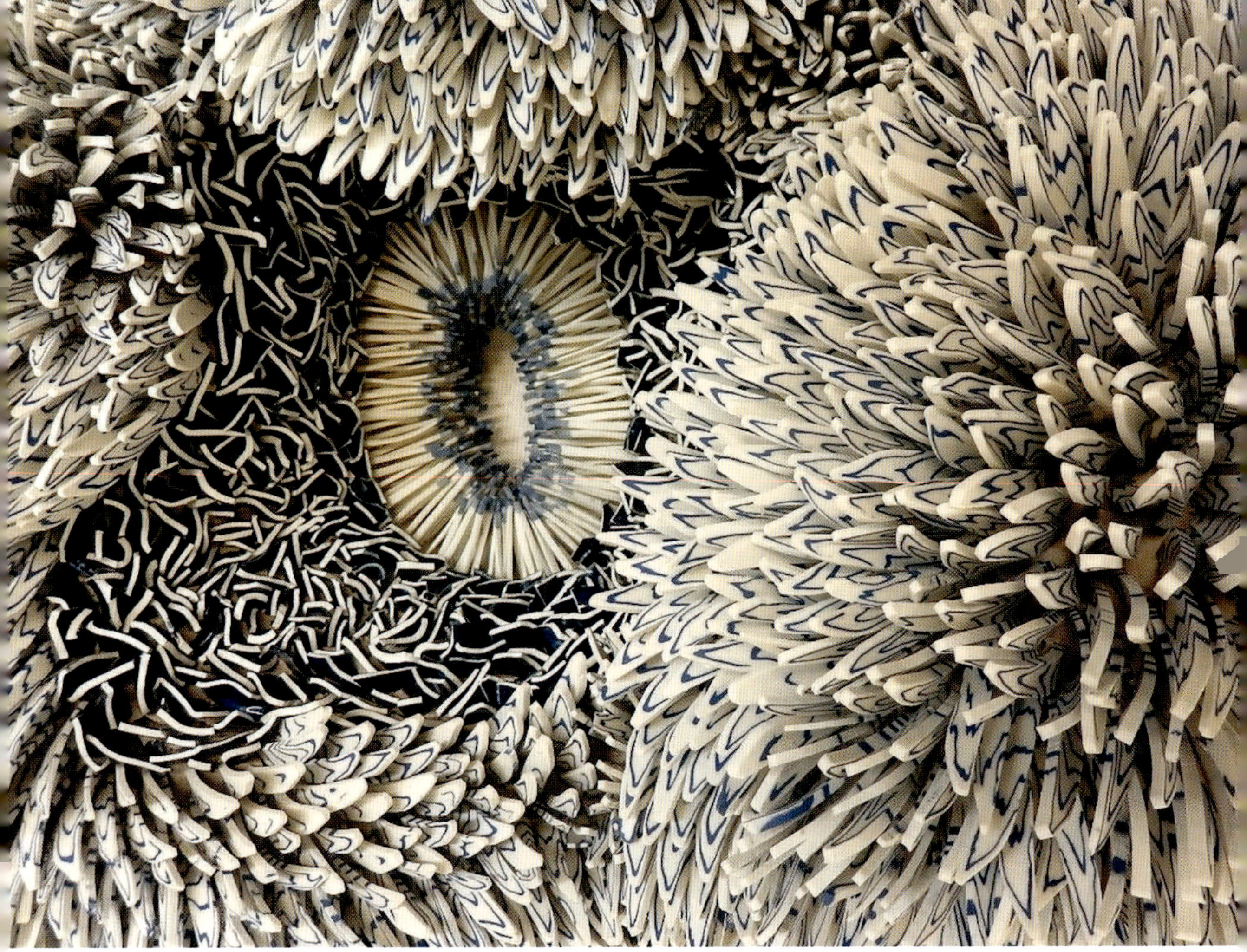

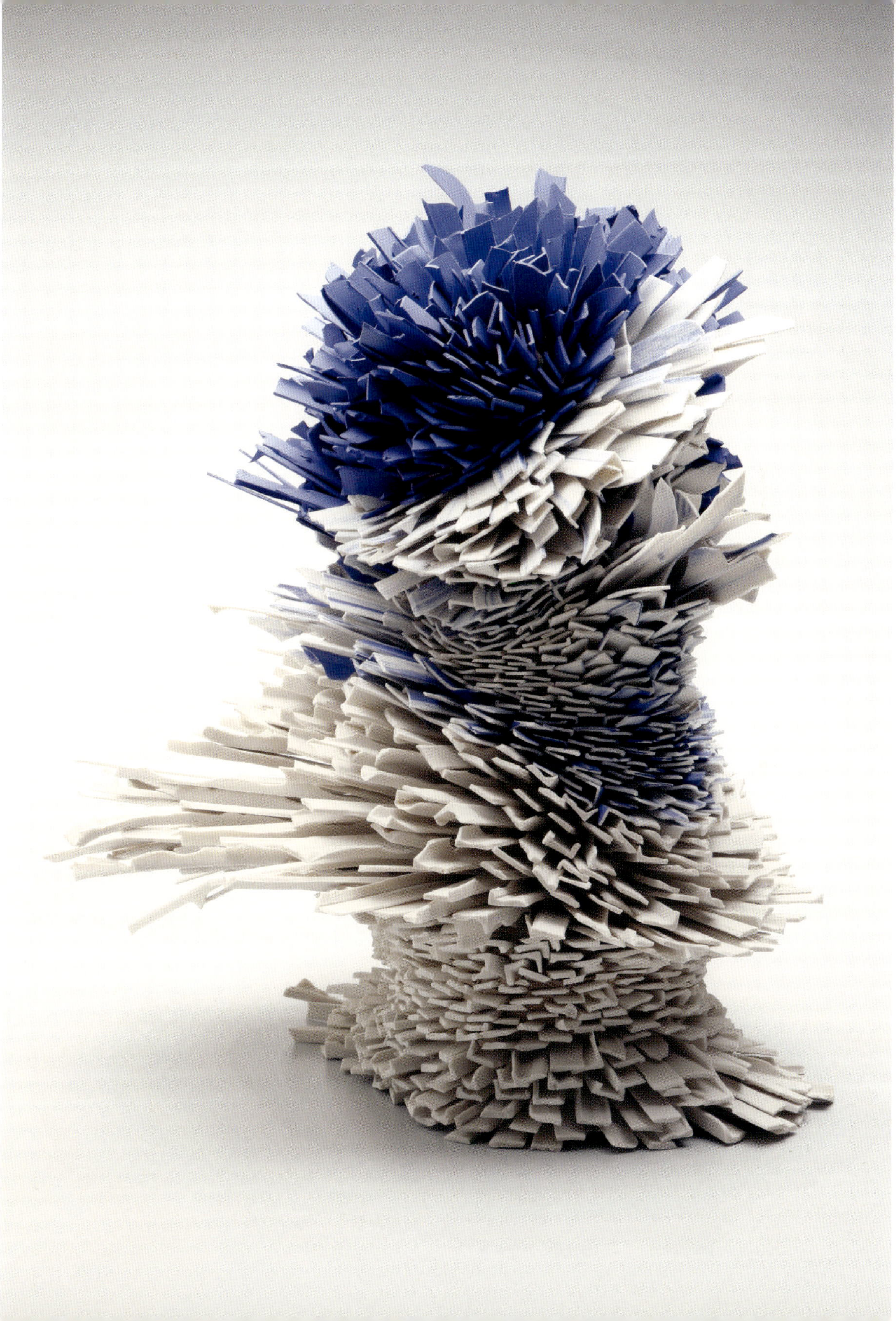

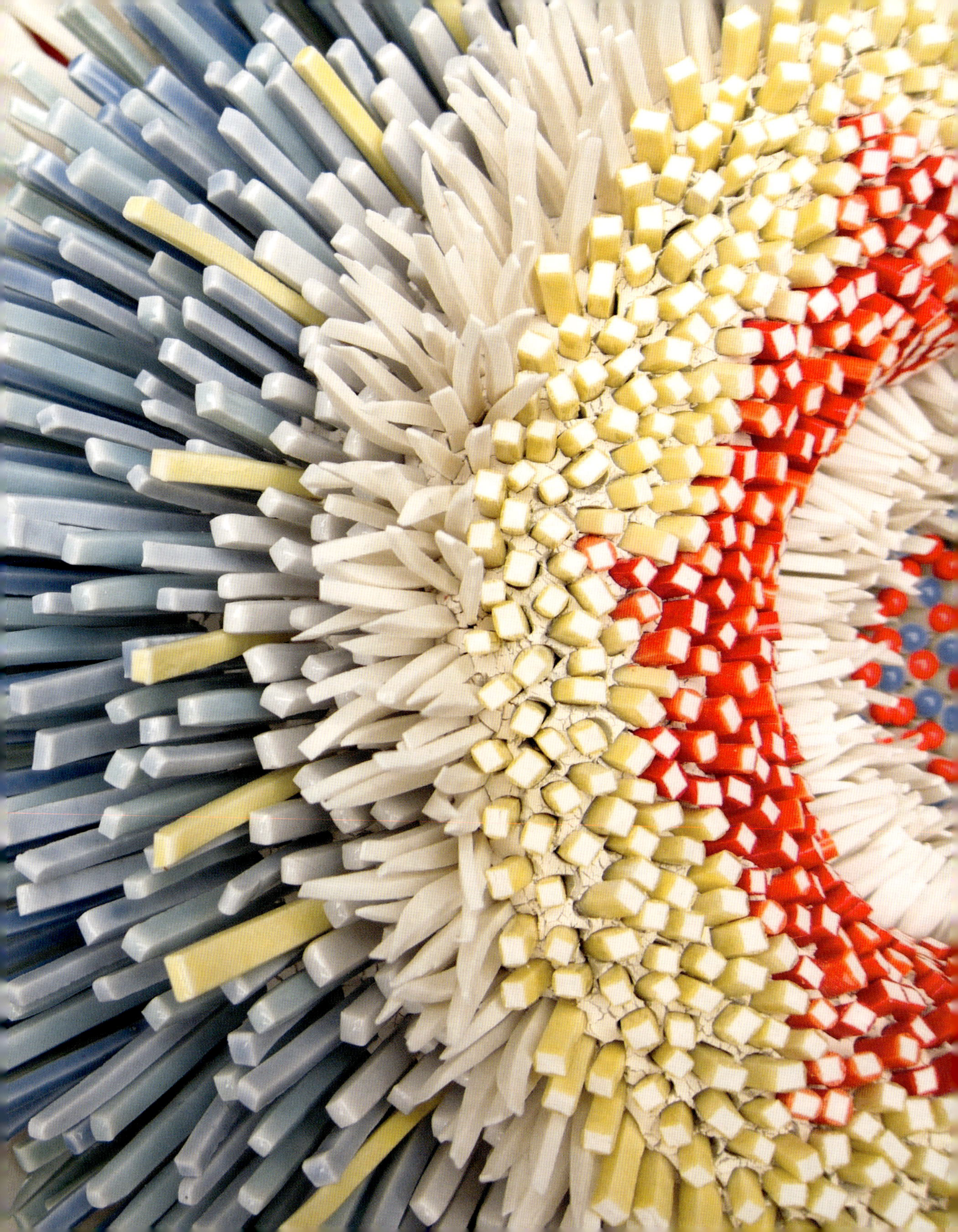

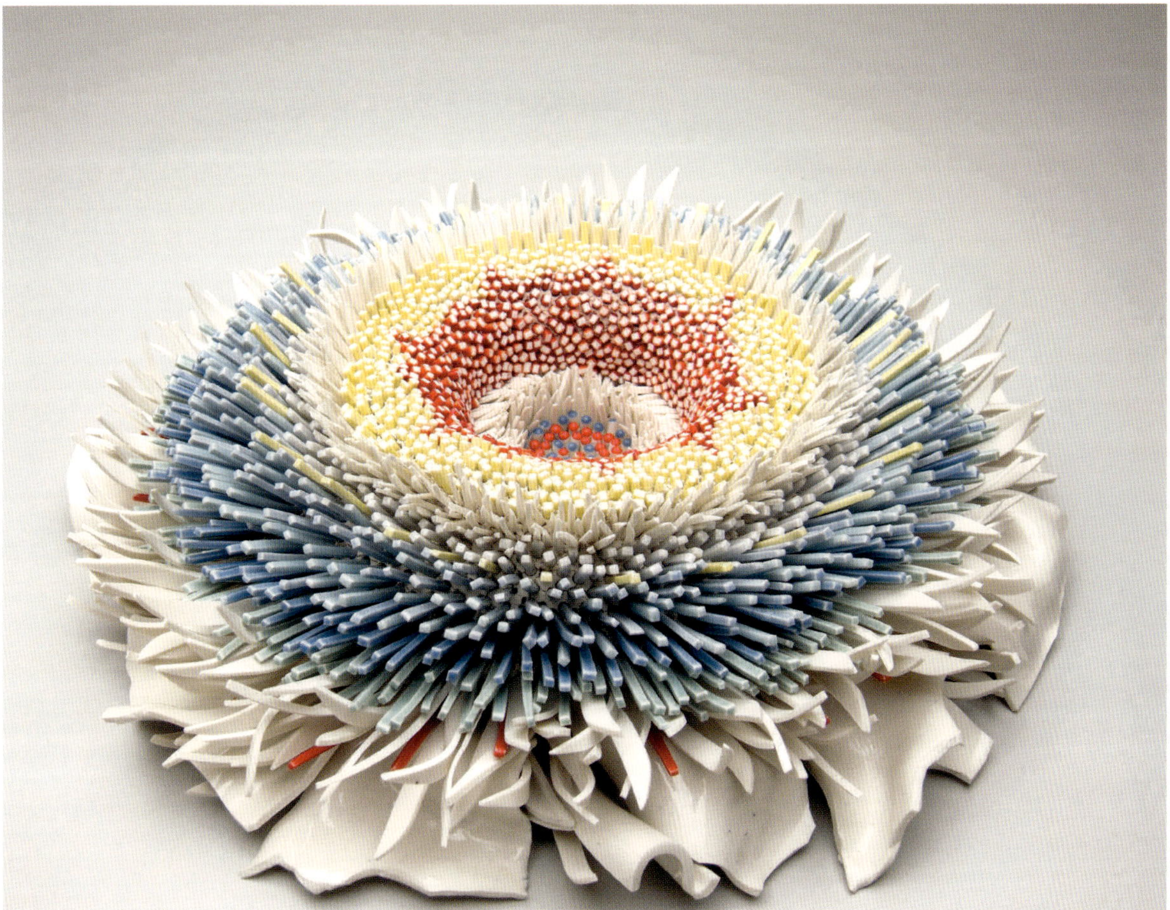

BRENDAN MONROE

LOCATION Oakland, CA
YEAR OF BIRTH 1980
MATERIALS Hand built and cast ceramics
EXPERIENCE 2 years

'I think it's important to constantly challenge oneself with new ideas and new mediums. My interpretations of the world are mostly rooted in science, then executed through painting and sculpting. These are the best ways for me to communicate, but I always enjoy making other things as well.

I tend to focus on the movement and activity of a surface. I've always gone between different types of mediums and transitioned to ceramic through wood sculpture and plaster mold making. The newest body of work incorporates attention to the gesture of the abstract animal shape and the activity of the glazes at their melting temperatures.'

Brendan Monroe is a painter, sculptor, and illustrator living and working in Oakland, CA with his wife Evah. Monroe completed a series of works with Heath Ceramics.

Photos courtesy of Heath Ceramics and the artist

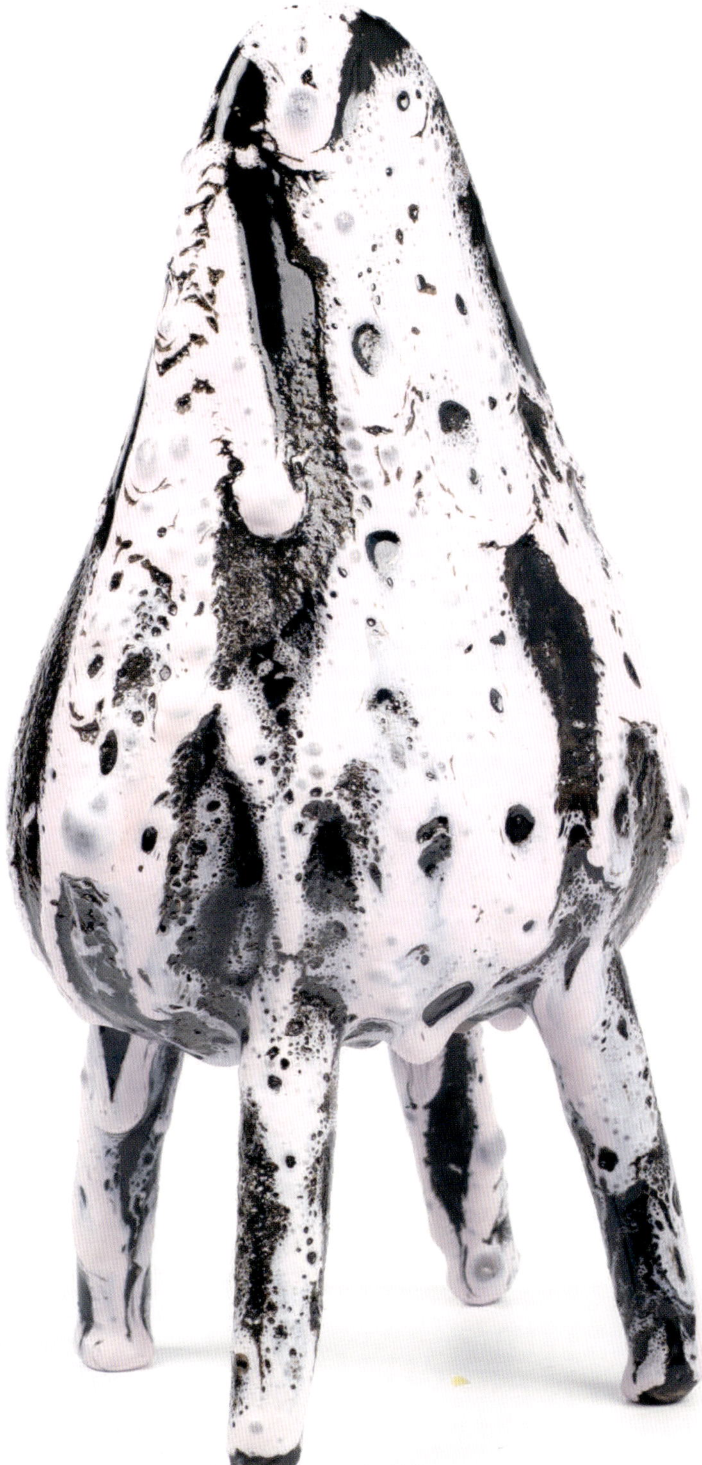

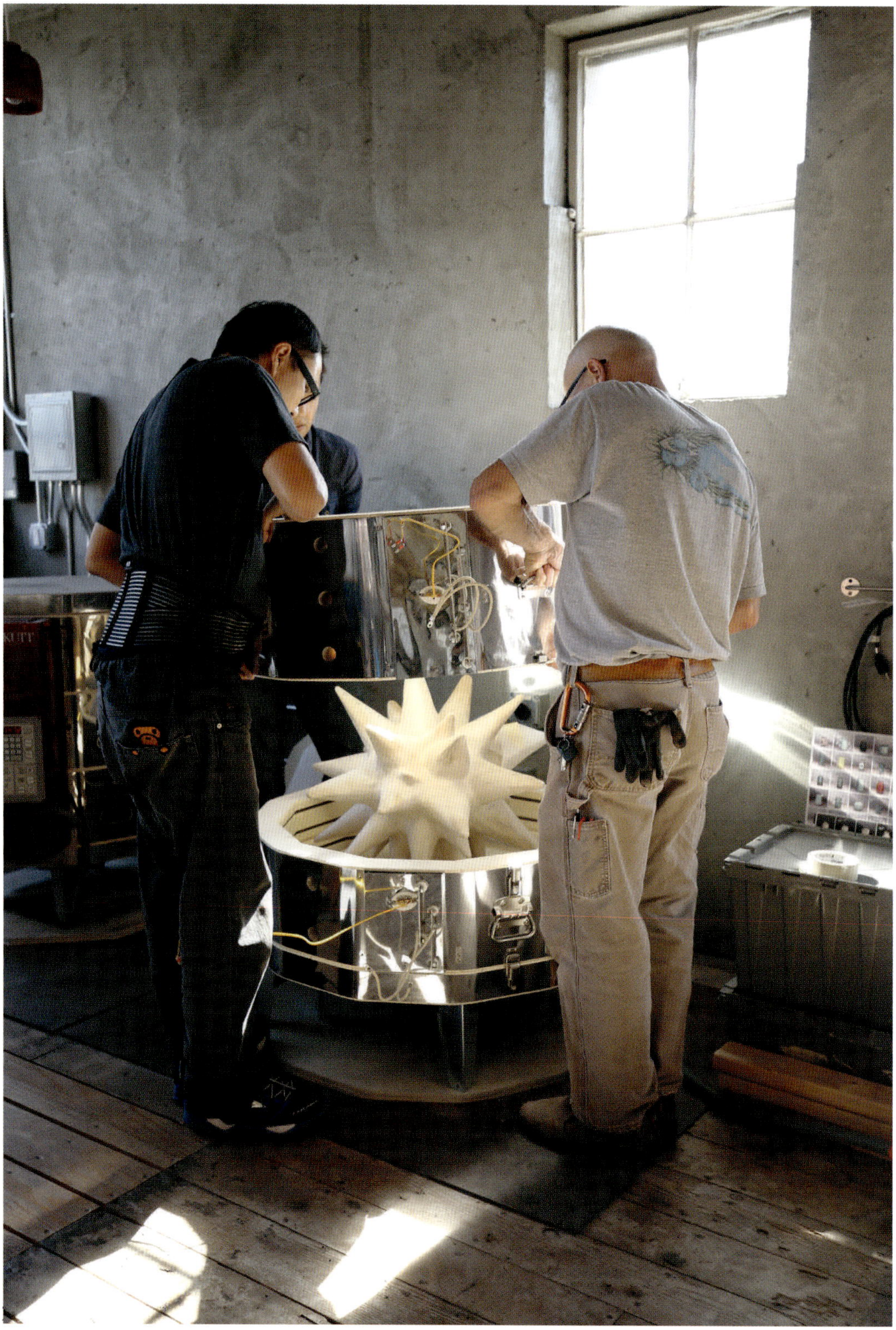

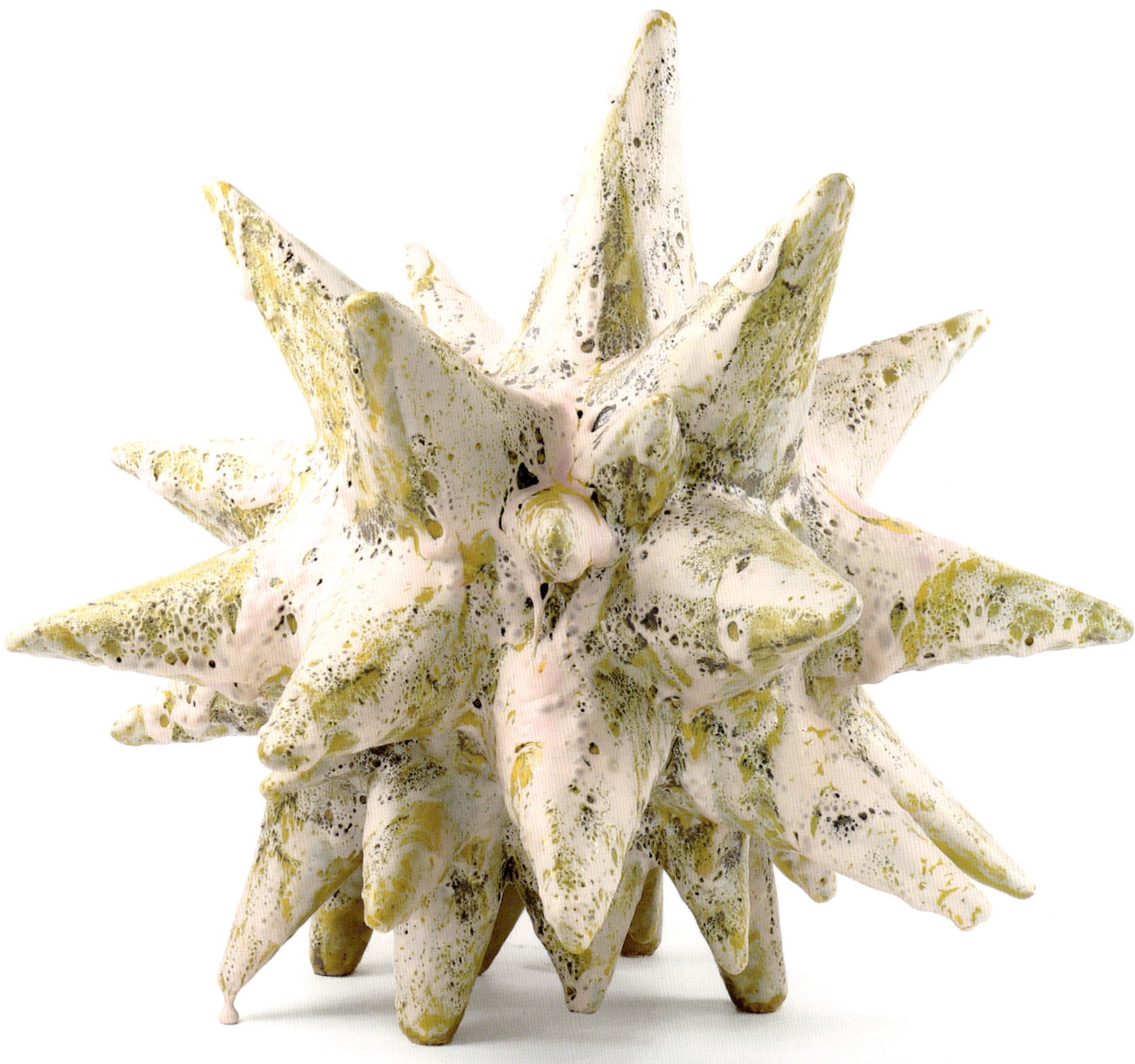

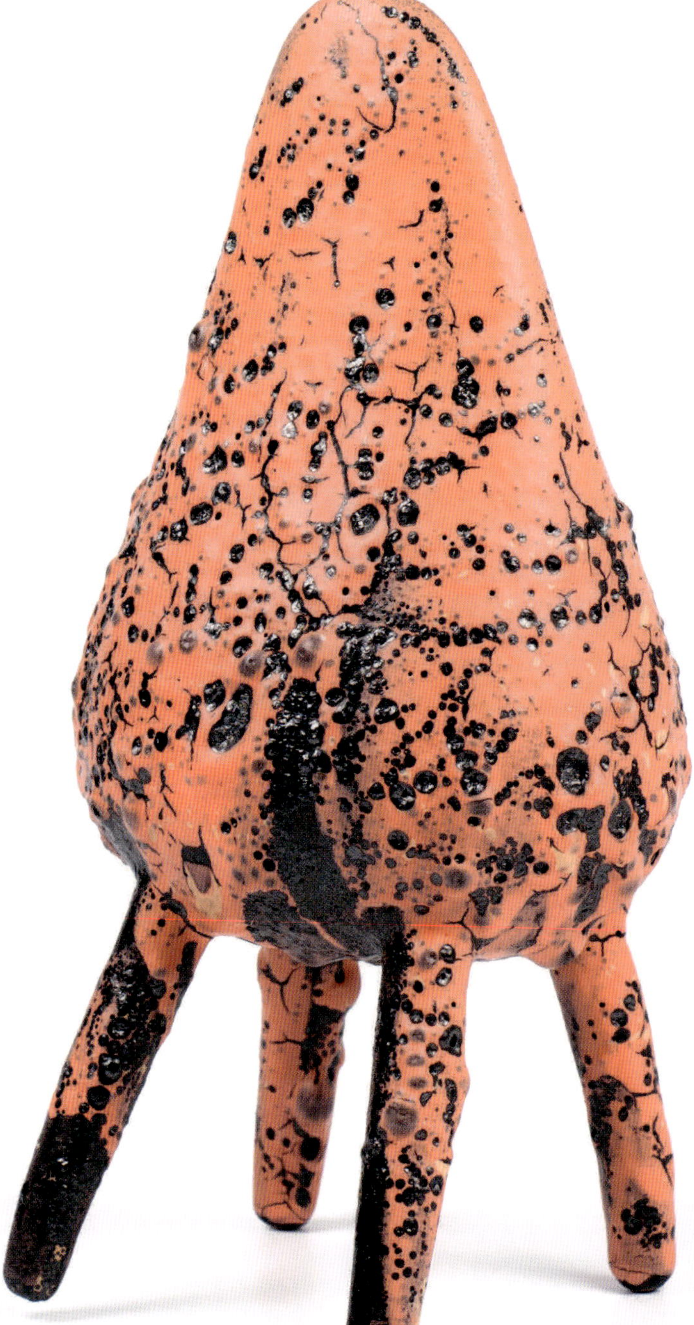

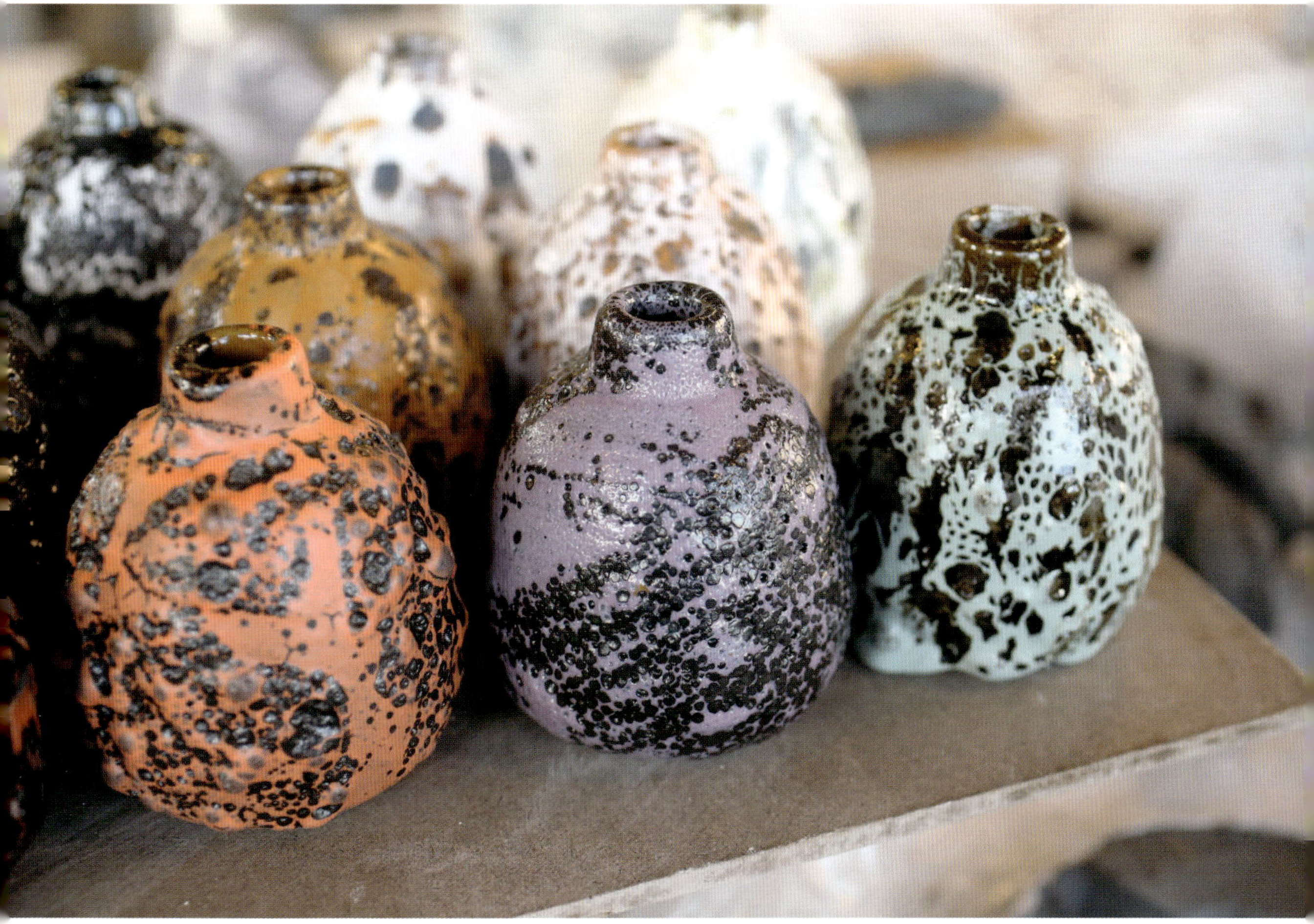

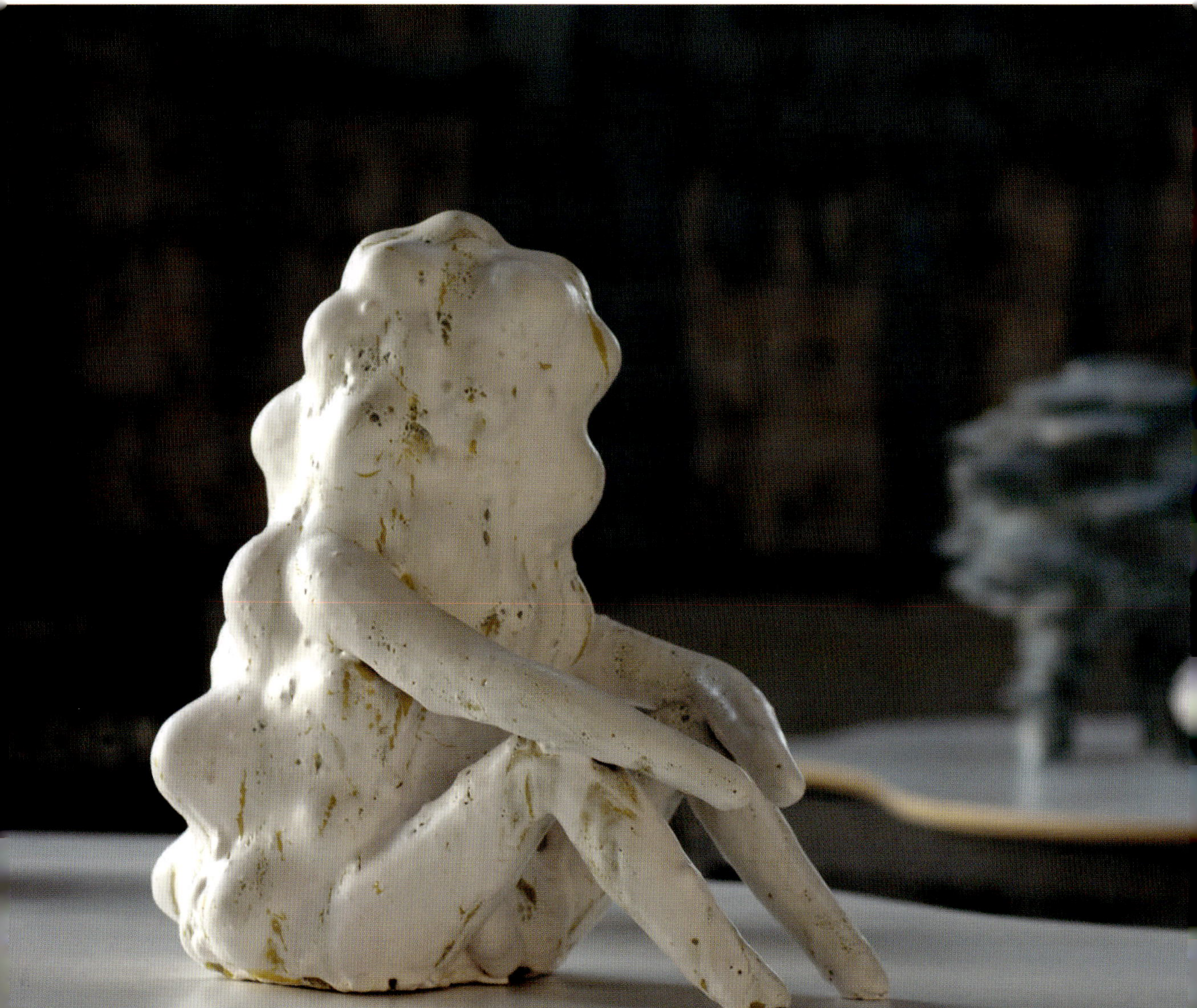

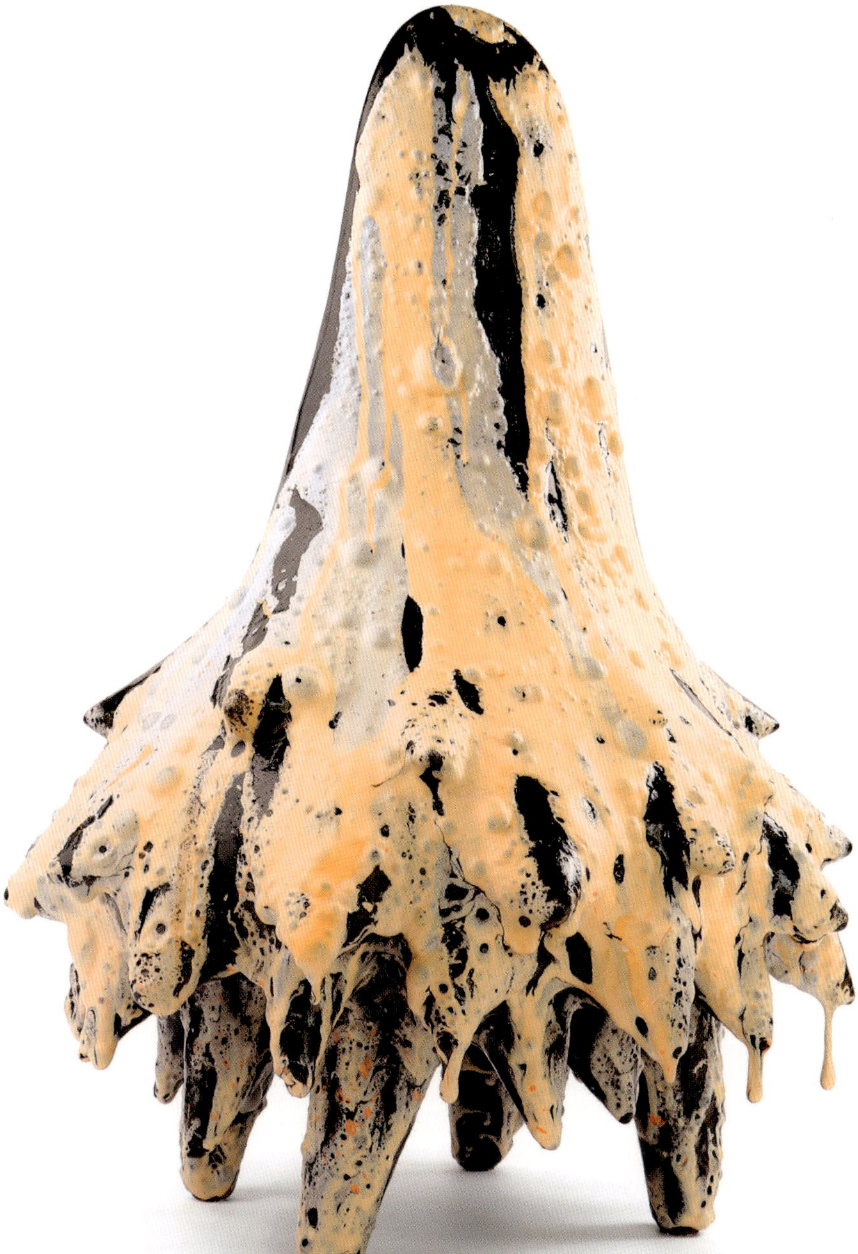

LAUREN SHAPIRO

LOCATION Miami, FL
YEAR OF BIRTH 1984
MATERIALS Slip-cast and hand built ceramics
EXPERIENCE 10 years

Lauren Shapiro's art is founded upon groupings of geometric, dimensional forms to collectively achieve a balance between growth, rhythm, and repetition, and evoking imagery like visual poems or musical compositions. Symbolically, Lauren perceives the foundation shapes as an accumulation of her time in minutes, aggregated in hours and days. Her chosen medium of ceramics permits Lauren to dually explore the contradictory ideas of strength versus fragility while simultaneously conveying the functional geometry of organic elements. The end result is a defined form, frozen forever like cave structures in a neutral white. The choice of white is intentional, blurring the transformative fabrication line between the paper and the porcelain and suggesting to the viewer the illusion of either or both materials.

'I use folded paper forms to create plaster molds for slip-casting the ceramic pieces, which become the raw material from which I assemble my sculptures.'

Shapiro is a South Florida born-and-raised sculptor working almost exclusively with ceramics. After graduating with a B.F.A. in Studio Arts from Florida Atlantic University in 2009, Shapiro commenced studies toward her Master of Fine Arts degree at the University of Miami, Florida, concurrent with a Graduate Teaching Fellowship; she is a 2016 MFA Candidate.

Photos courtesy of Heath Ceramics and the artist

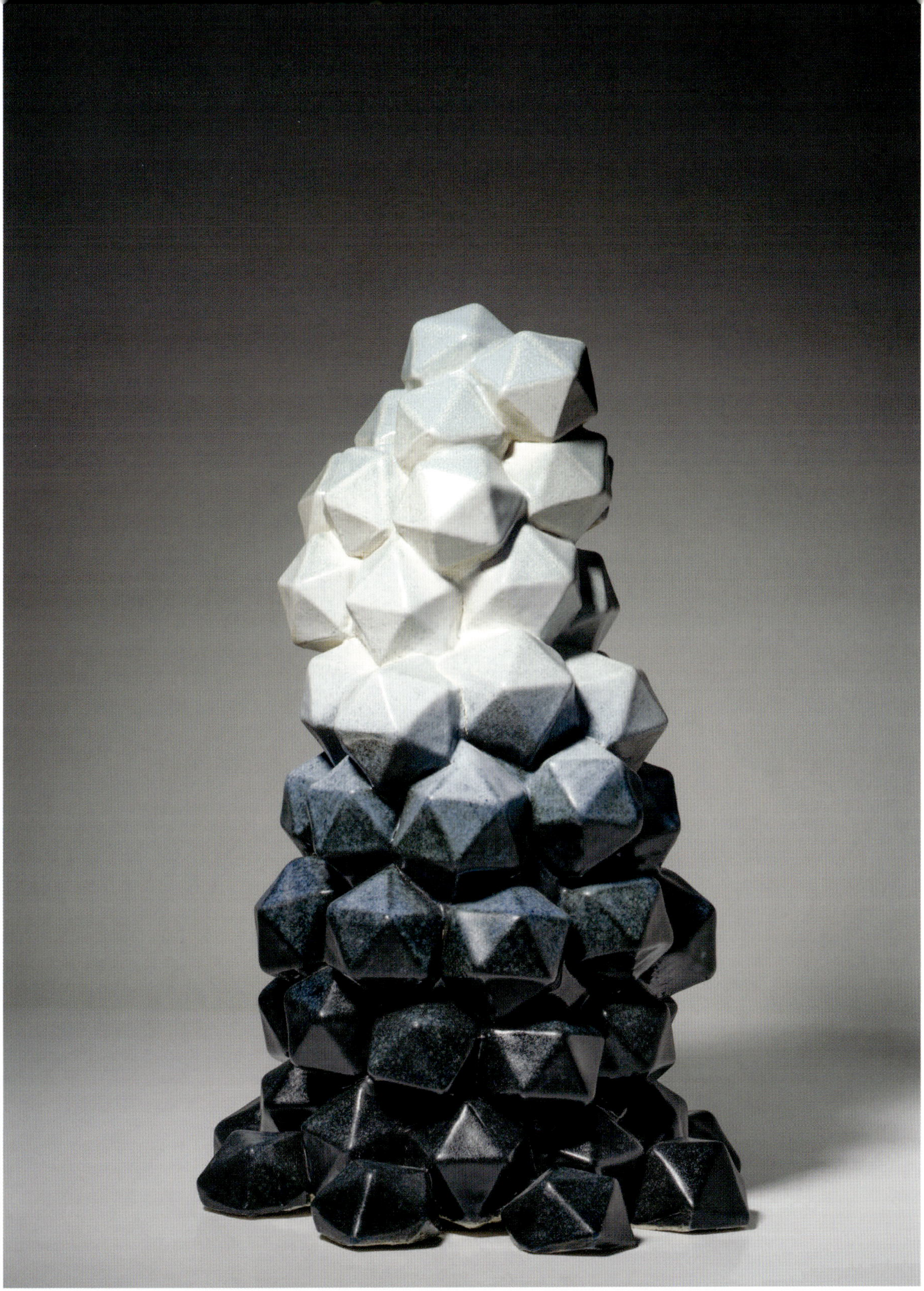

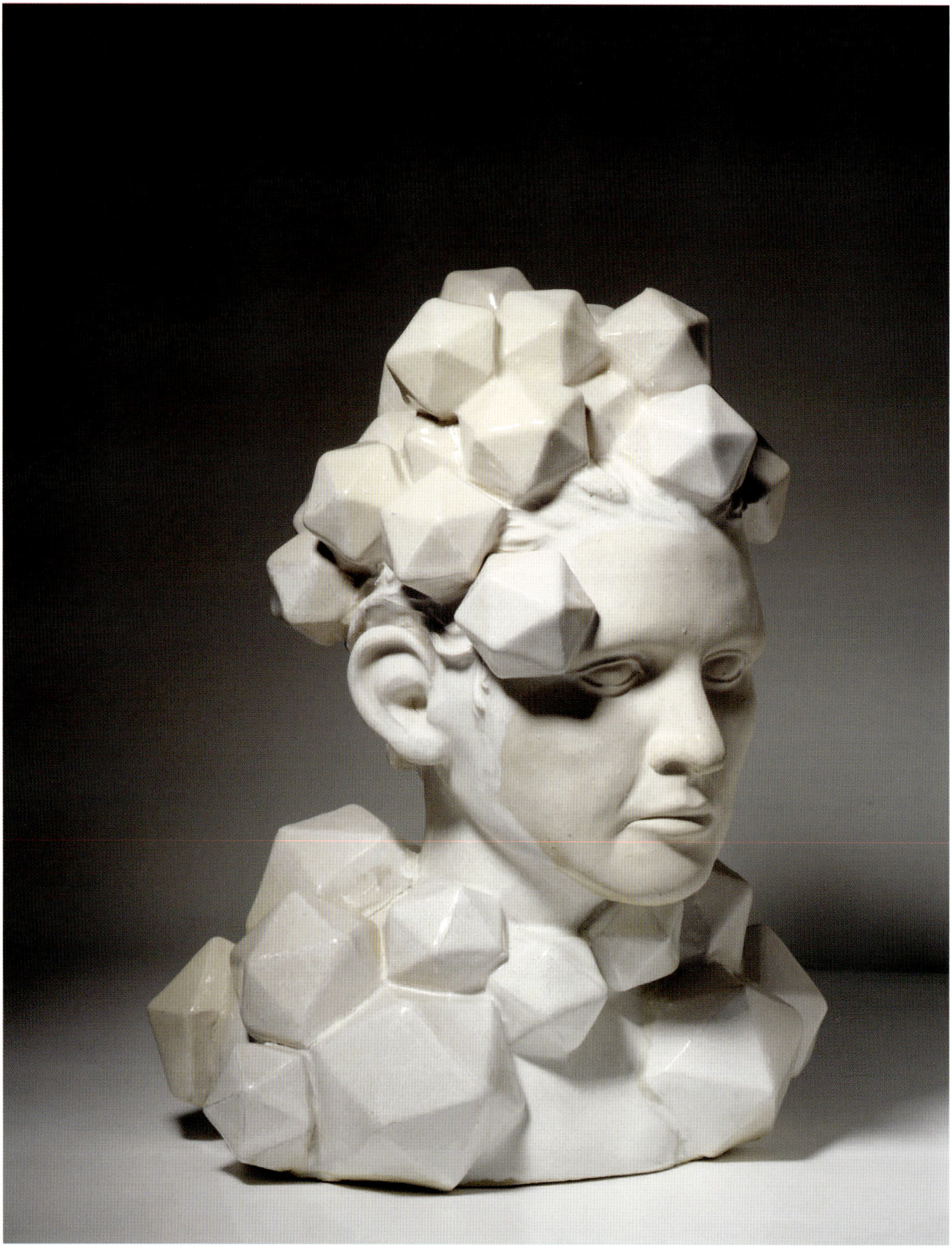

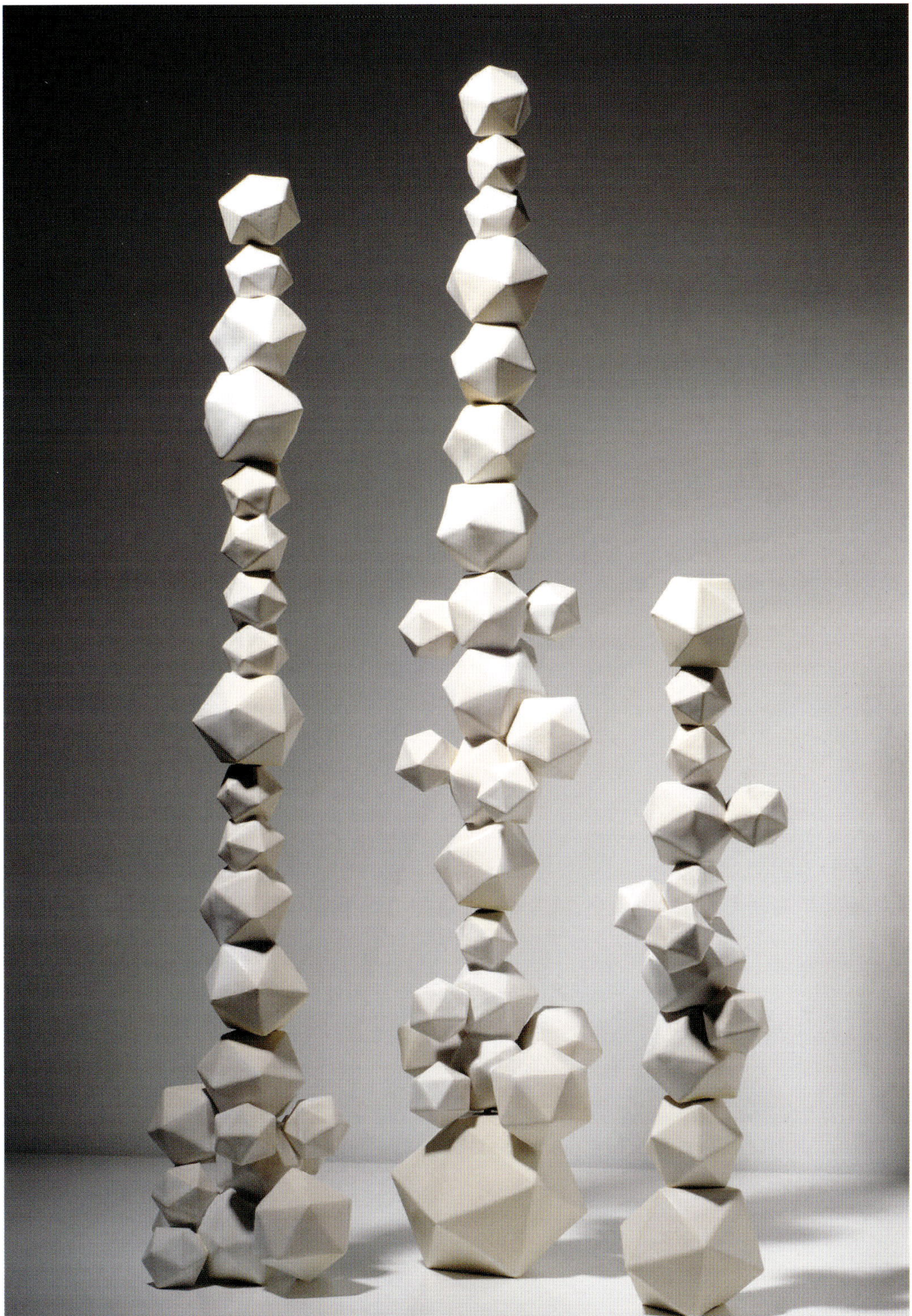

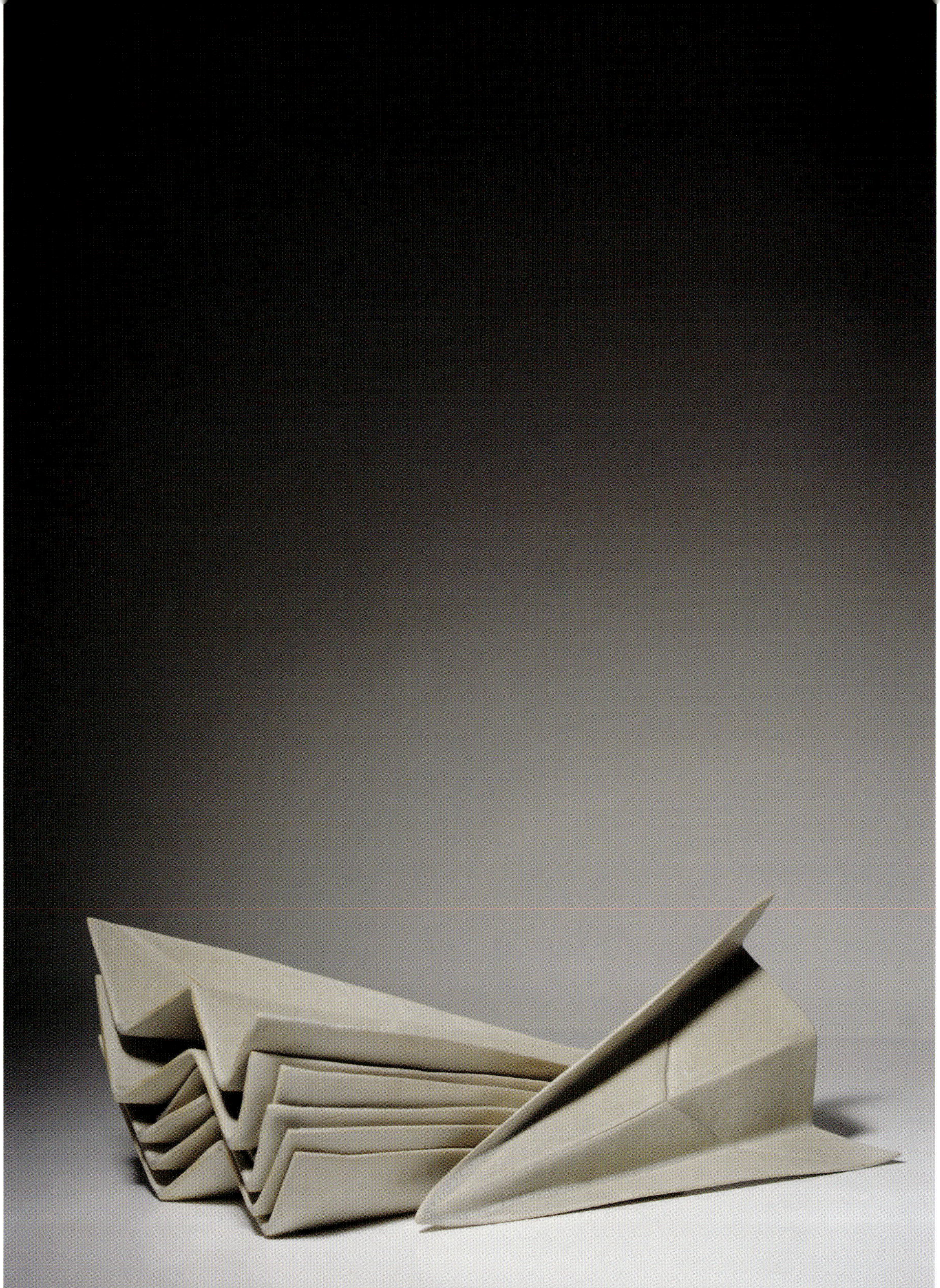

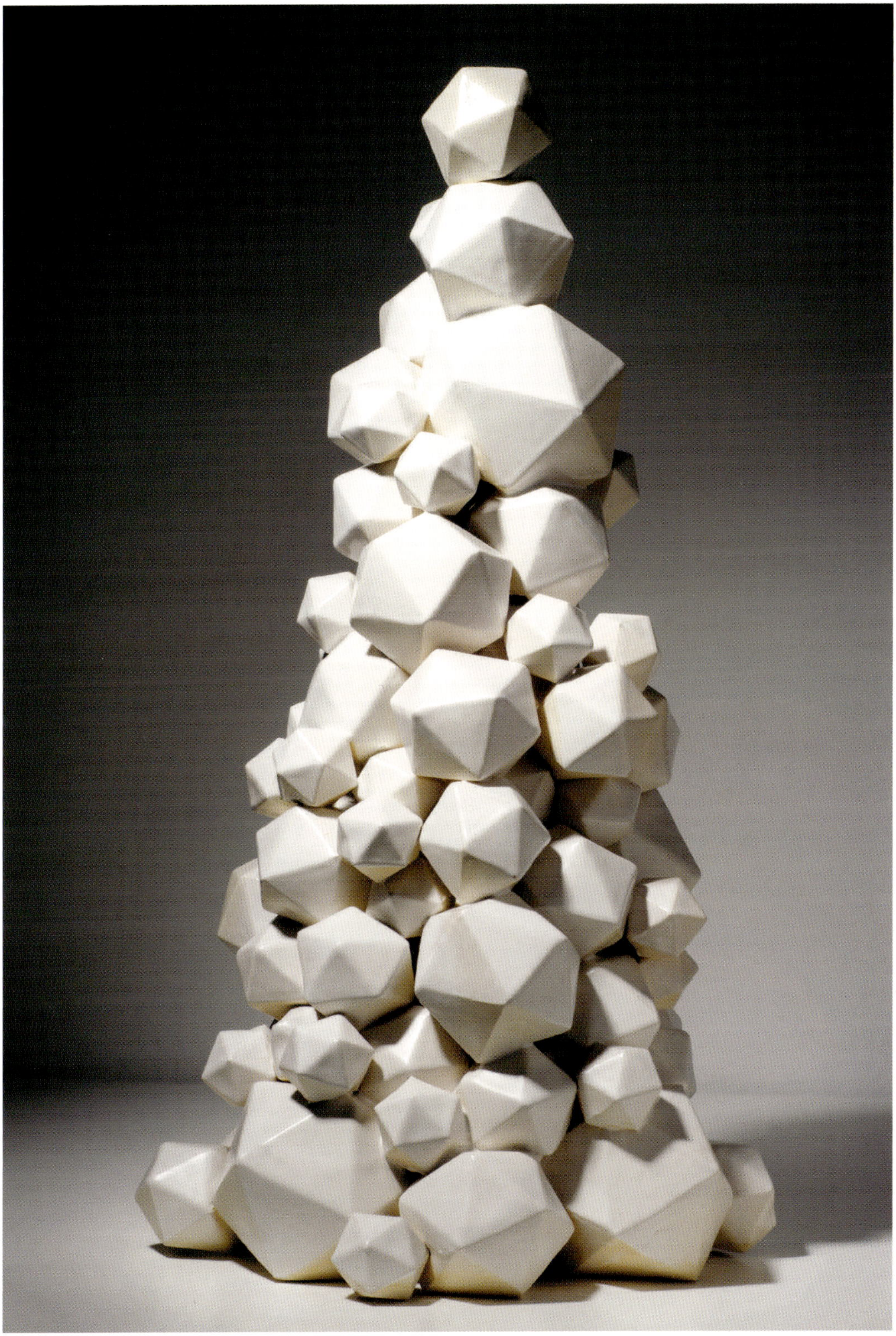

NICK MAKANNA

LOCATION San Francisco, CA
YEAR OF BIRTH 1988
MATERIALS Low-fire clay and glazes, slabs and coils and slipping everything together
EXPERIENCE 2 years

'My work centers around the ruin as a means of exploring and understanding place in relation to my hometown of San Francisco. The abandoned industrial relics that dotted the city, and more specifically the surrounding flats of Potrero Hill where I grew up, served as important sites of lawlessness; a shambled utopia welcoming of any and all pursuits. As much of this fragile landscape has fallen at the hands of new development, abstracted interpretations of previous structures begin to populate an alternate reality within my own work. Built in brightly glazed ceramic and saturated oil paint, a world of dilapidated and collapsing forms unearths connections to longer lineages of Bay Area artistic tradition: abstract-expressionist ceramics, fluorescent 70's funk, and sculptural 60's figuration to name a few. The appreciation of the ruin rejects modernist principles of technological progress and innovation — often arrogant and misguided notions that guide much of the change and displacement sweeping the Bay Area. Through the translation, abstraction, and depiction of these once-vibrant areas I am able, momentarily, to assuage my own anxieties of a quickly changing San Francisco, all in one fickle attempt to make the ephemeral eternal.'

Photos courtesy of Laura Rokas and the artist

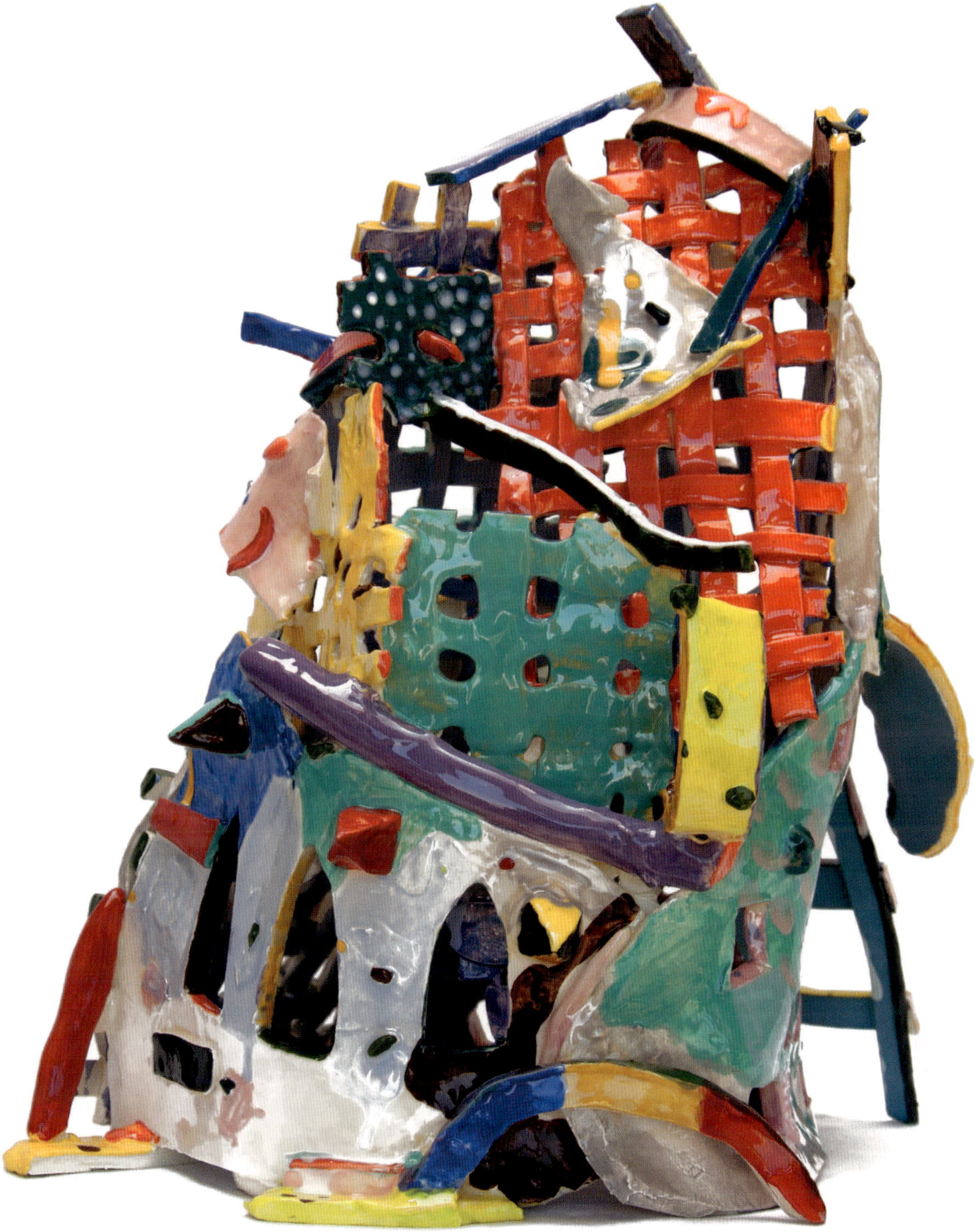

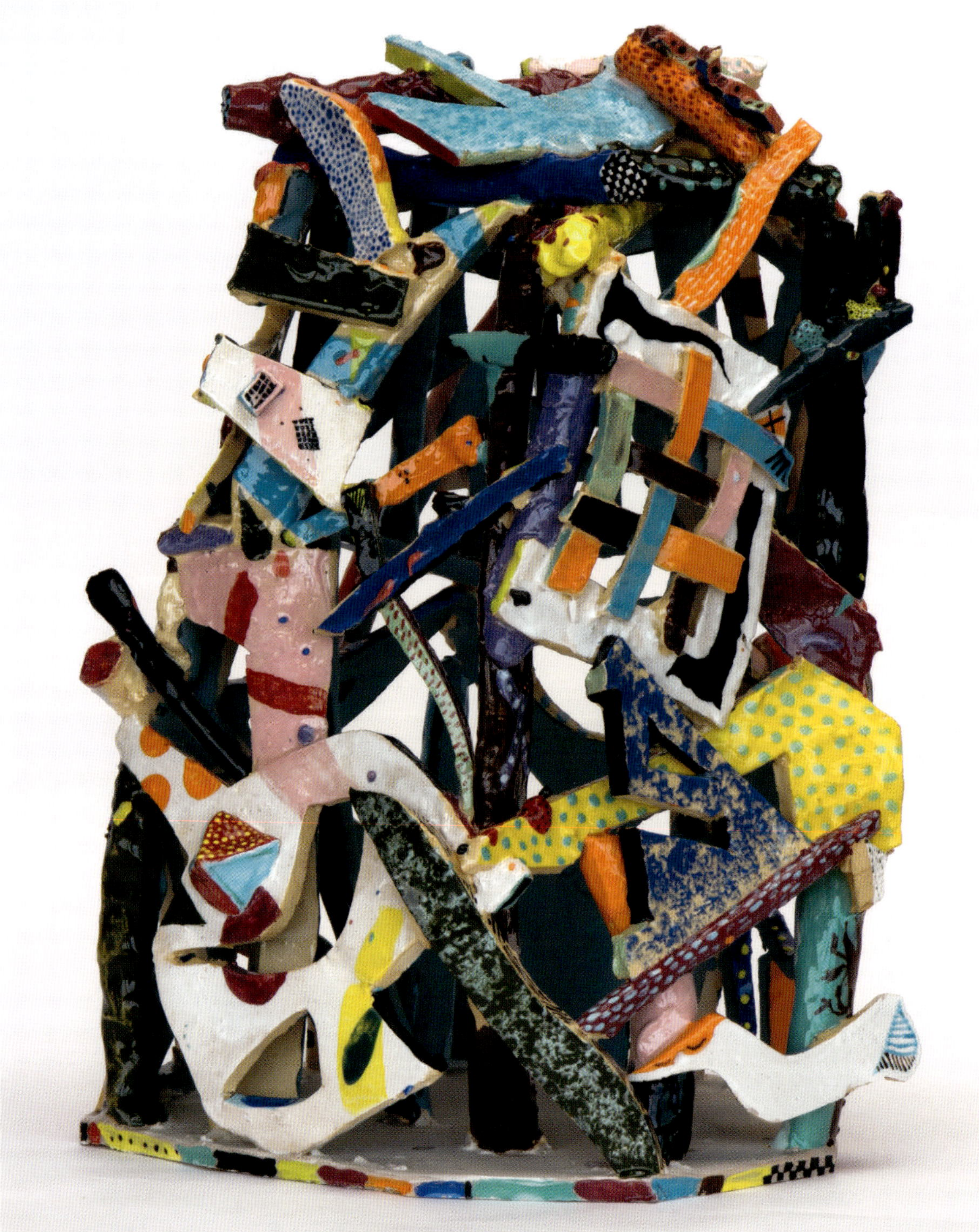

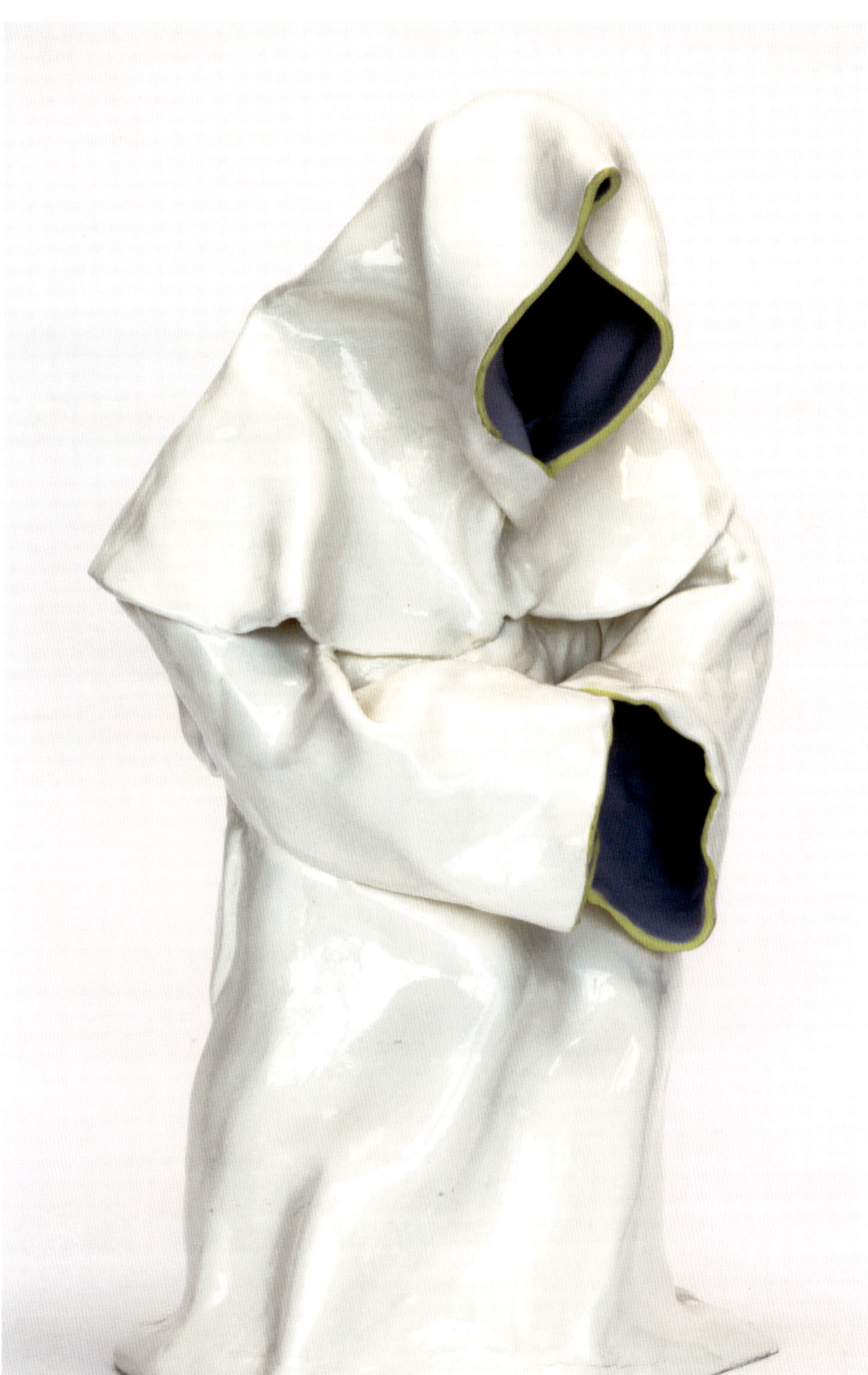

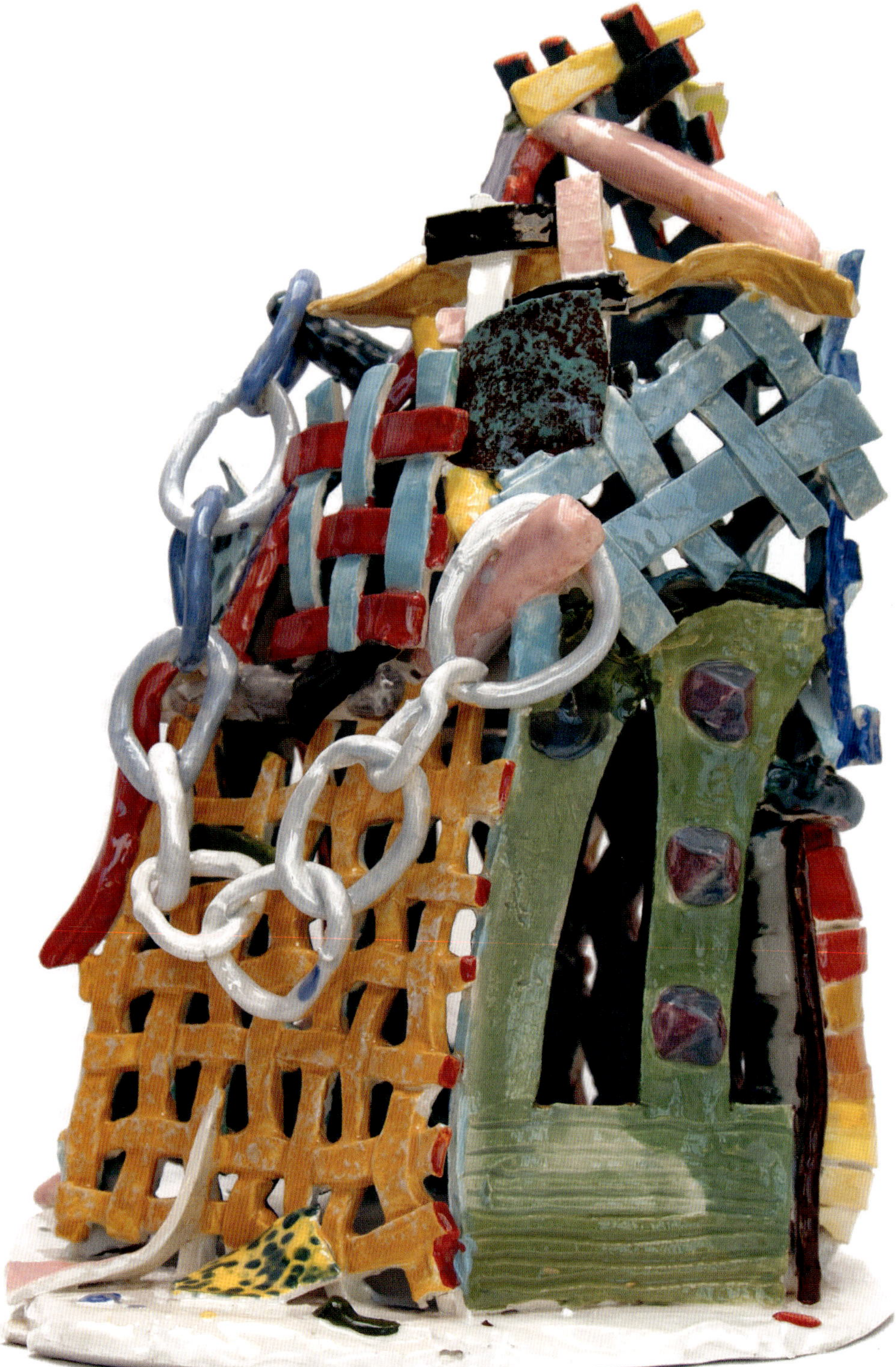

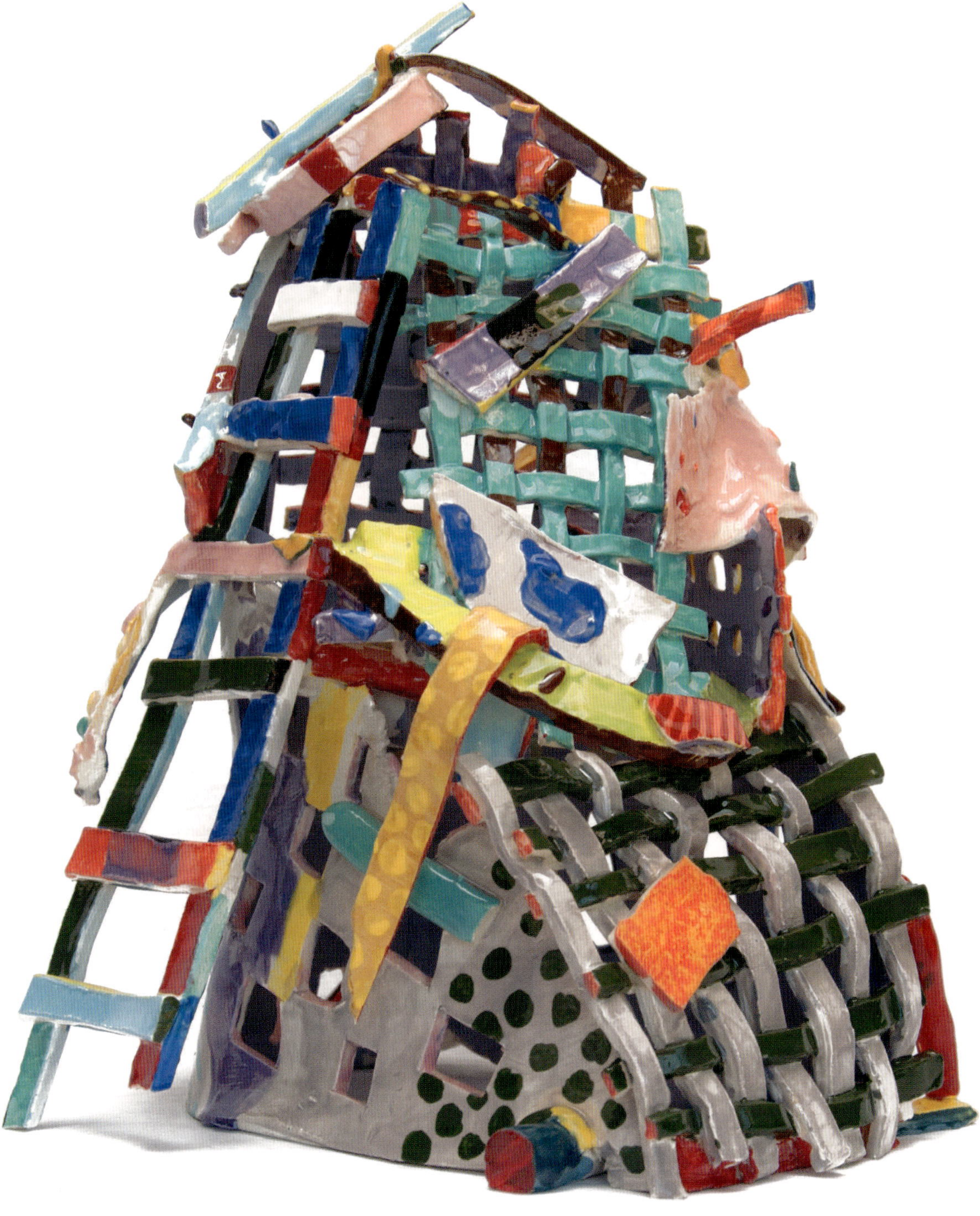

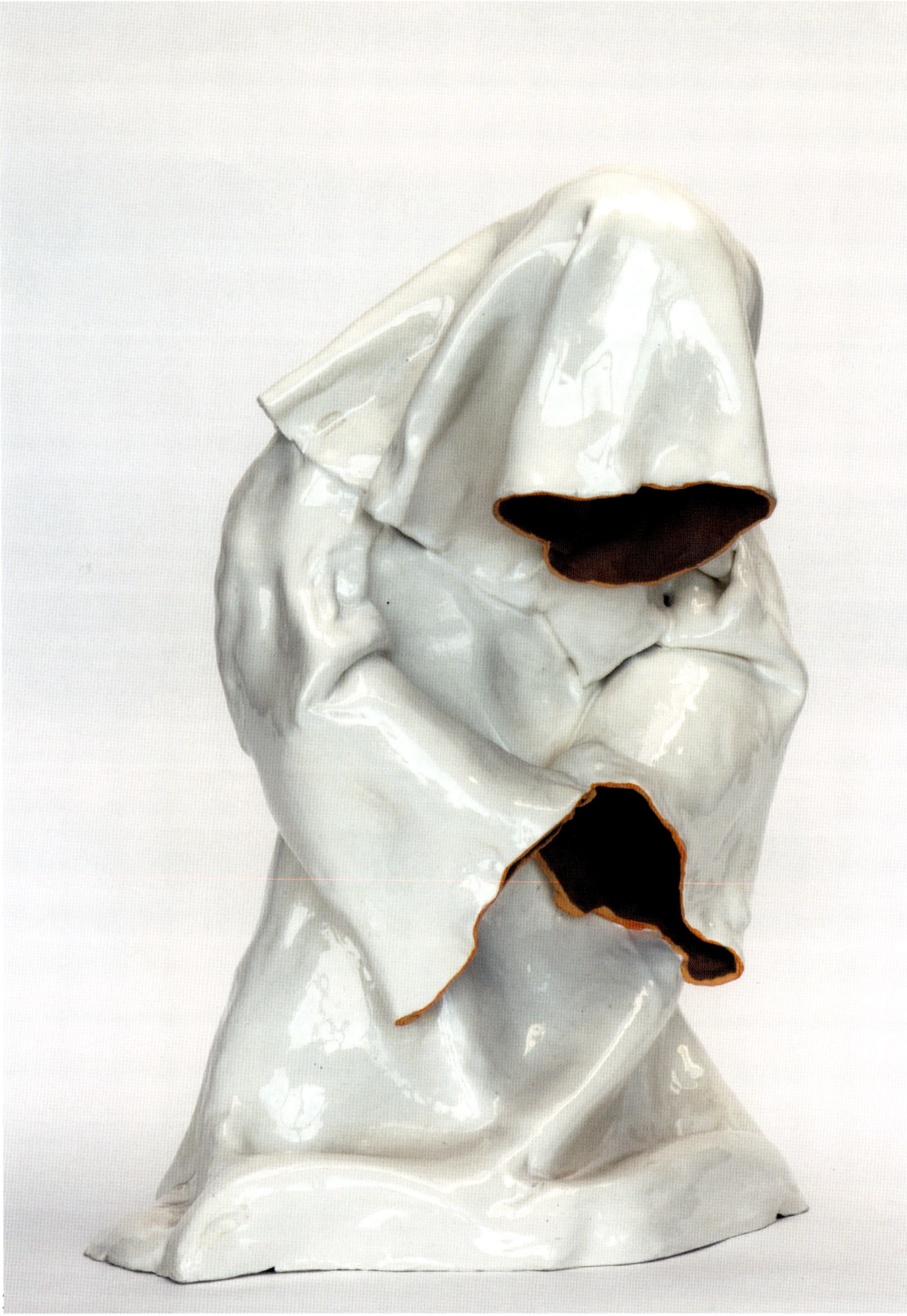

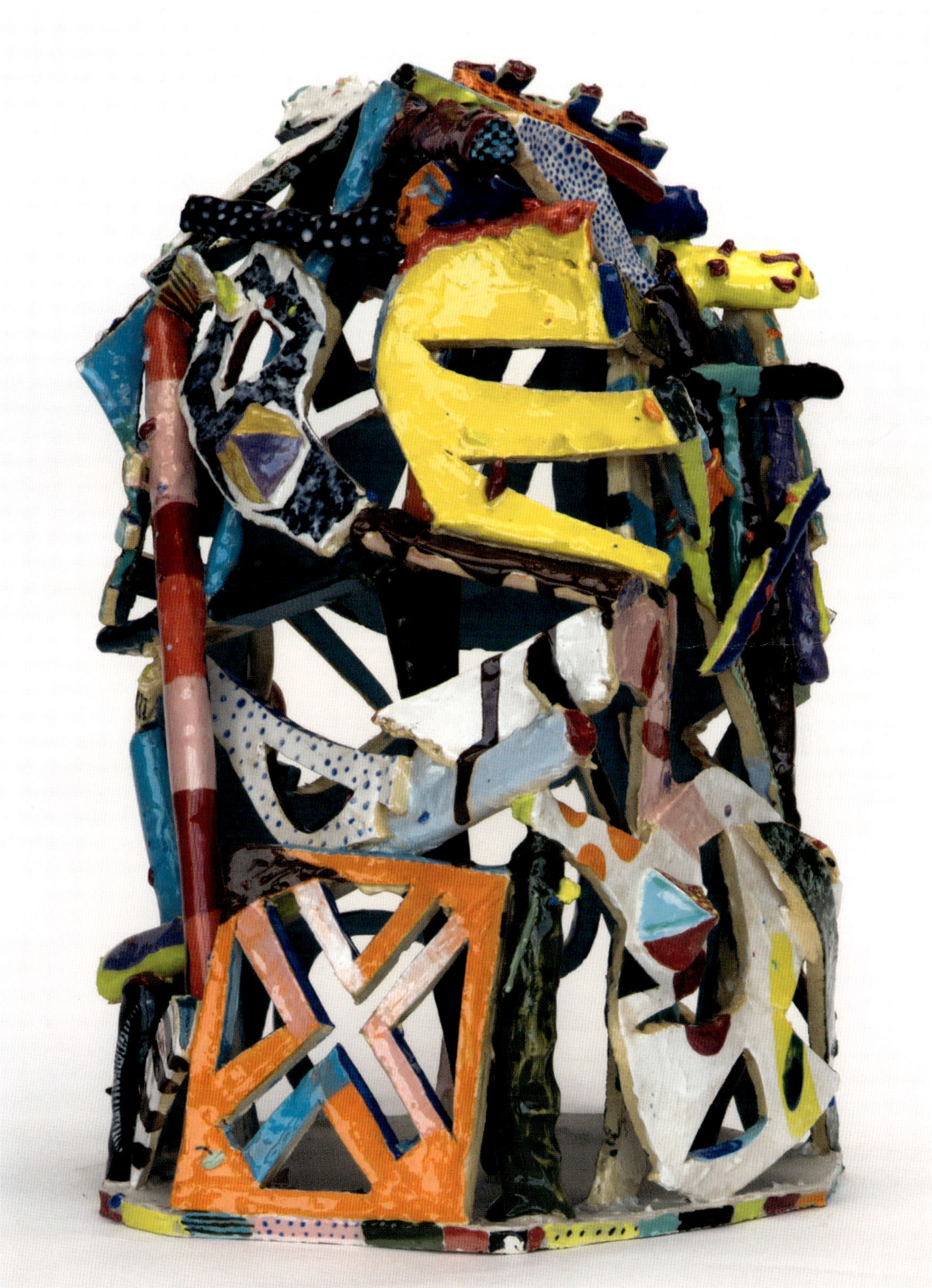

FIGURATIVE WORKS

TANYA BATURA

JESS RIVA COOPER

CRYSTAL MOREY

KATE MACDOWELL

CLAIRE PARTINGTON

JESSICA HARRISON

TANYA BATURA

LOCATION Los Angeles, CA
YEAR OF BIRTH 1974
MATERIALS Earthenware
EXPERIENCE 23 years

Tanya Batura is an artist with a predisposition toward the absurd and incongruous. Working in clay, her large-scale heads defy even their own materiality, and are at once monumental and delicate, geometric and languorously sensual in their fluidity, starkly devoid of color yet strangely shadowed from within. Pushing both material and content, her agenda is less about pleasing any perceived "viewer," and much more about absorption into her own process. Batura was born in Hartford, Connecticut in 1974. She received a BFA from the University of Washington, Seattle in 1998 and received her MFA from the University of California, Los Angeles in 2003. Tanya's work has been widely exhibited, and several of her sculptures reside in museum collections, including The Nerman Museum of Contemporary Art in Kansas, The Portland Art Museum, and The Arizona State Art Museum. Tanya Lives and works in Los Angeles, CA.

'I use large, flat, extruded coils to build my work. After the form is complete, I will often turn it in space, and/or create facets with slabs. I also experiment with piercing the form in a variety of ways, while the work is wet or dry. After the work has dried I sand it, then bisque it. After firing I will either glaze or paint with acrylics. I use an airbrush to paint.'

Photos courtesy of Alan Shaffer, Ryan Van Der Hout and the artist

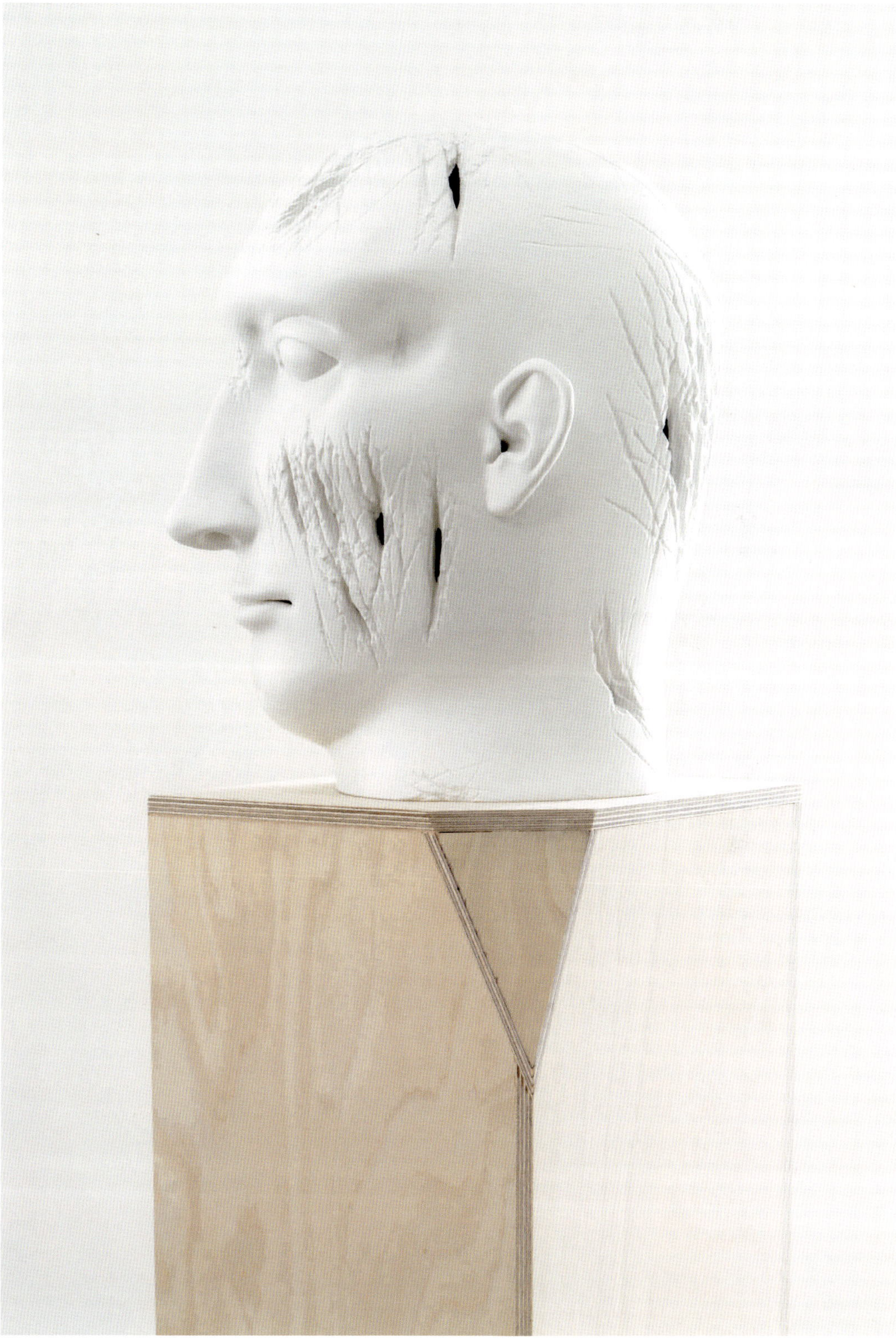

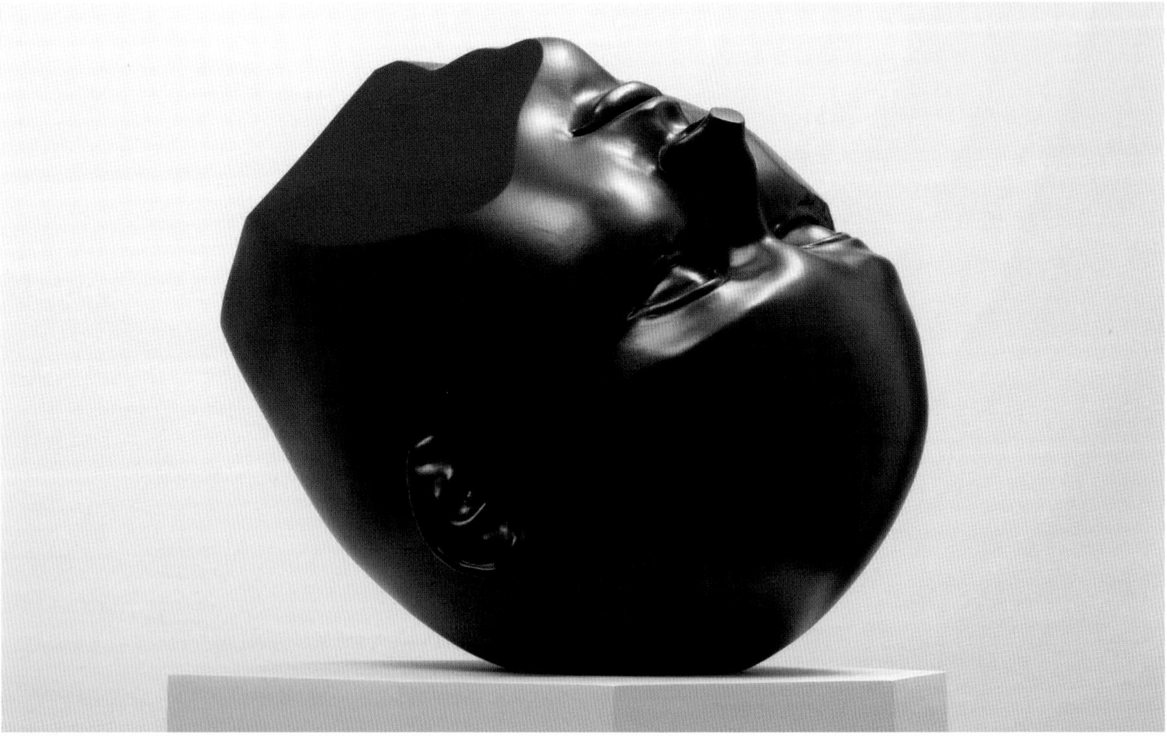

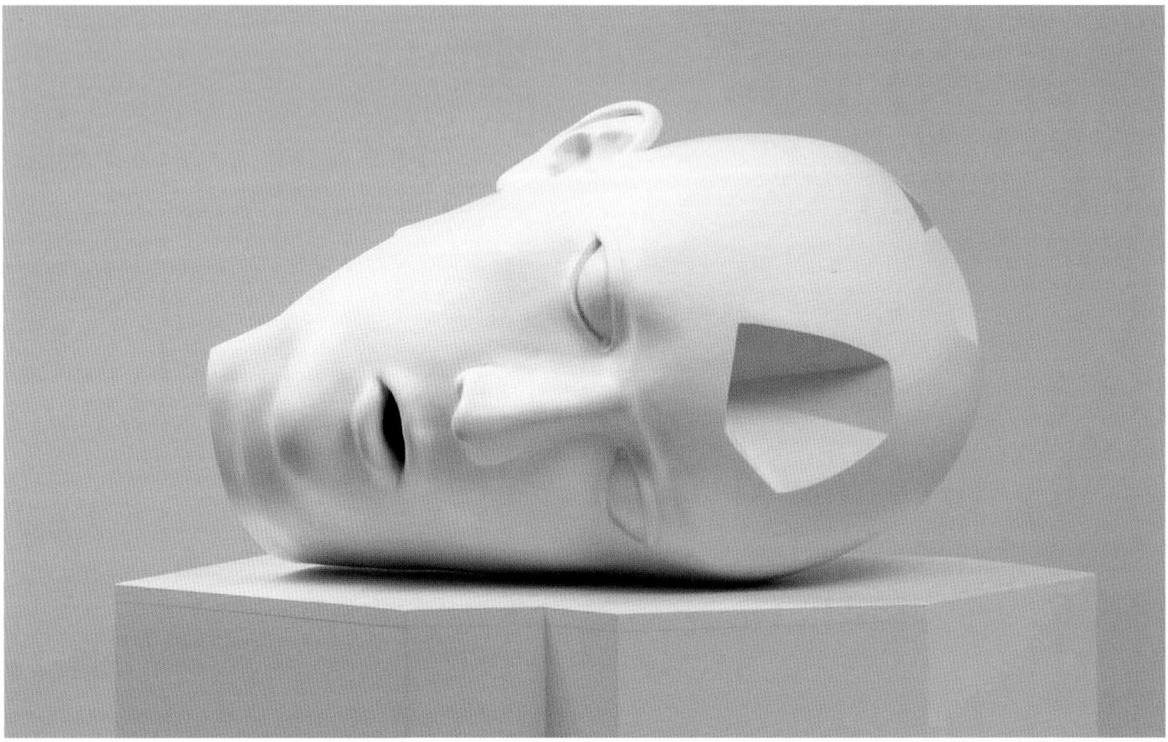

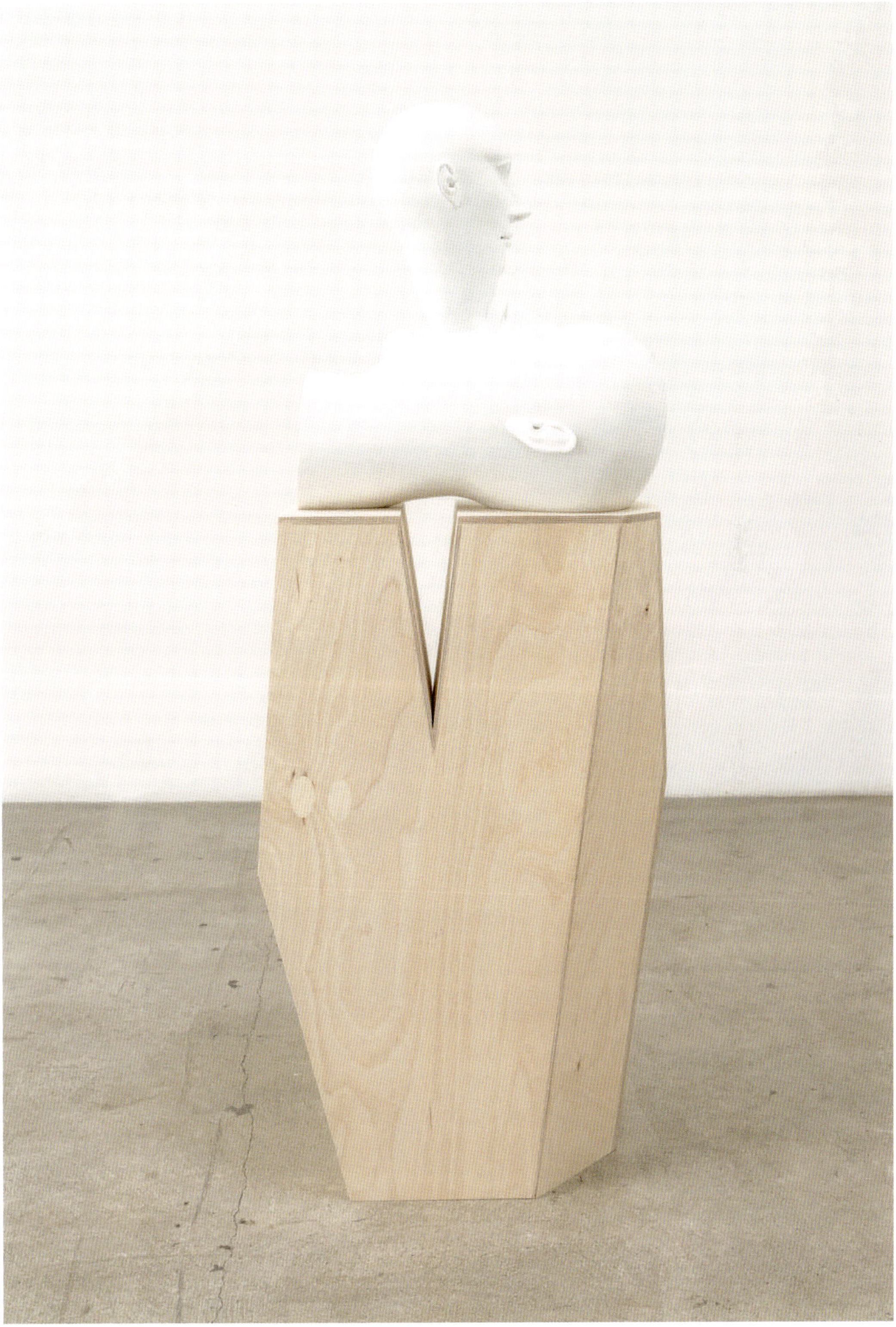

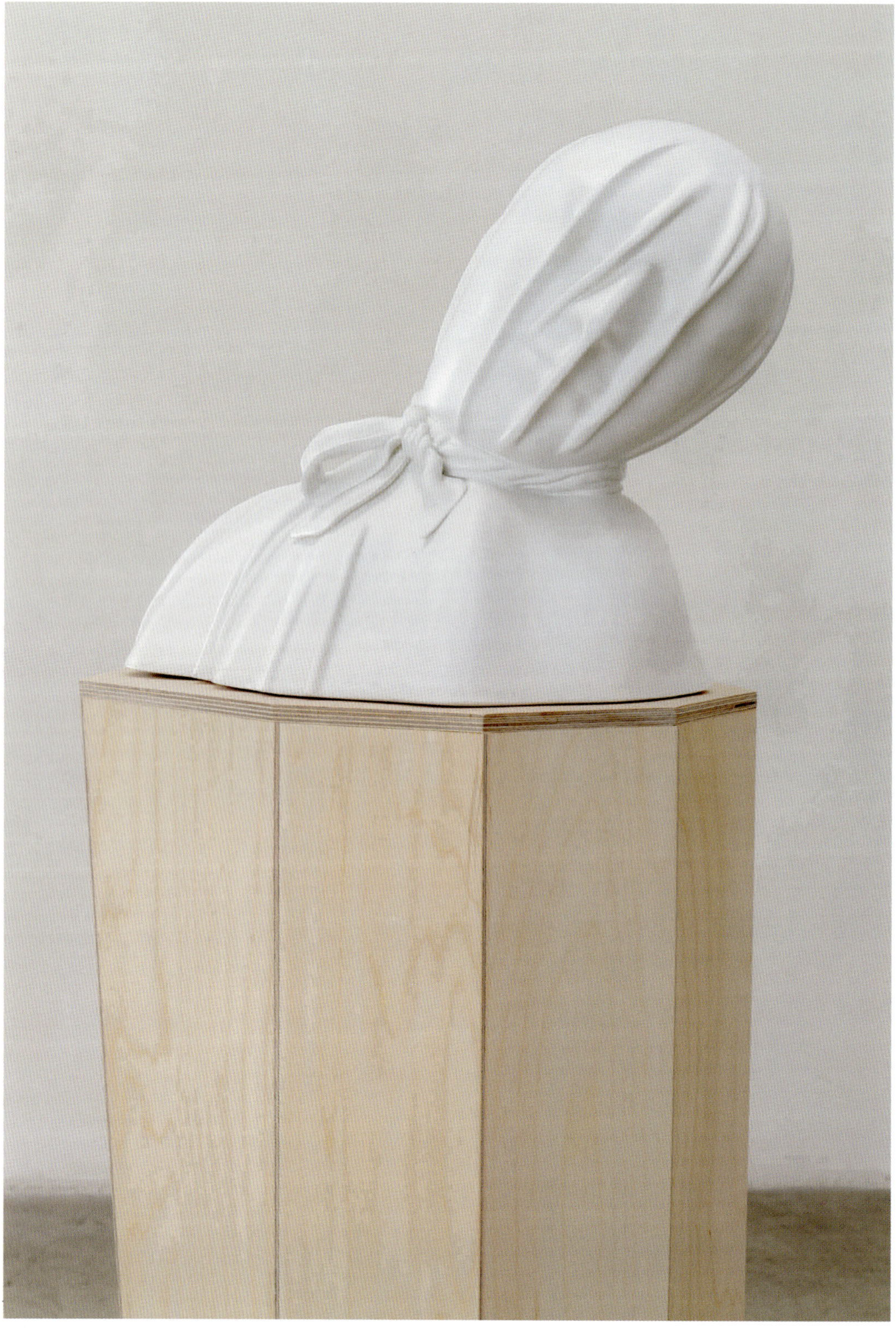

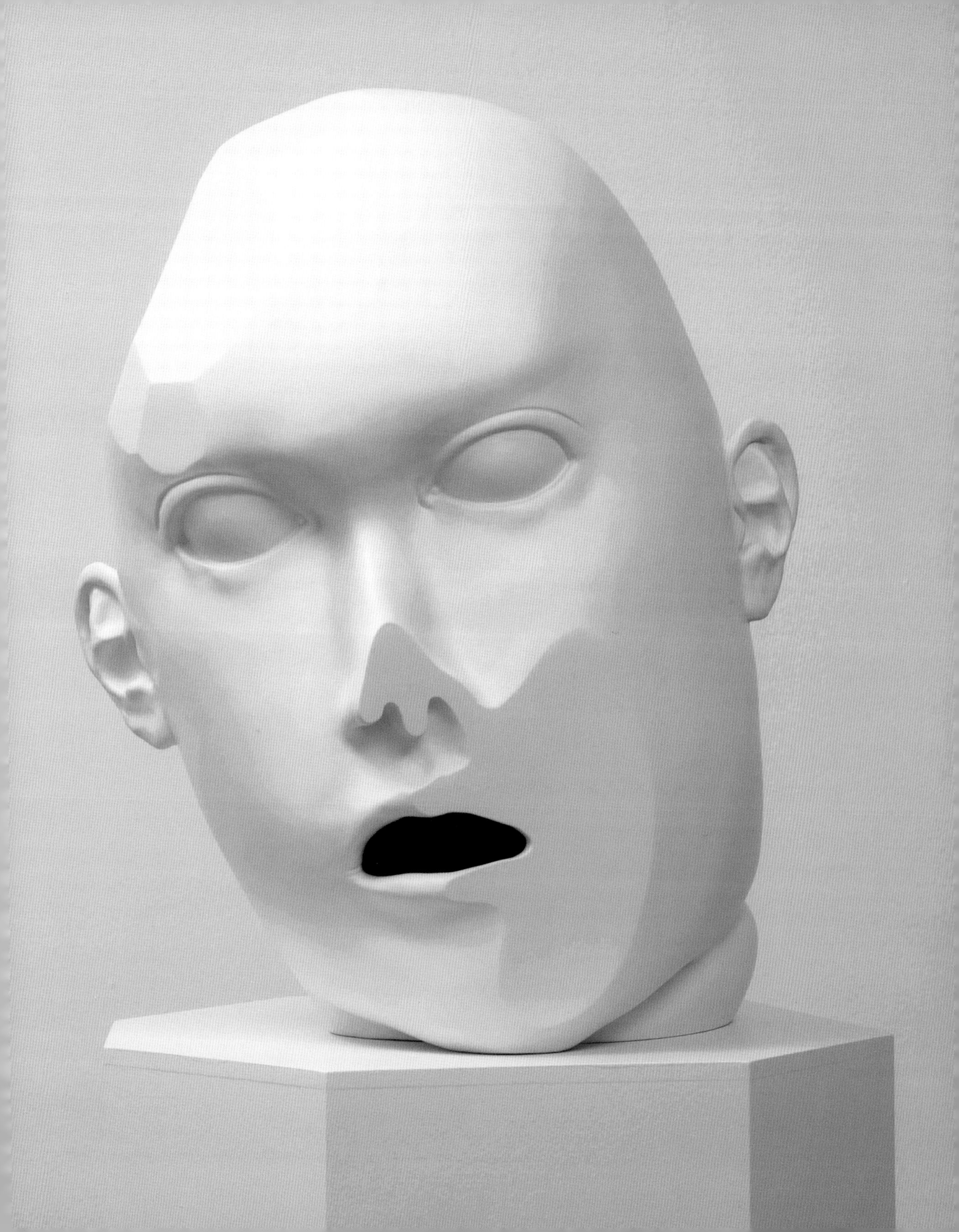

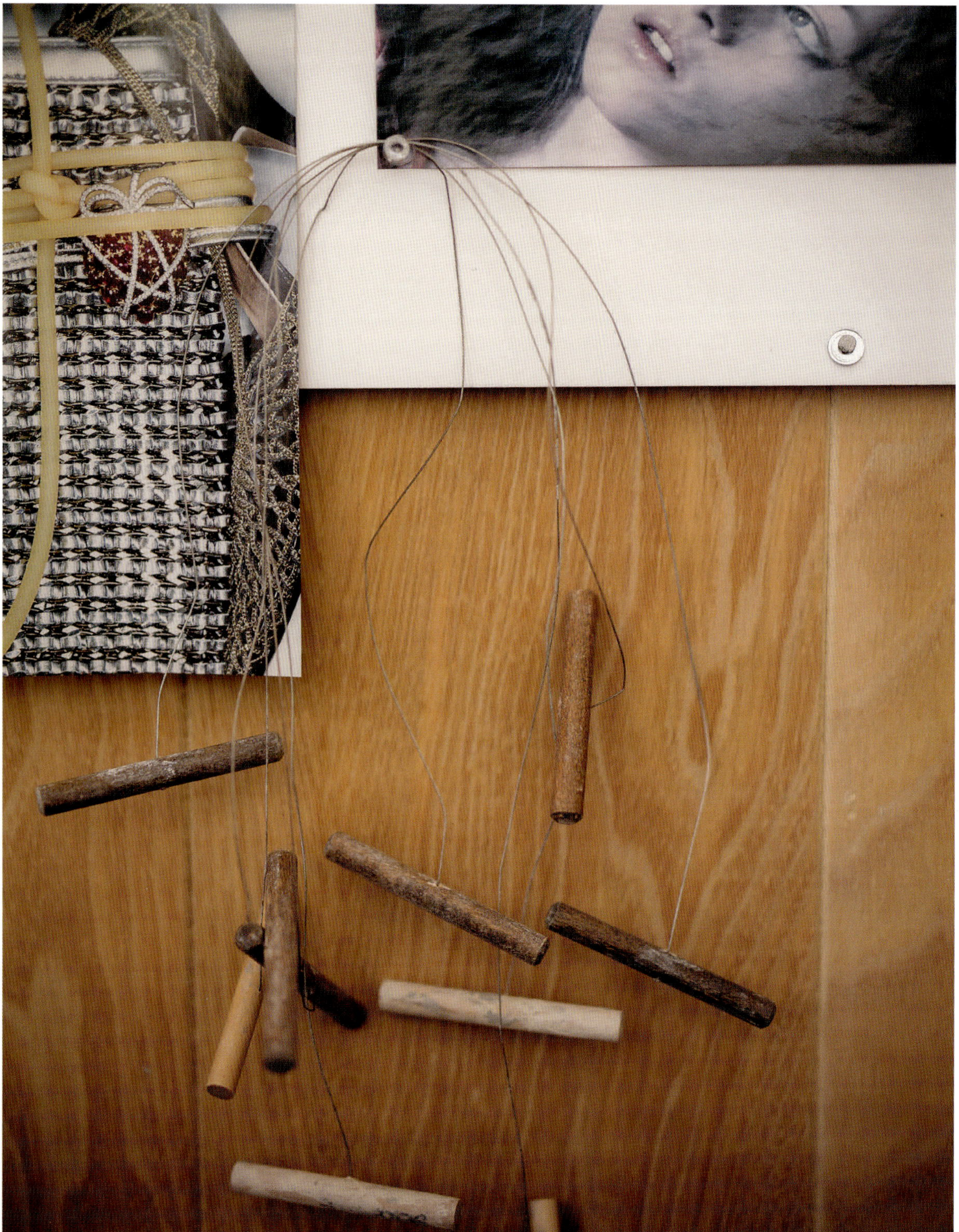

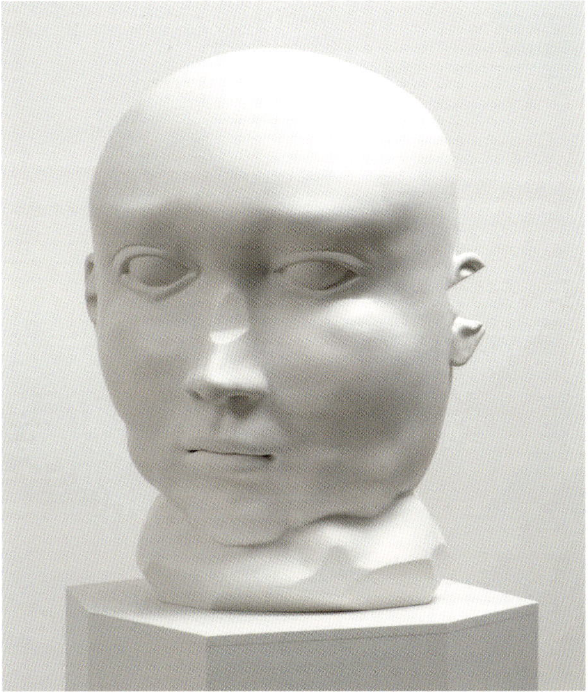
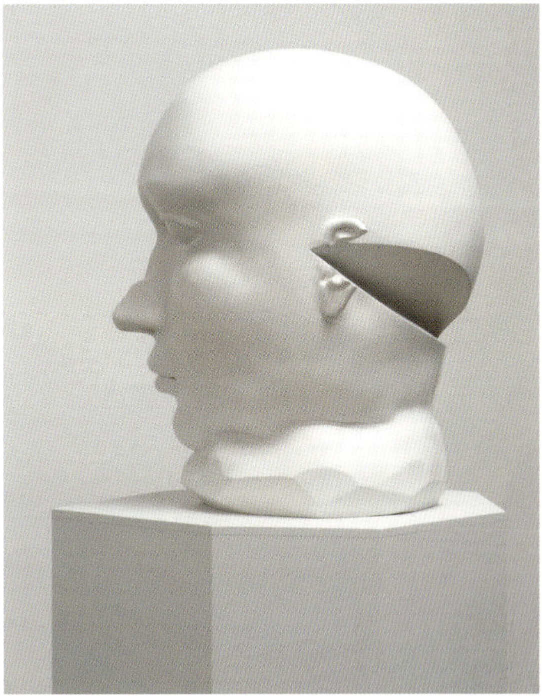
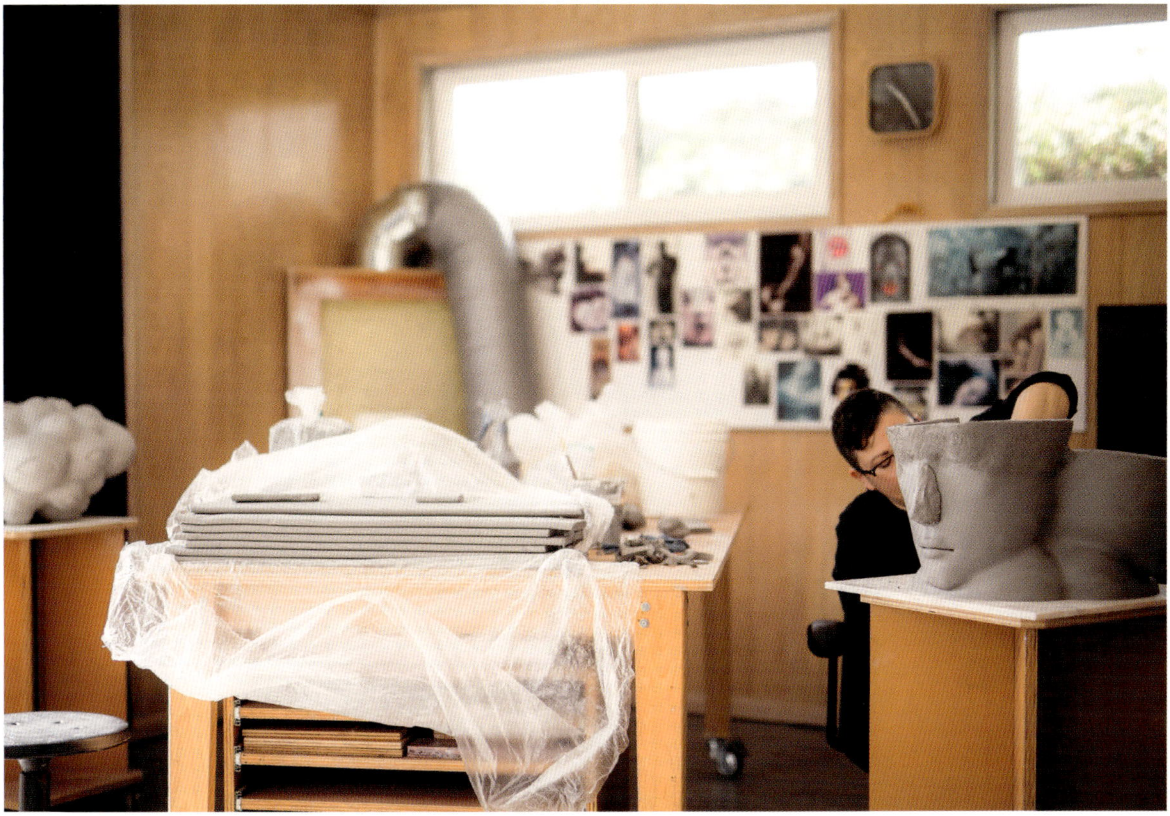

JESS RIVA COOPER

LOCATION Toronto, Ontario, Canada
YEAR OF BIRTH 1981
MATERIALS Cone 6 porcelain, mixed media
EXPERIENCE 15 years

'In my sculptures, the world sprouts plant matter. Color and form burst forth from quiet gardens and bring chaos to ordered spaces. Nature undergoes a reclamation process by creeping over structures, subverting past states and creating a preternatural transformation. I see a direct parallel between my interest in insidious plant life and the malevolent Dybbuk spirit of Yiddish folklore, which takes over the human body. In both situations a loss of control is suffered as the parasitic entity subsumes the host.

I use Cone 6 slip cast and plastic porcelaneous clay, often with the addition of paper or nylon fibre. I create surface decoration with decals, overglaze, and other multiple firing techniques. I use many other materials for larger scale sculptures and installations such as pvc tubing, welded steel, and airline cable.'

Jess Riva Cooper is a ceramic artist and educator currently based in Toronto, Canada. Cooper received her MFA in ceramics from the Rhode Island School of Design.

Photos courtesy of the artist

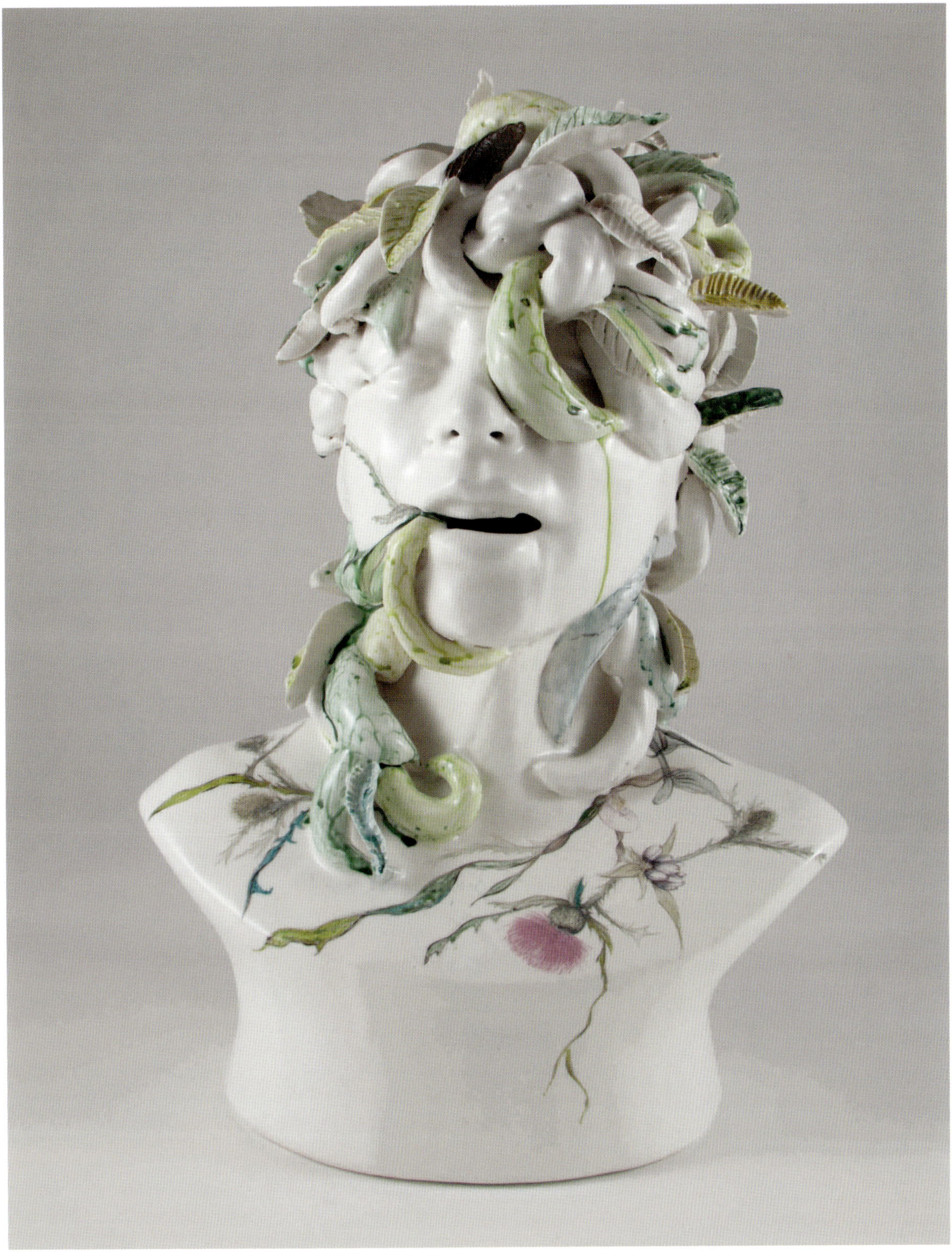

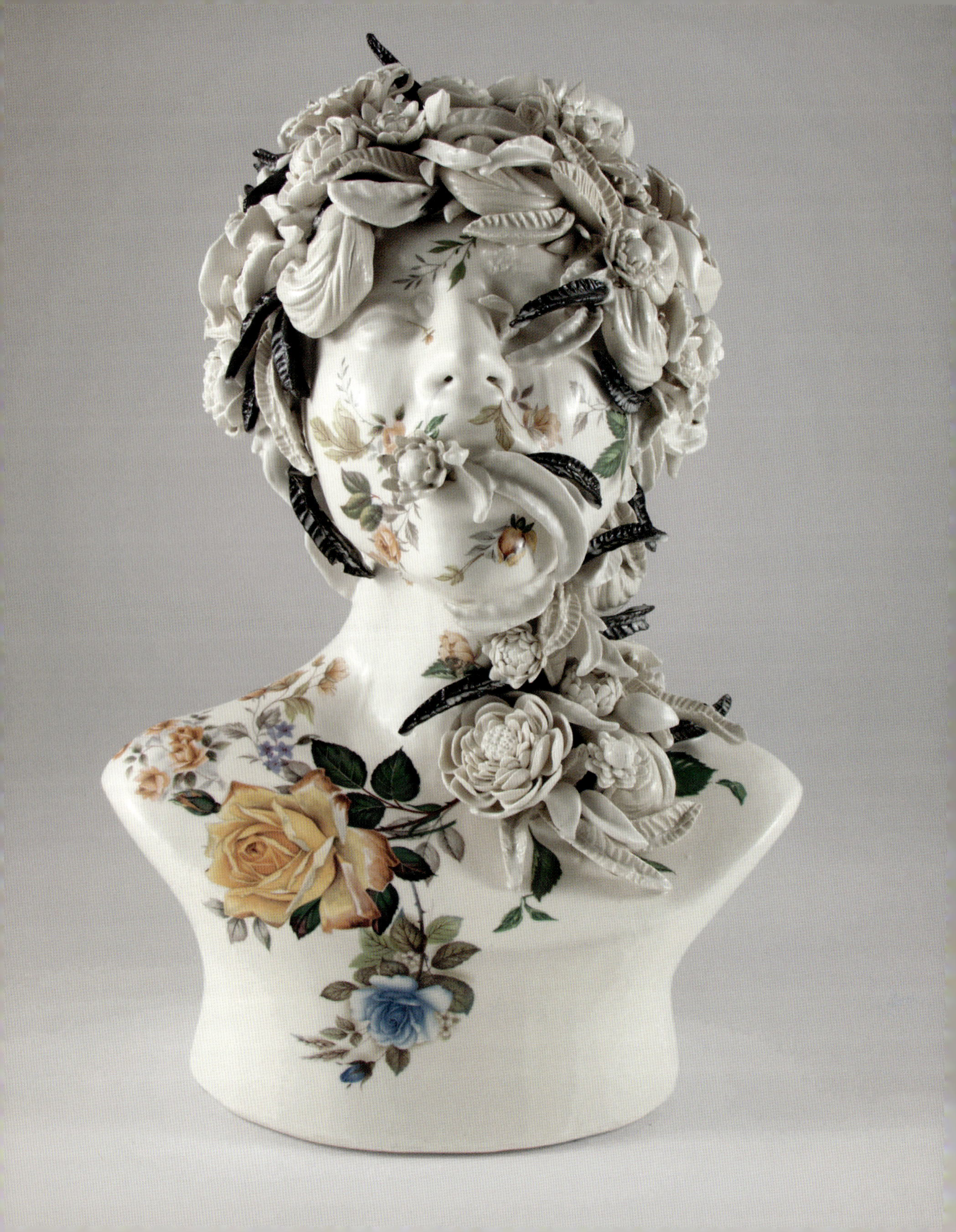

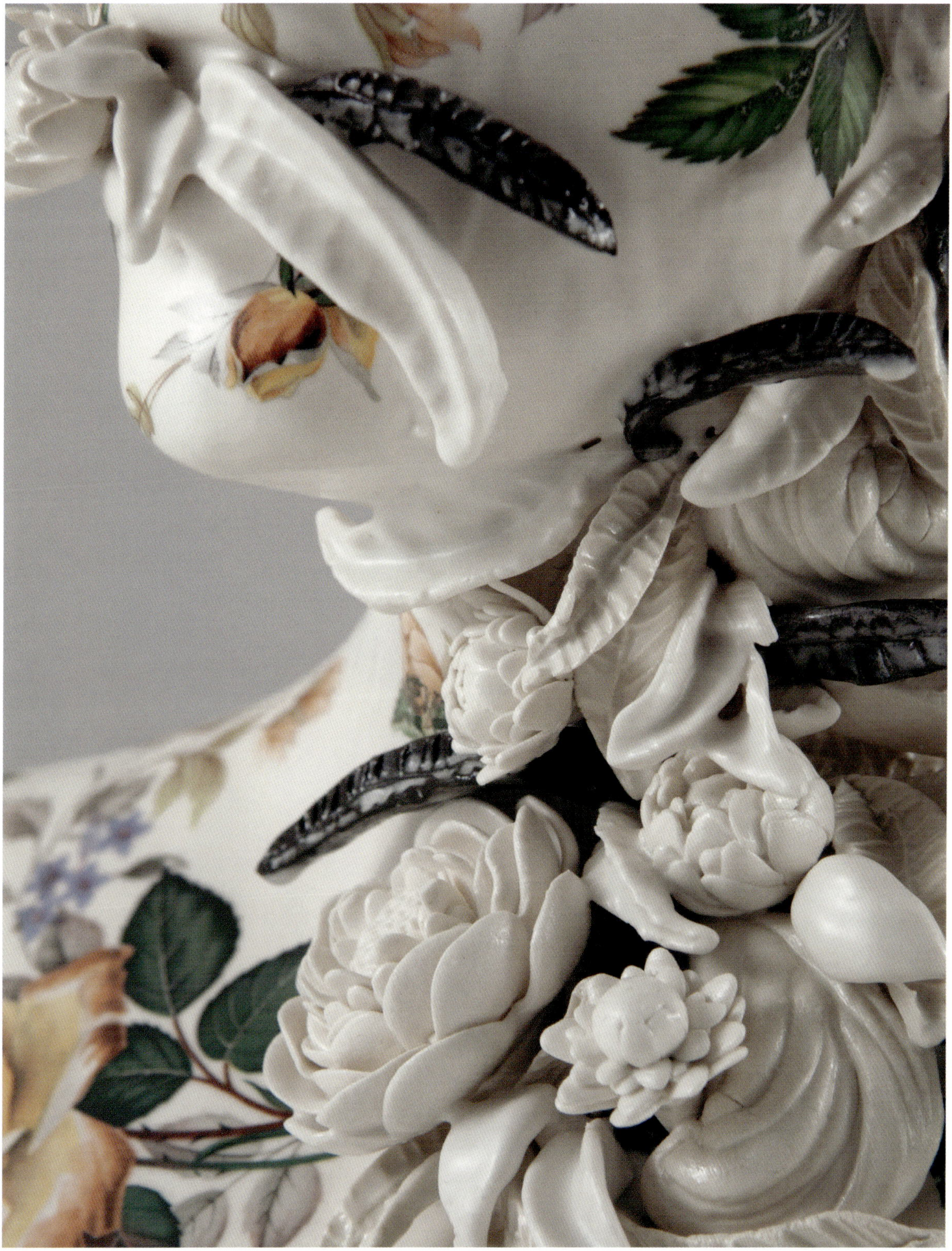

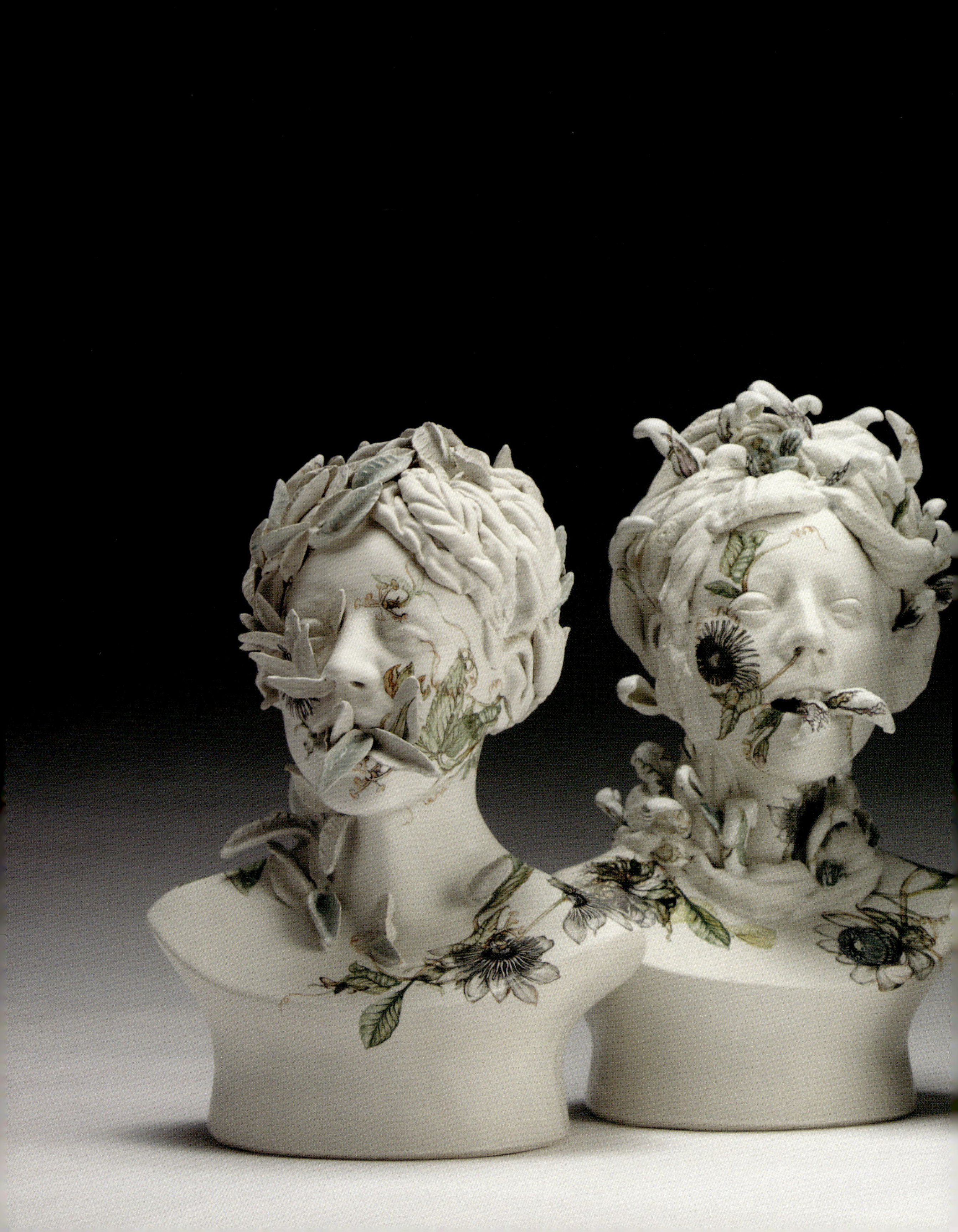

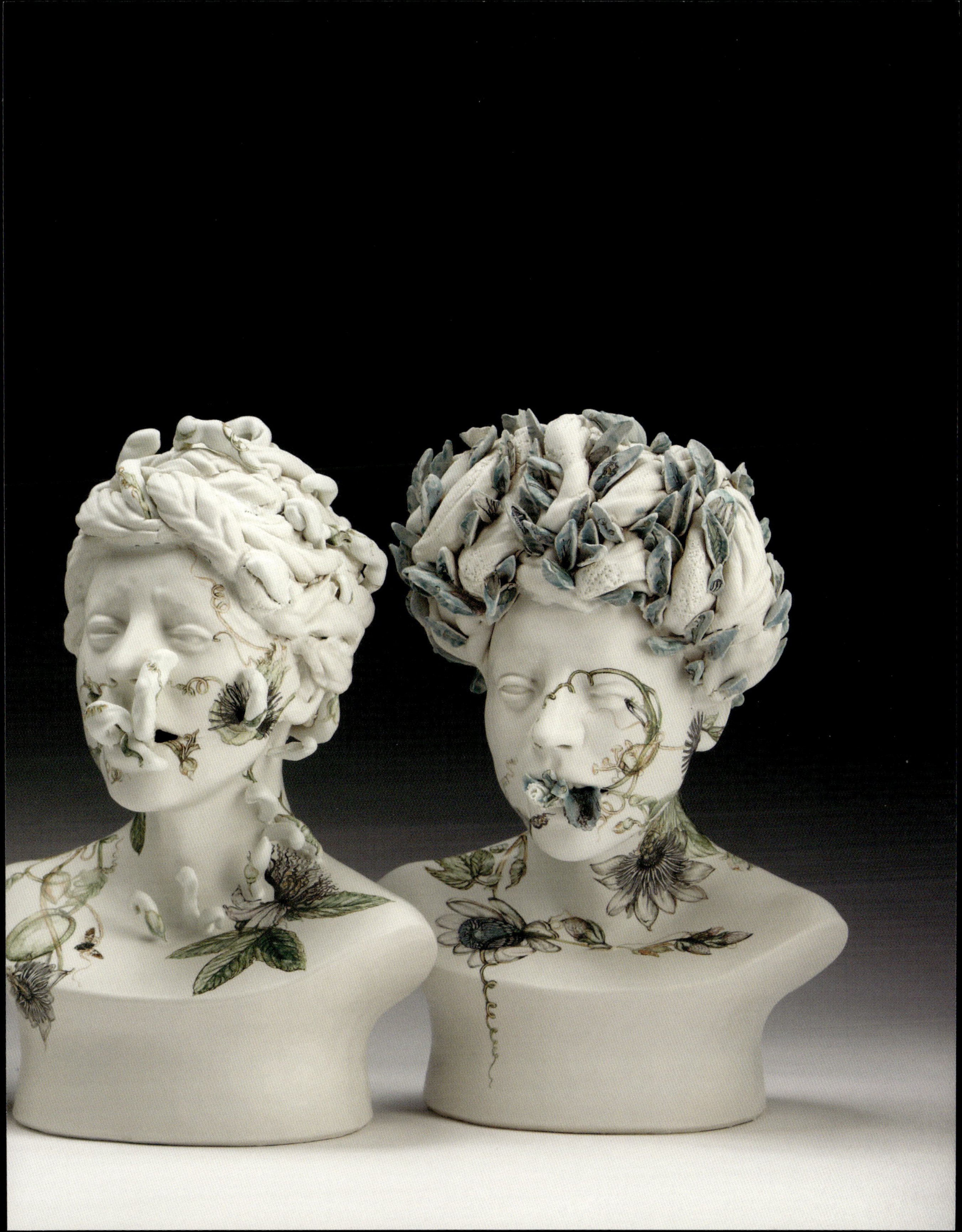

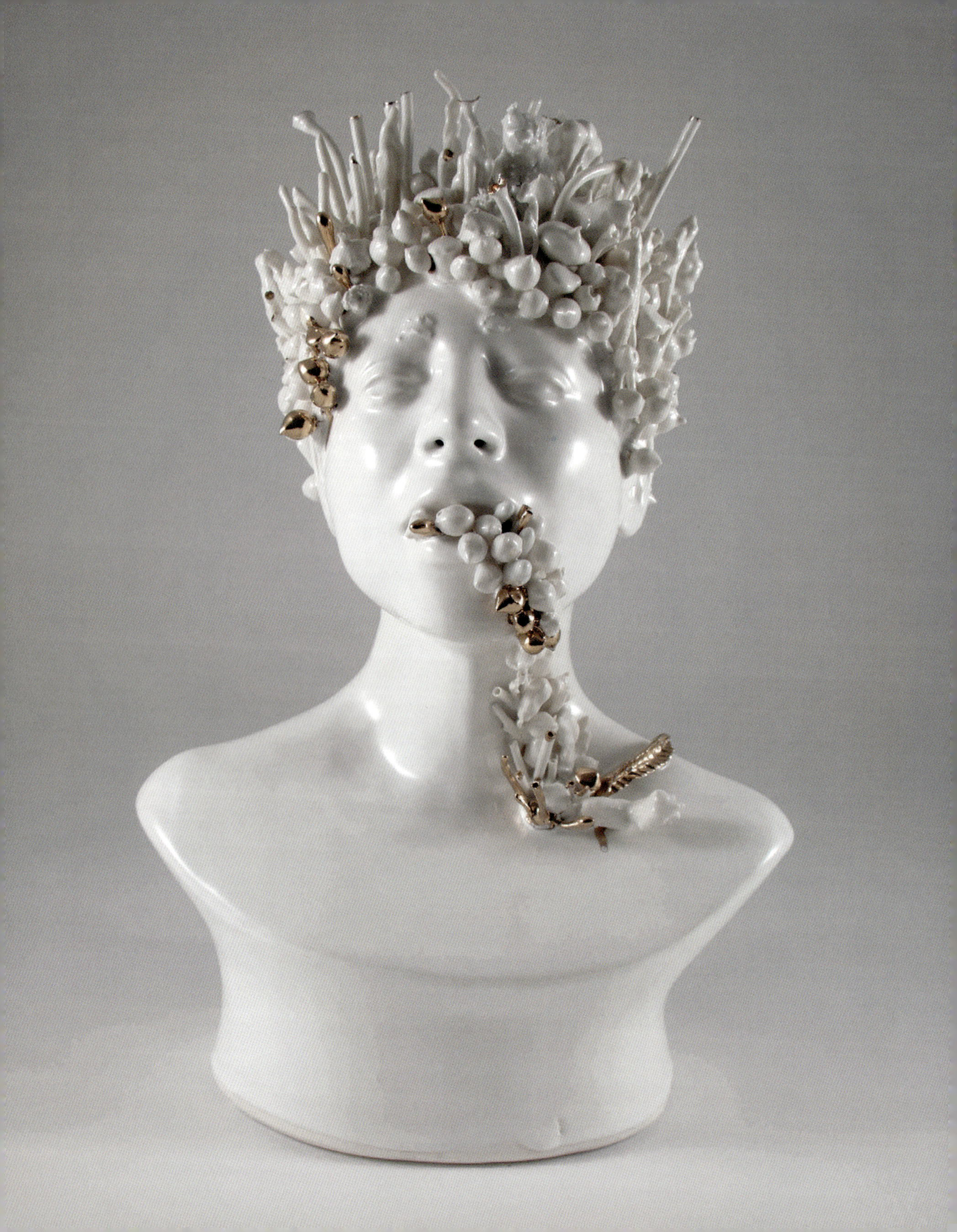

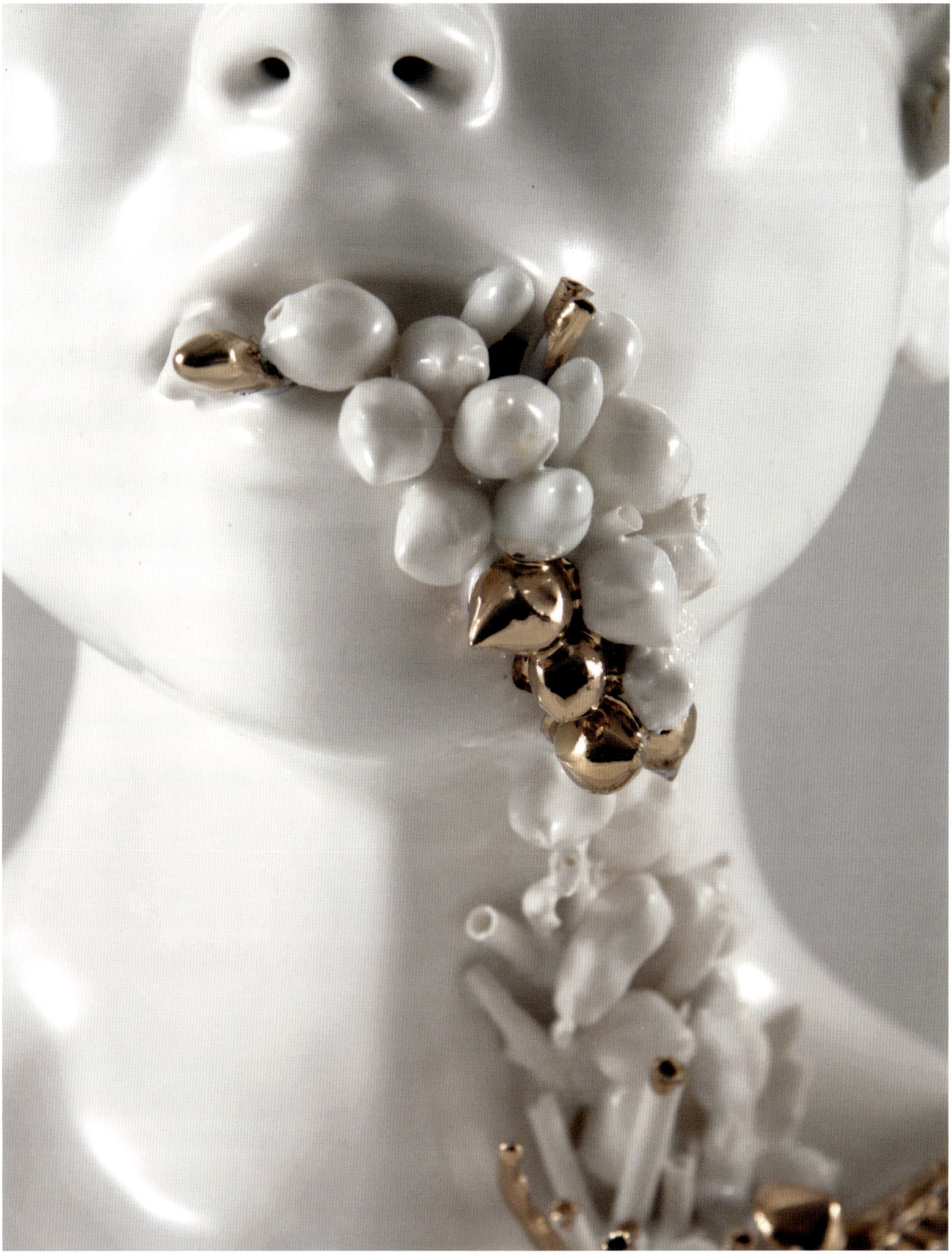

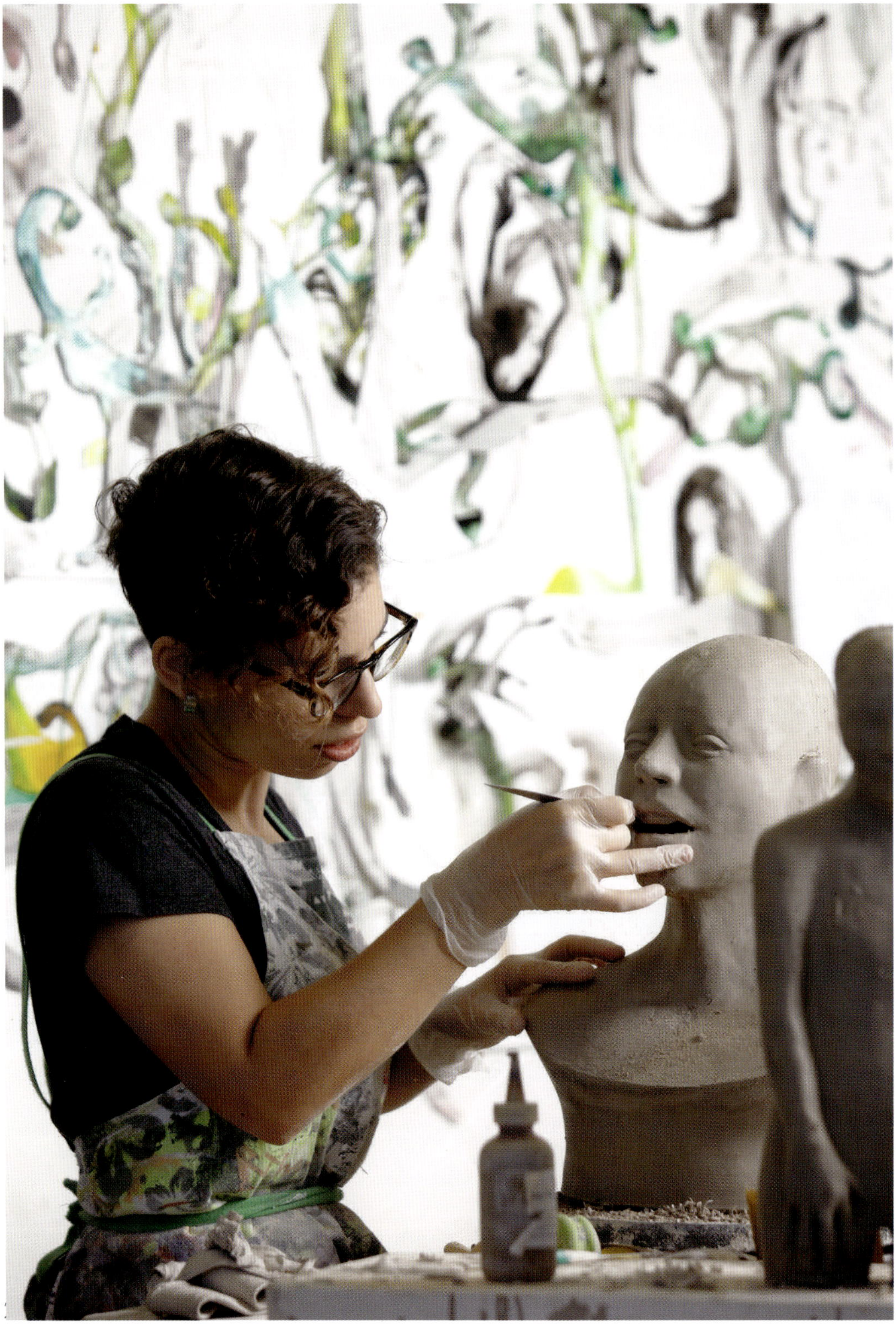

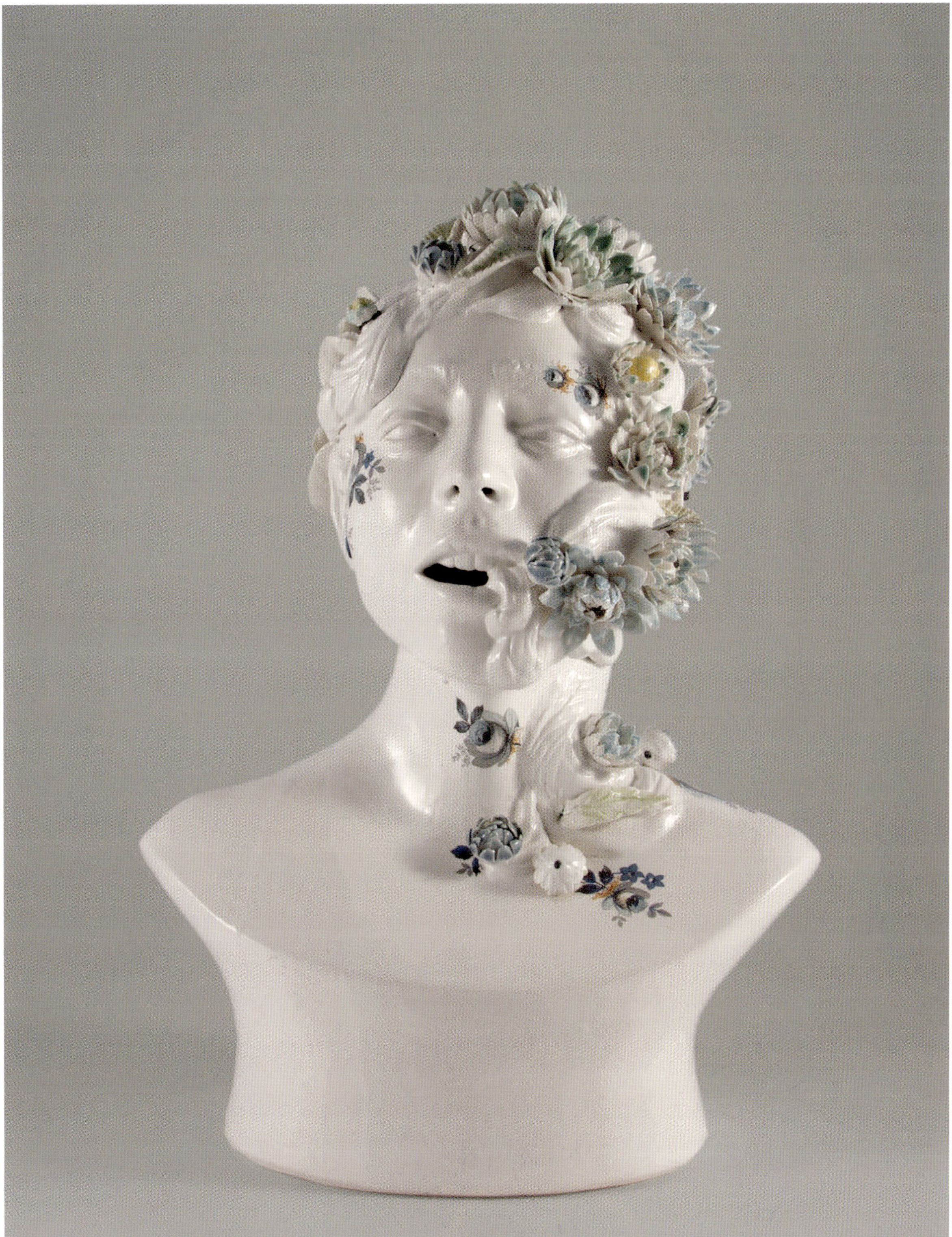

CRYSTAL MOREY

LOCATION Oakland, CA
YEAR OF BIRTH 1983
MATERIALS Porcelain
EXPERIENCE 14 years

As a visual artist, Crystal Morey takes inspiration from an alternative upbringing where she closely connected with the natural landscape around her. Living in rural Northern California shaped her perspective on nature and how humans interact with land, animals, and each other. Now living in an urban environment, Morey aims to show our relationship to the world around us through the fragile medium of porcelain. With this delicate material she creates a heightened sense of urgency and stress, commenting on our human evolutionary path.

Morey received her BFA in Ceramic Sculpture from the California College of the Arts and her MFA in Spatial Art from San Jose State University. Her work has been included in many publications and online periodicals such as 500 Figures in Clay, Juxtapoz, Hi-Fructose, Beautiful Surface, and Beautiful Decay. Morey has been an artist in residence at both Watershed Center for the Ceramic Art and Penland School of Craft. Morey currently lives in Oakland, California where she maintains an active studio practice exhibiting her work in galleries and museums regionally, nationally, and internationally.

'I create hand sculpted, finely detailed anthropomorphic sculptures with the fragile medium of porcelain. I aim to create a tenuous connection between parts that shows emotion and struggle in our current environmental state.'

Photos courtesy of the artist

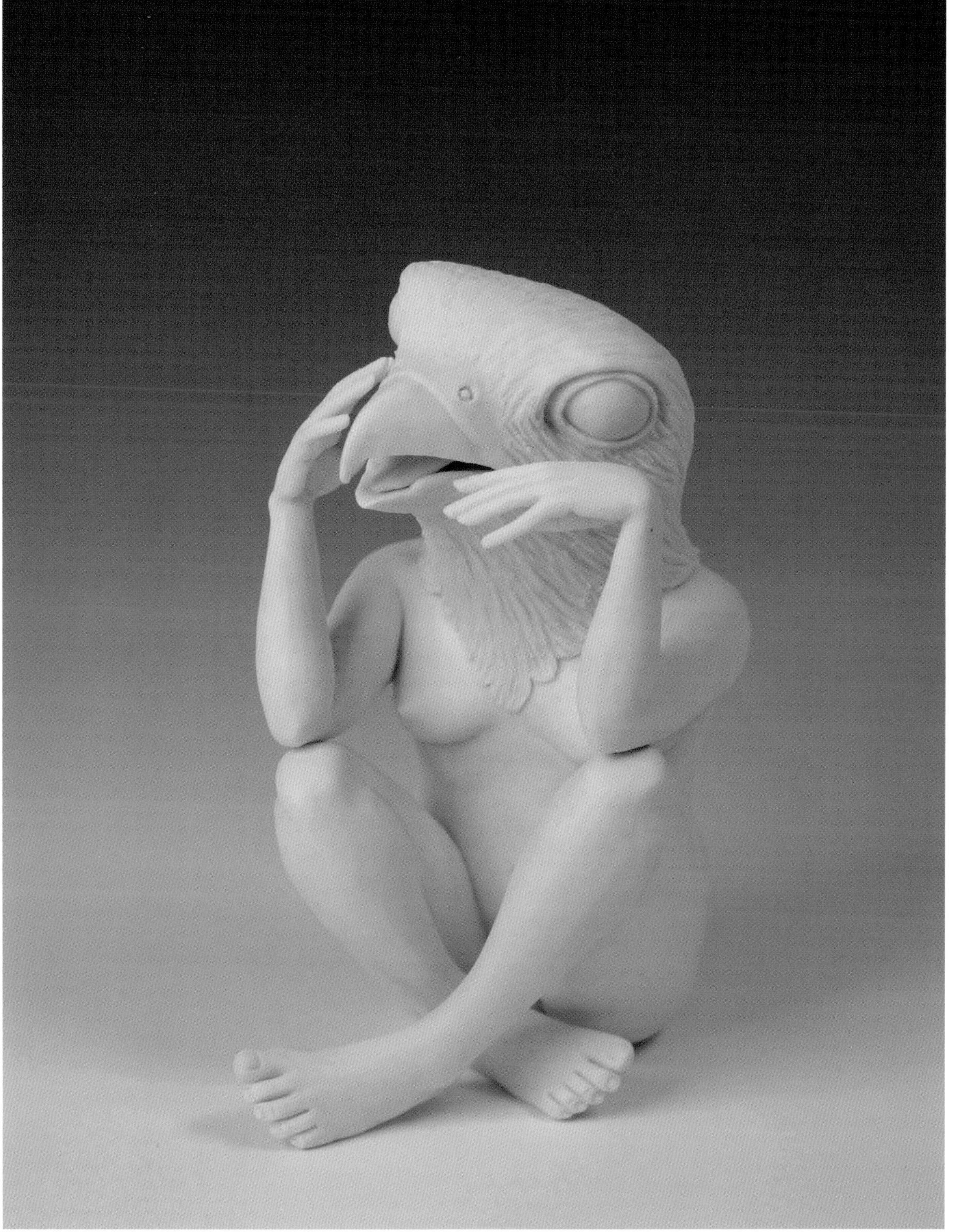

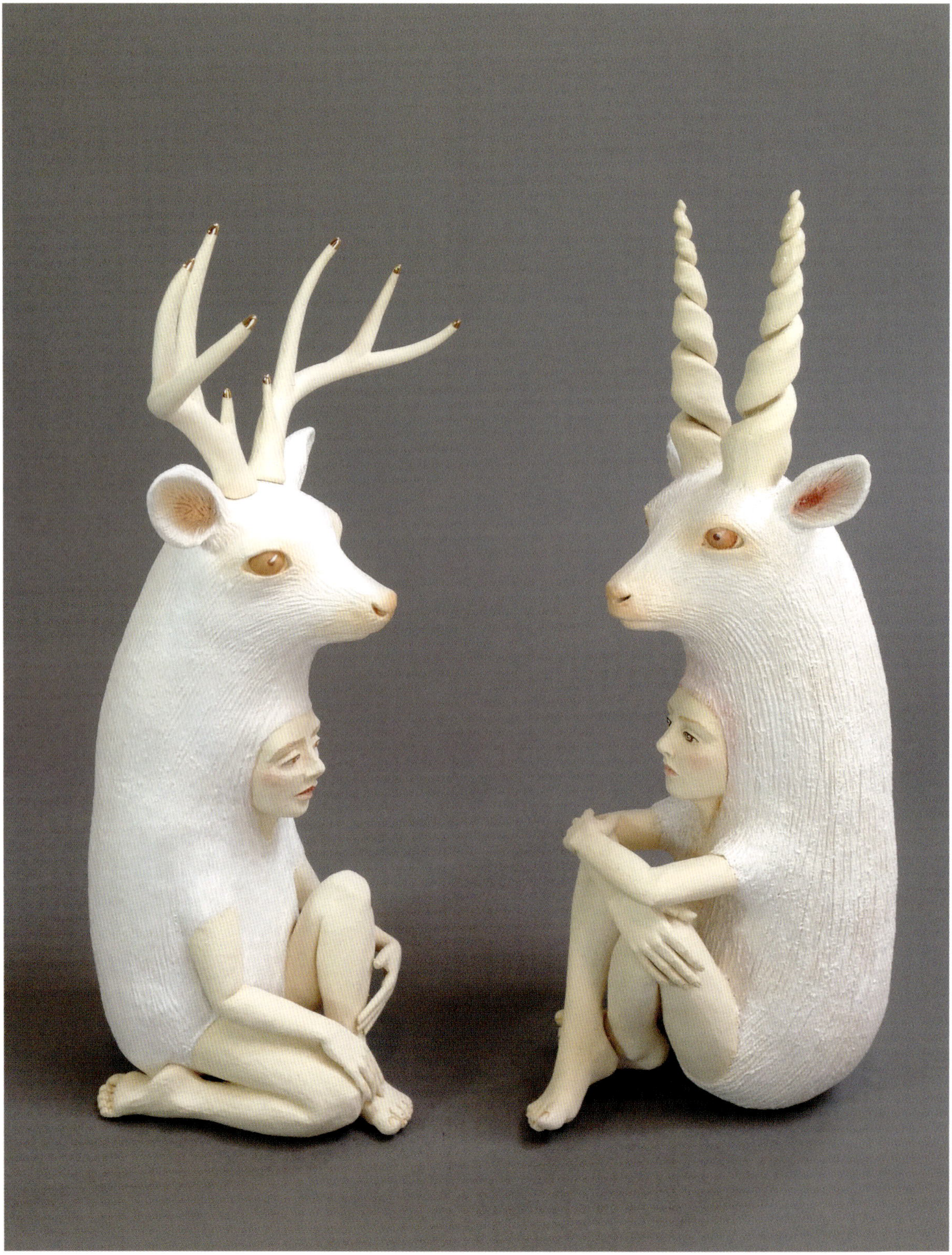

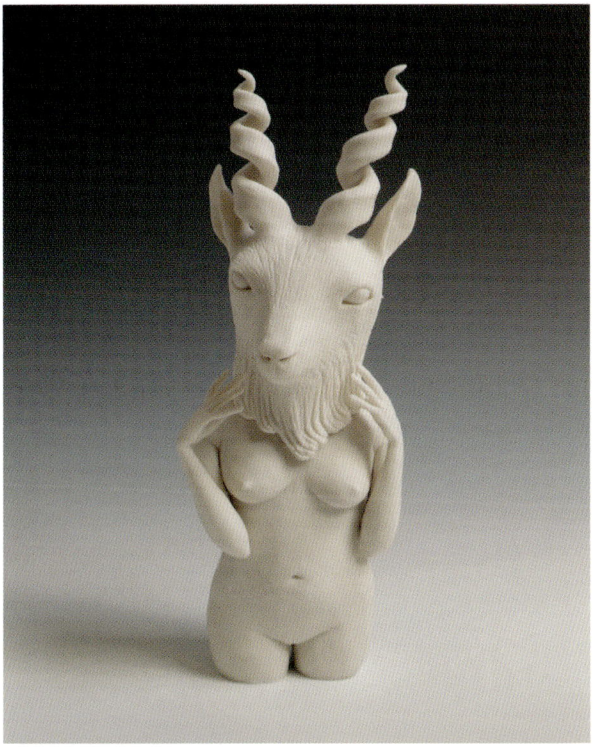
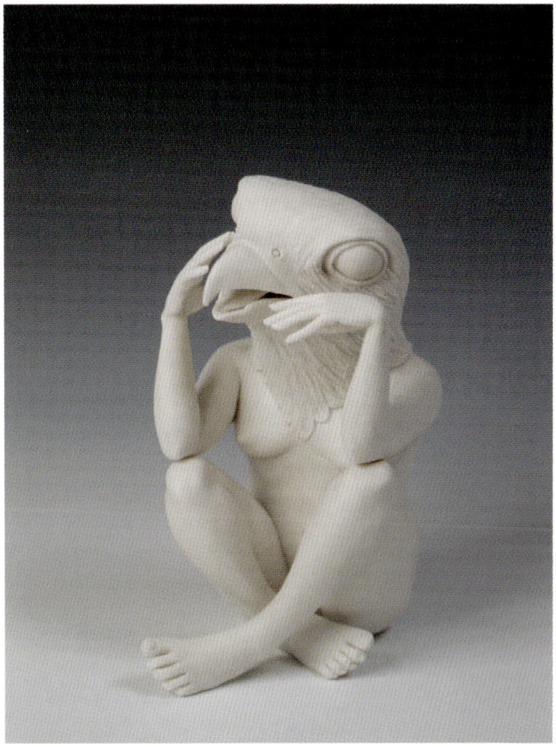
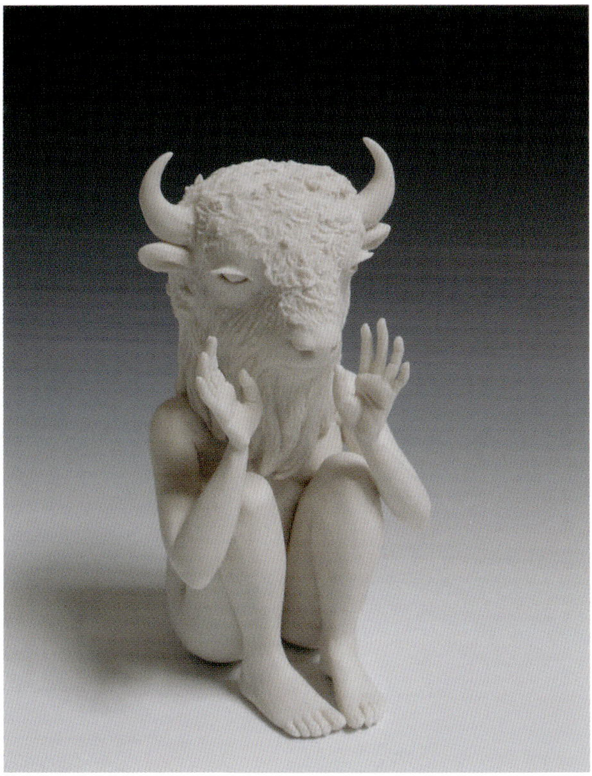
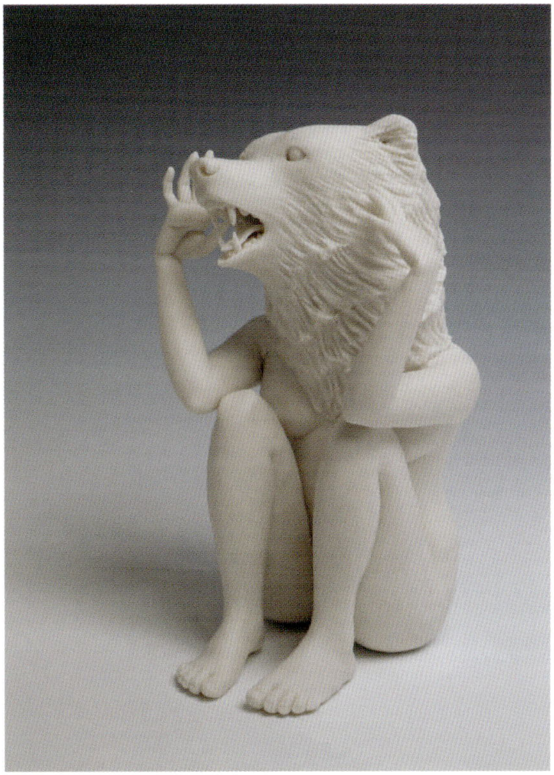

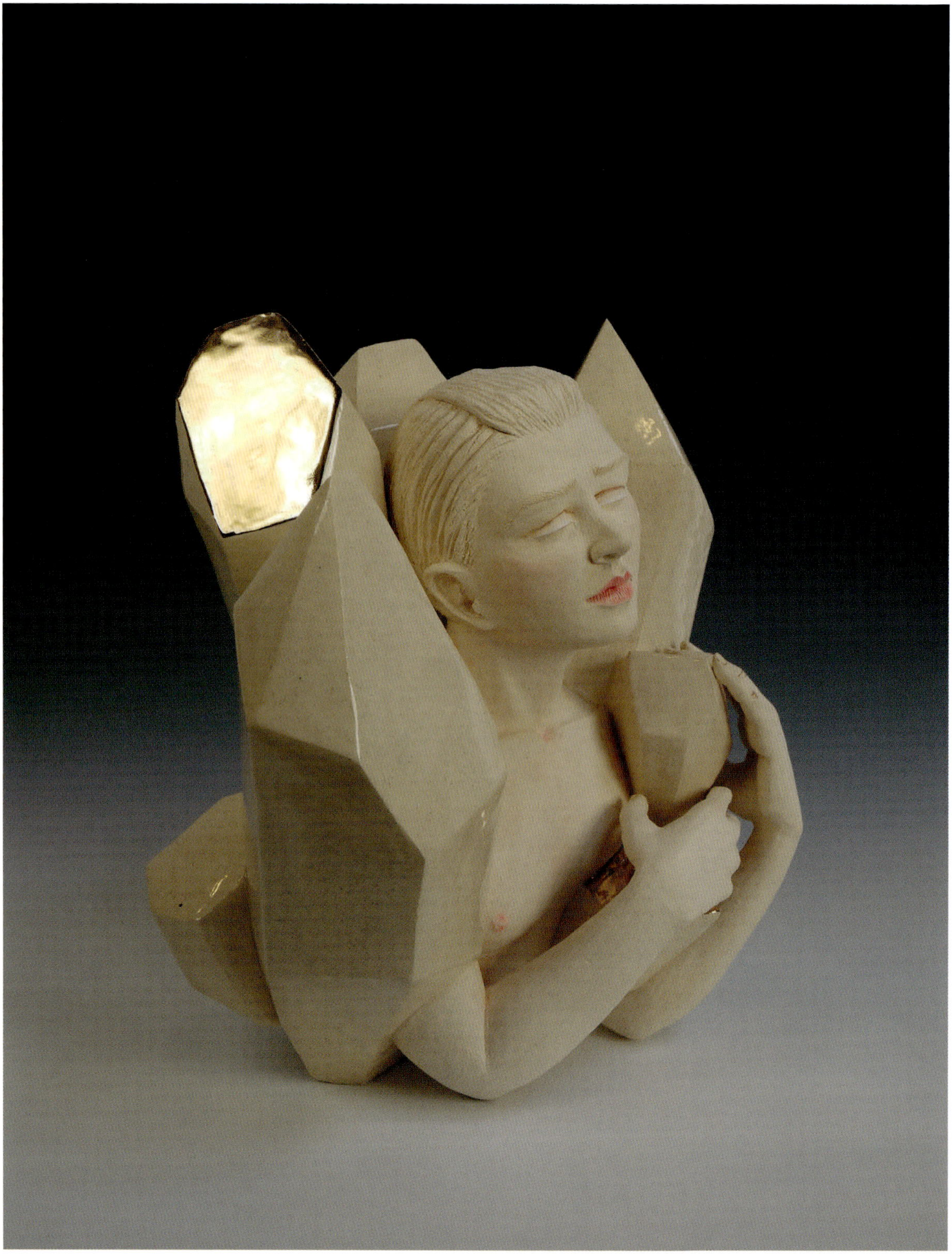

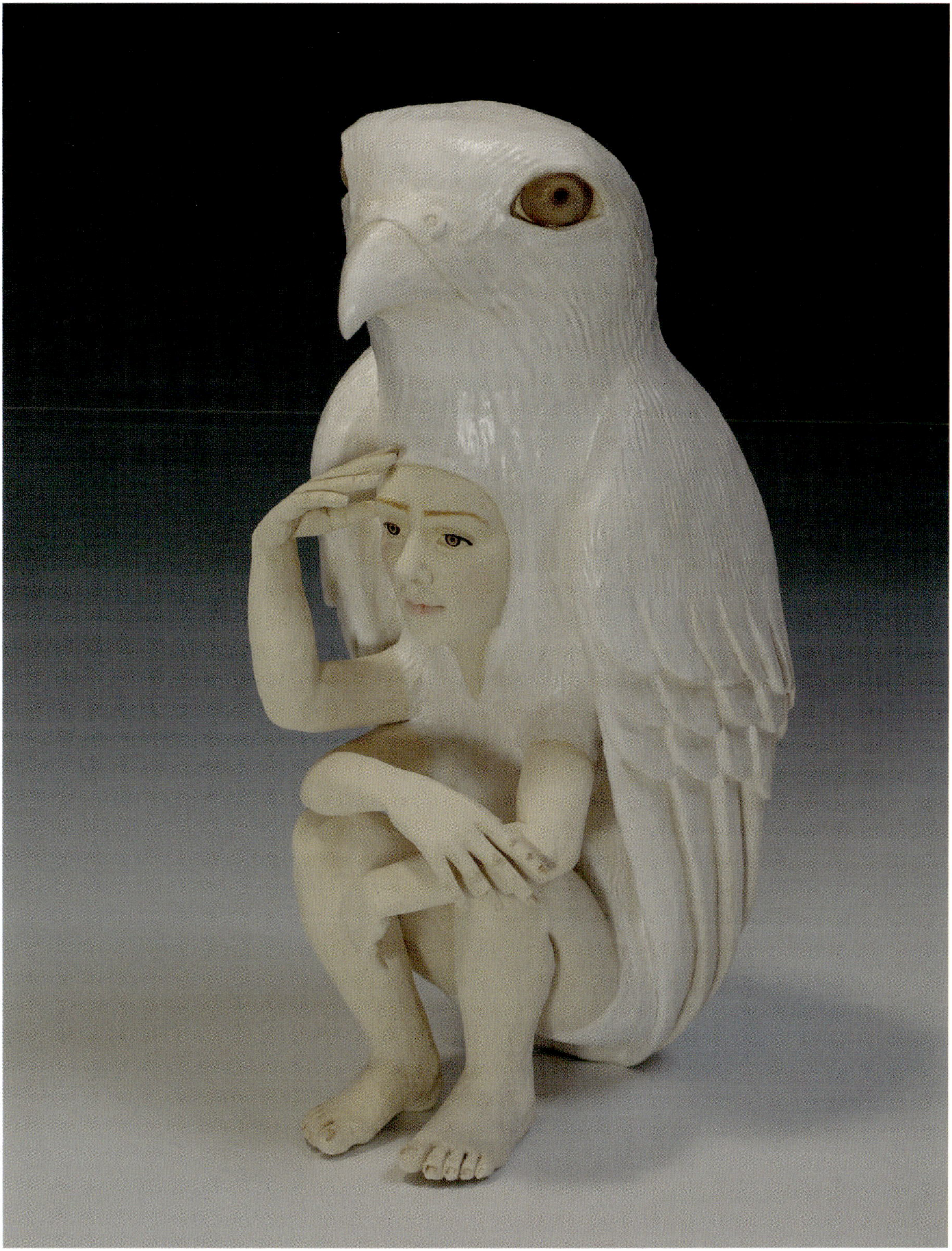

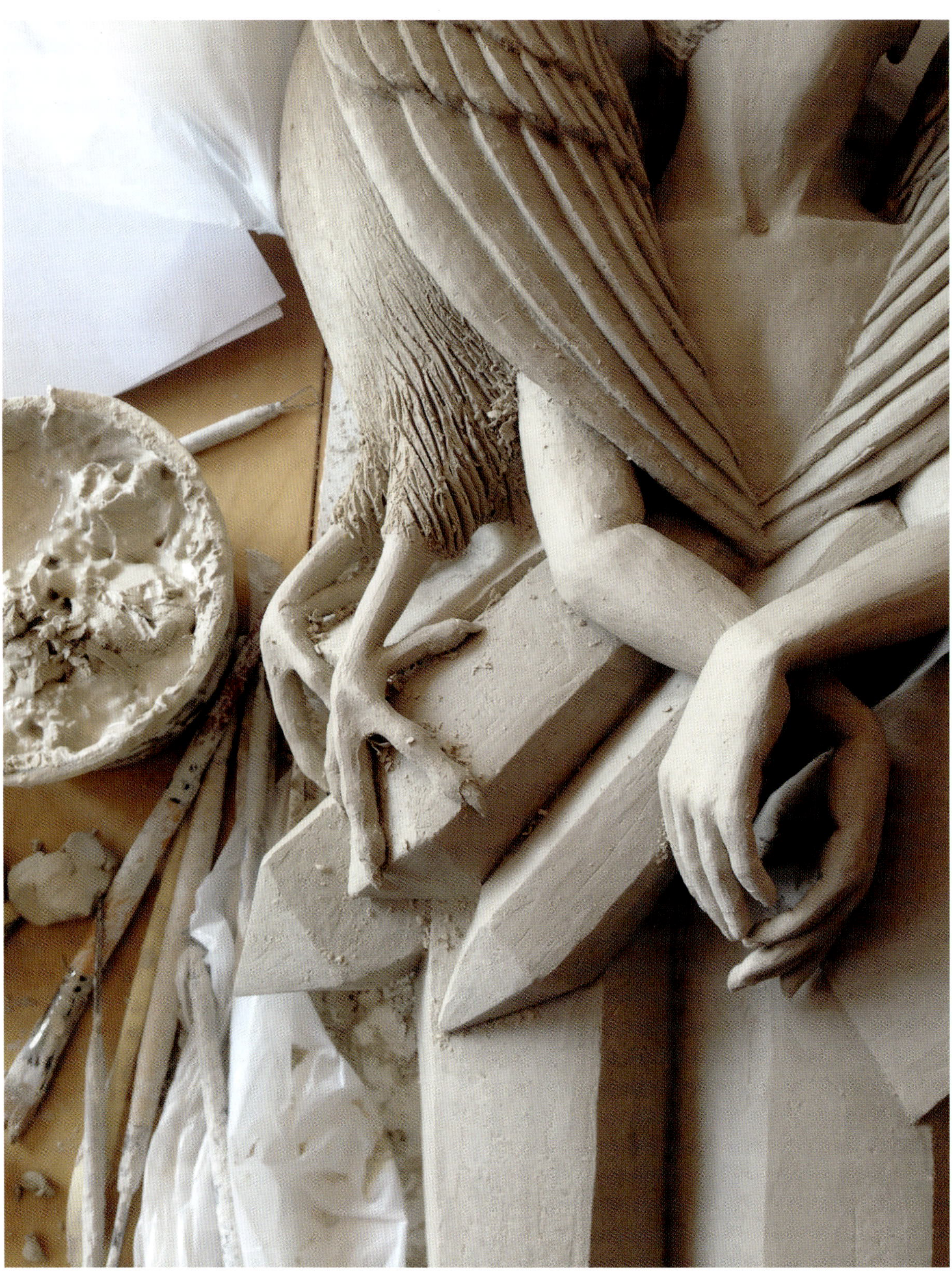

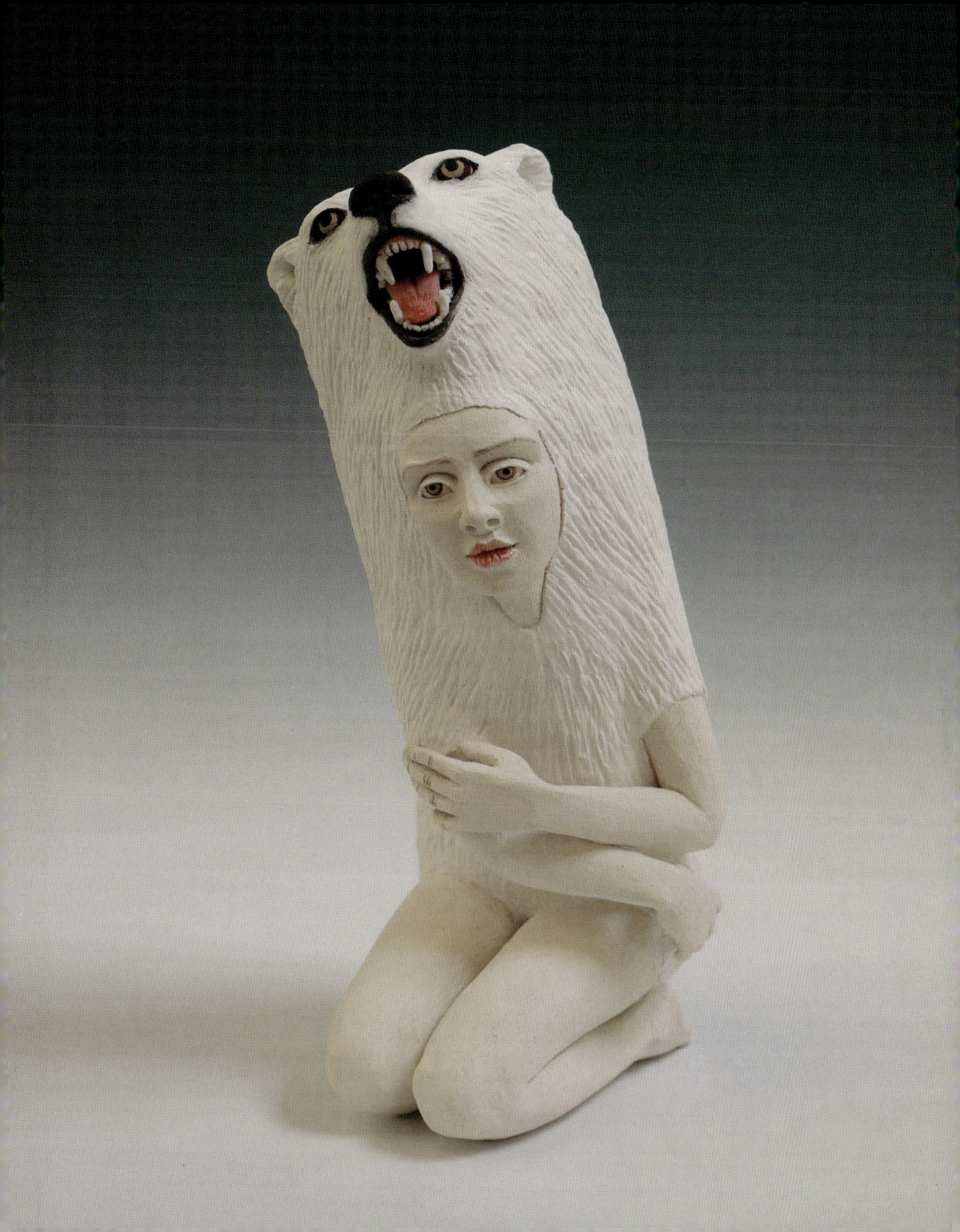

KATE MACDOWELL

LOCATION Portland, OR
YEAR OF BIRTH 1972
MATERIALS Hand sculpted porcelain
EXPERIENCE 10 years

'In my work the romantic ideal of union with the natural world conflicts with our contemporary impact on the environment. These pieces are in part responses to environmental stressors, and also borrow from myth, art history, figures of speech, and other cultural touchstones. In some pieces aspects of the human figure stand in for ourselves and act out sometimes harrowing, sometimes humorous transformations which illustrate our current relationship with the natural world. In others, animals take on anthropomorphic qualities when they are given safety equipment to protect themselves from man-made environmental threats. The union between man and nature is shown to be one of friction and discomfort, with the disturbing implication that we too are vulnerable to being victimized by our destructive practices.

I hand sculpt each piece out of porcelain, often building a solid form and then hollowing it out. I chose porcelain for its luminous and ghostly qualities as well as its strength and ability to show fine texture. It highlights both the impermanence and fragility of natural forms in a dying ecosystem, while paradoxically being a material that can last for thousands of years. I see each piece as a captured and preserved specimen, a painstaking record of endangered natural forms and a commentary on our own culpability.'

Photos courtesy of Dan Kvitka and the artist

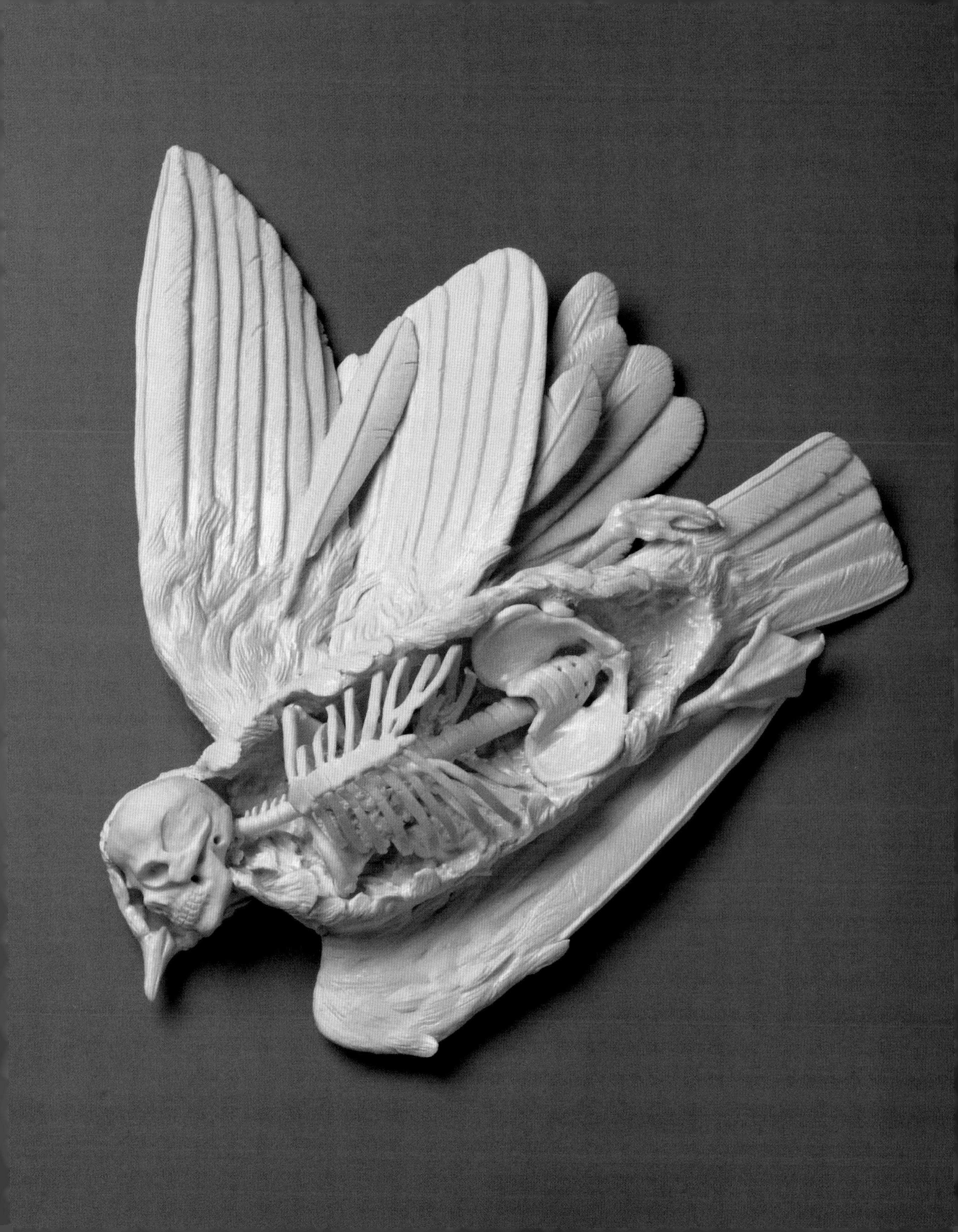

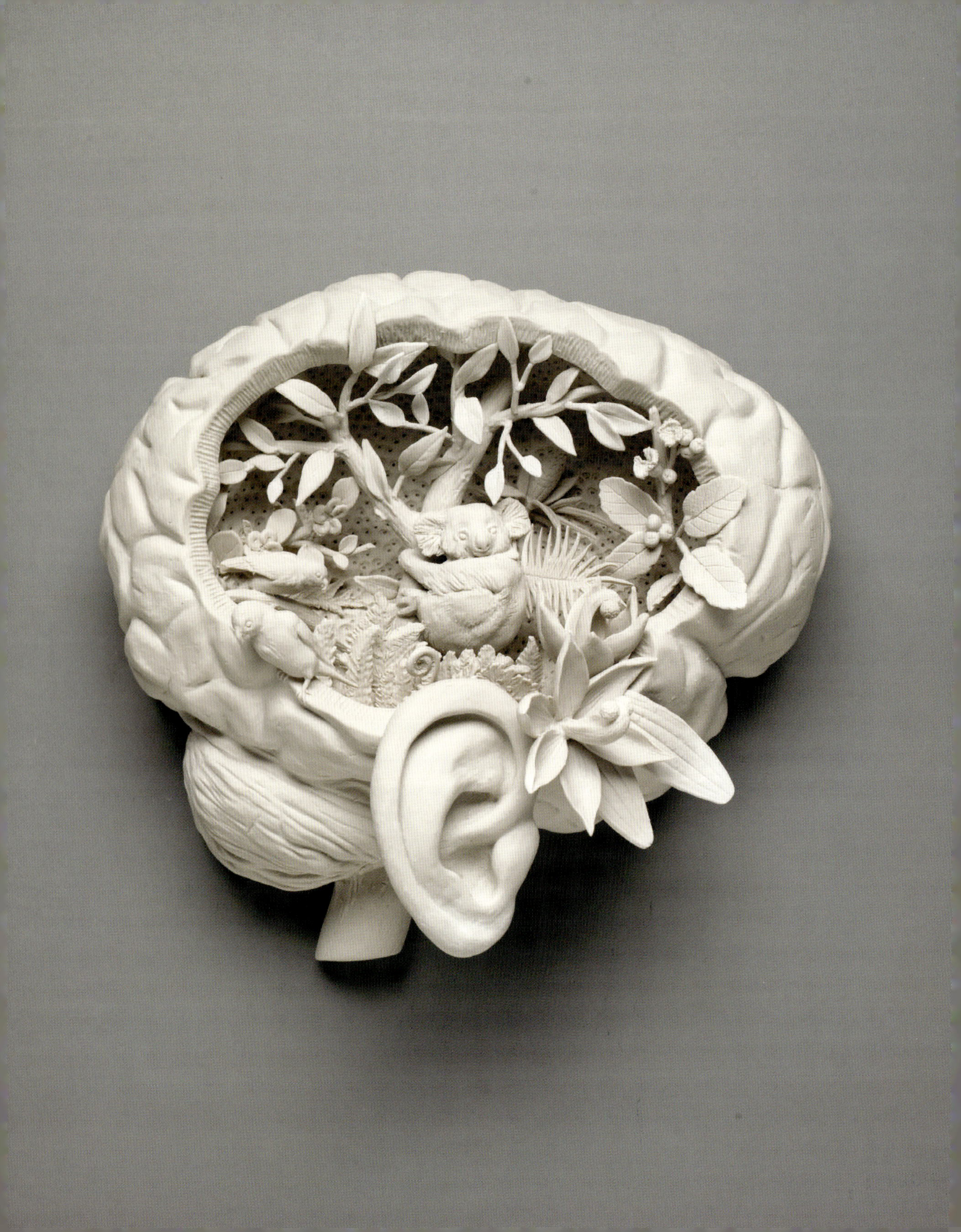

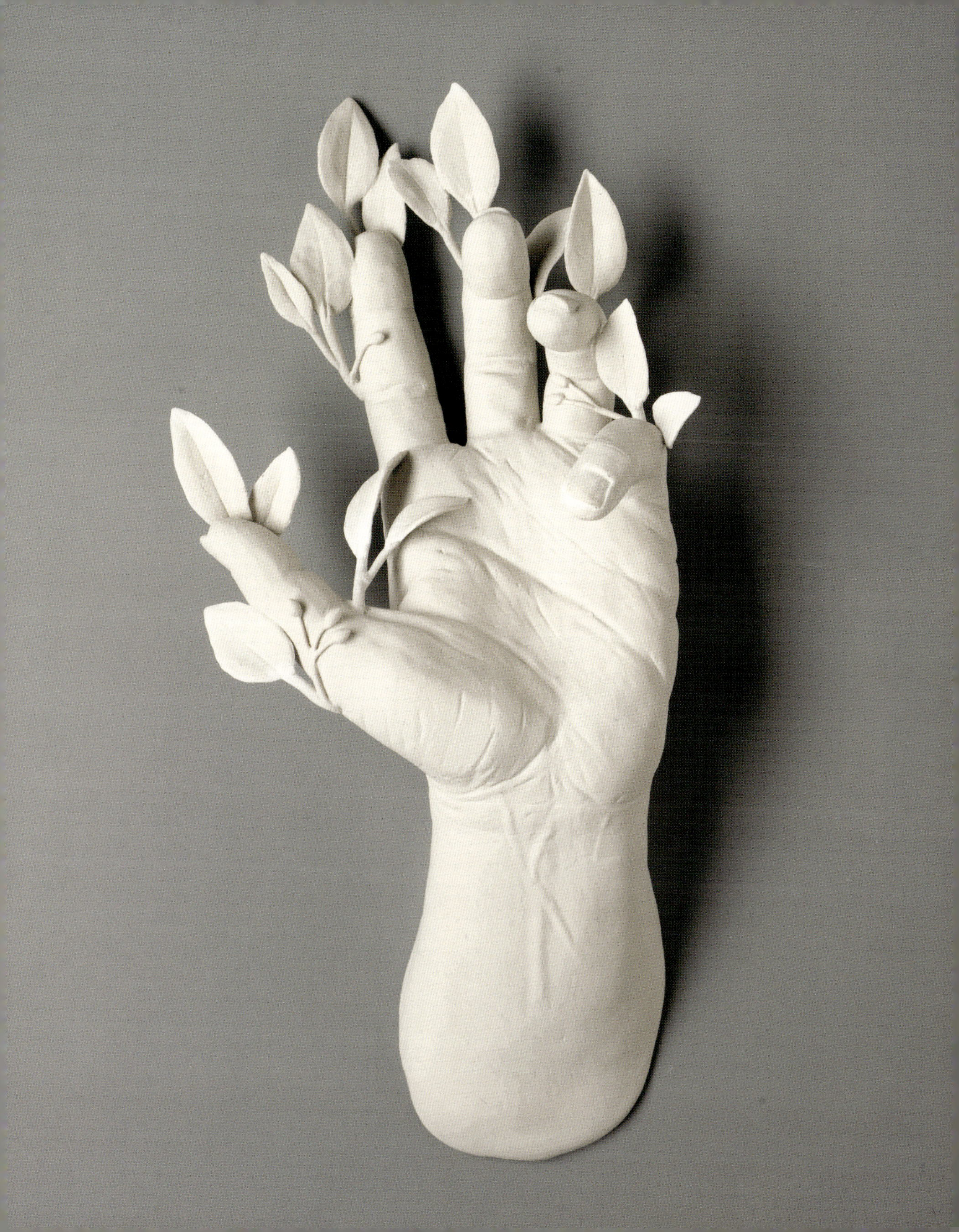

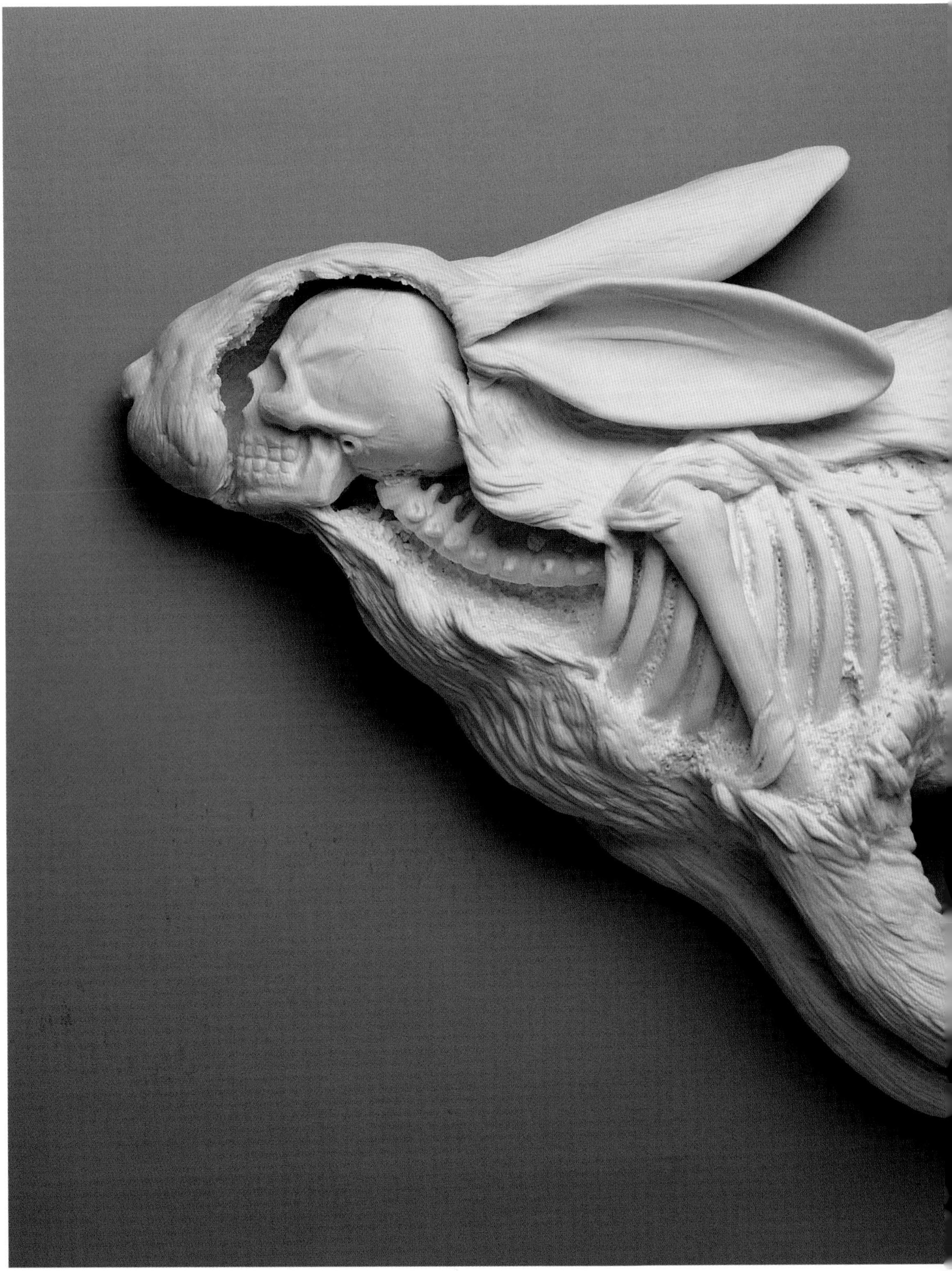

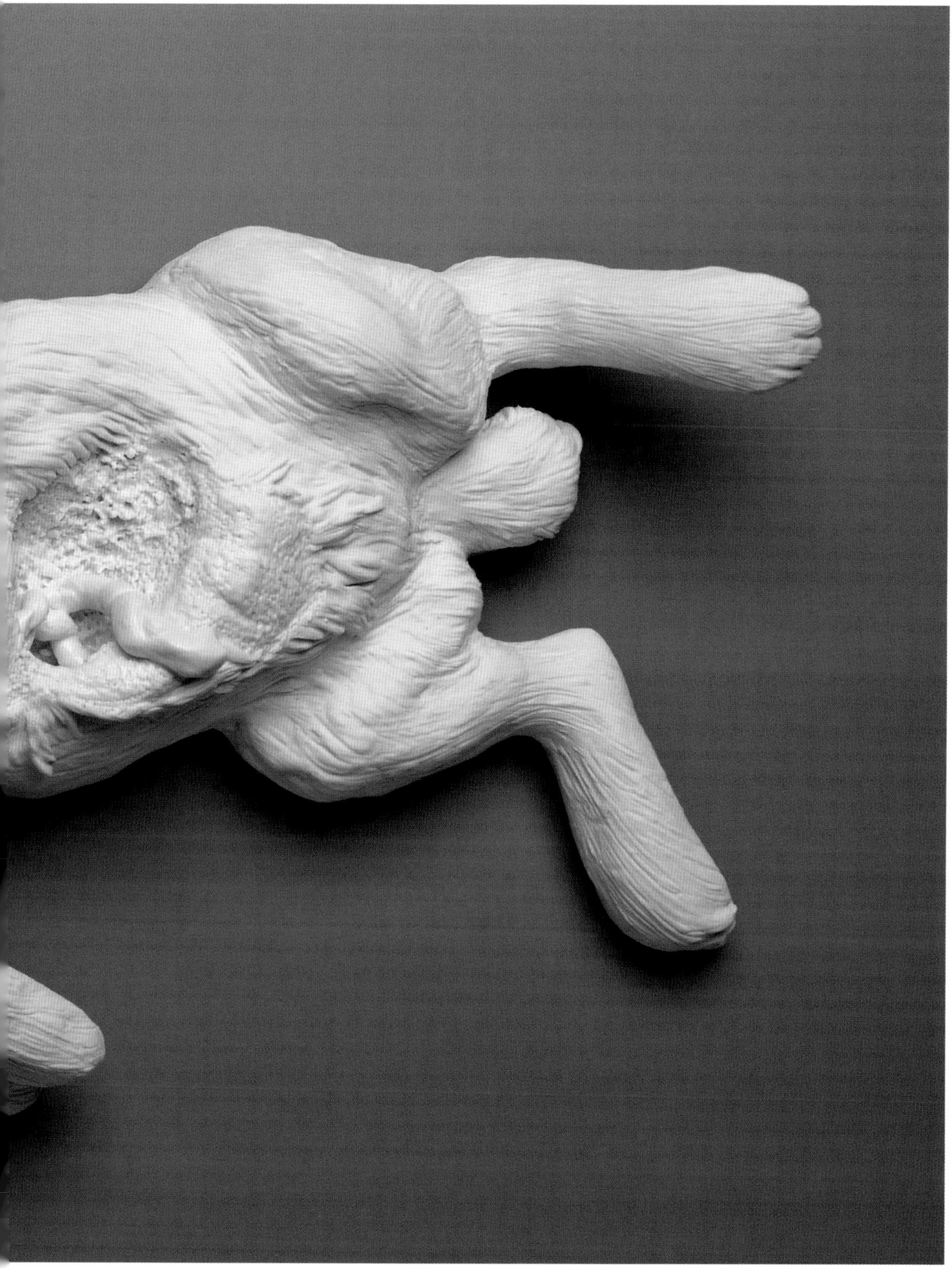

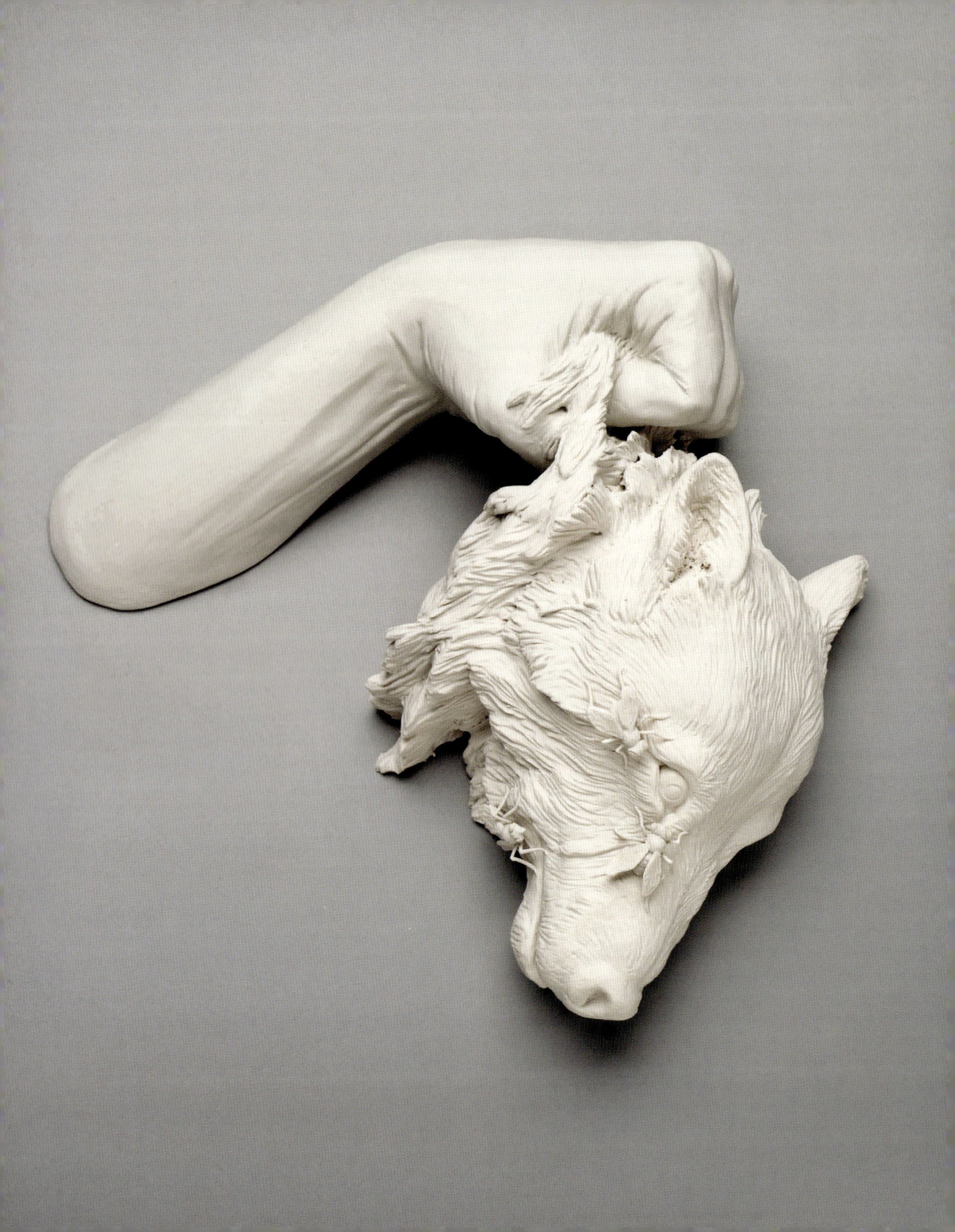

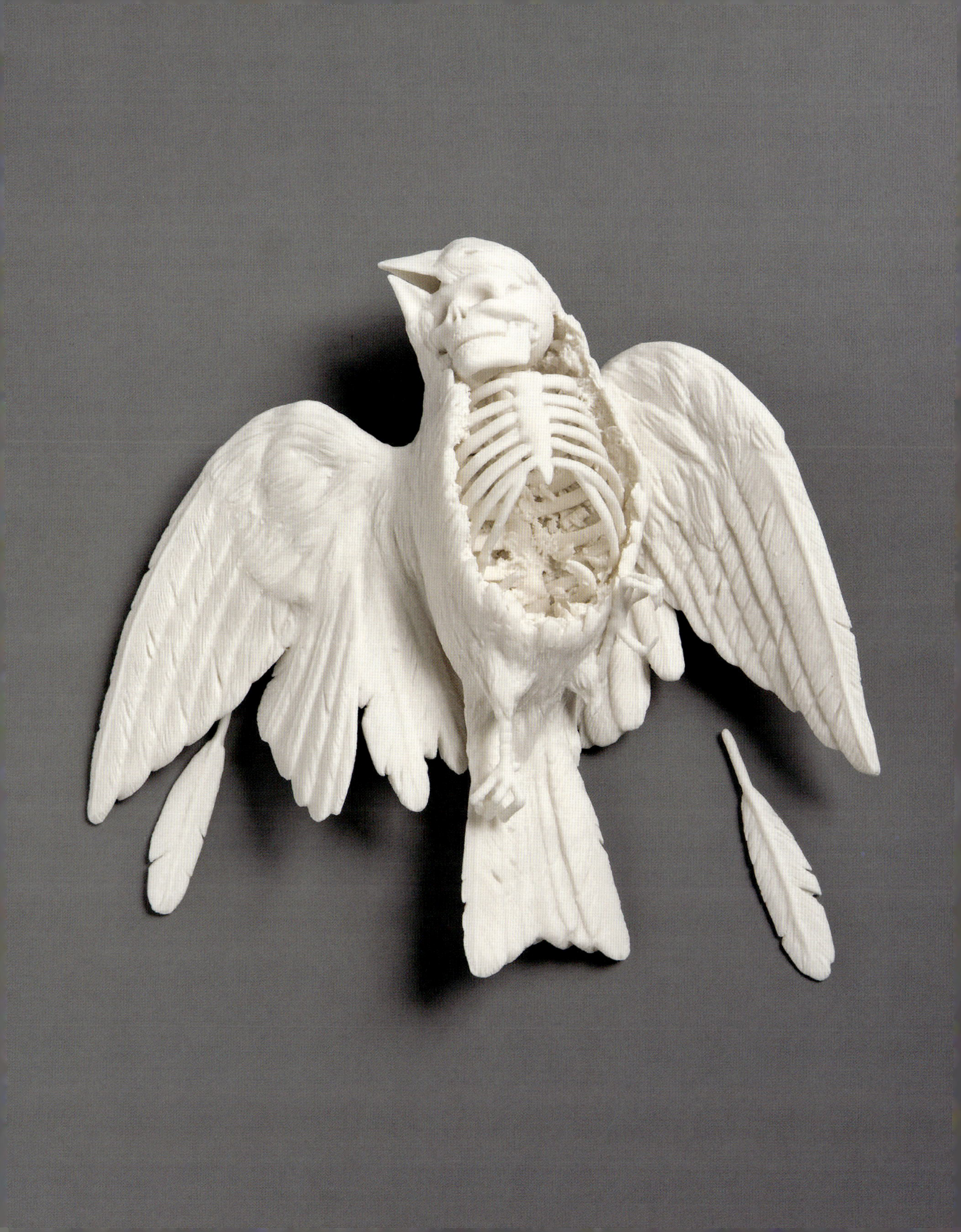

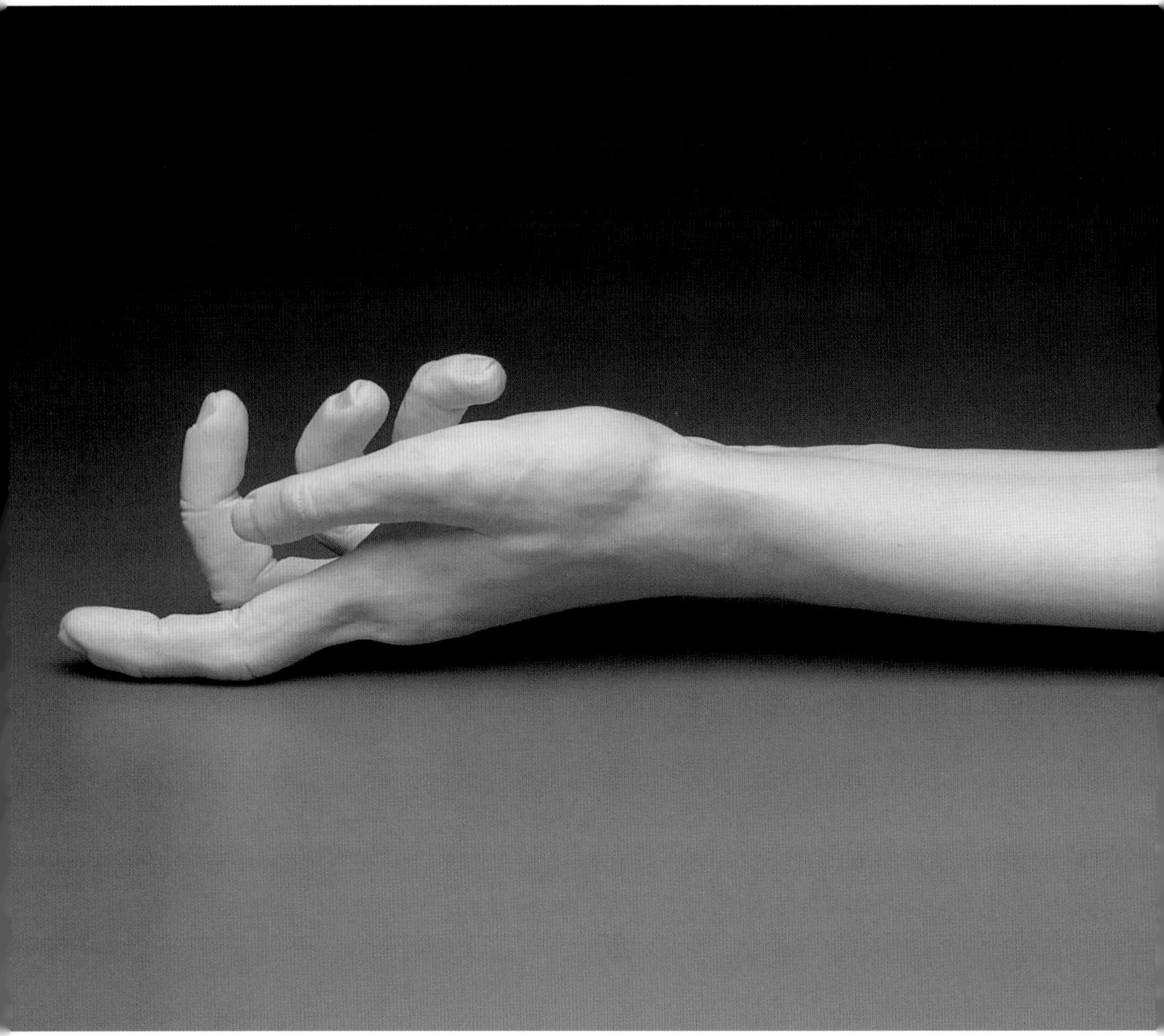

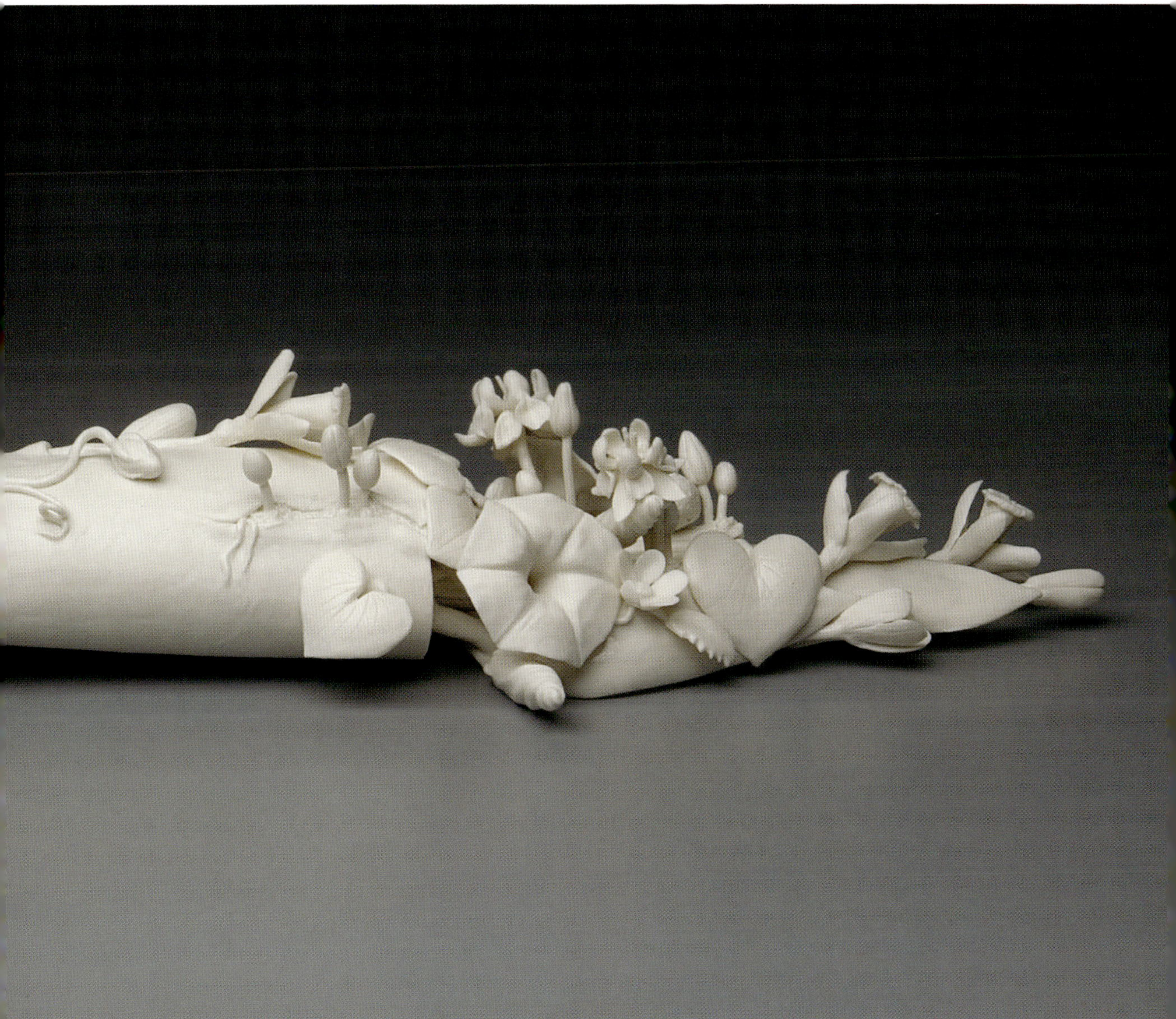

CLAIRE PARTINGTON

LOCATION London, UK
YEAR OF BIRTH 1973
MATERIALS Low fired earthenware, stoneware and porcelain
EXPERIENCE 13 years

'I build the pieces from the base up using coils of clay. The figures are then refined and details sculpted. The decoration is applied using a process called sprigging, where I cast objects in plaster and use the mold to create decorative reliefs. Once fired and glazed I use my own images as enamel decals on the surface and liberal amounts of gold lustre.'

Claire Partington studied Fine Art at Central Saint Martins. After a bit of a hiatus and some random jobs, she went on to work within some of the museums and galleries that inspired her enthusiasm for art and craft (most notably the V&A) and continued her post grad education in Museum Studies. In 2003 she started going to night school to make ceramics and stayed there for the next five years. She started producing work again, drawing on influences gained from her non-craft education and vocational experience working with National Collections. Her work is a result of her interests in art and social history, stories and the urban environment she lives in.

Photos courtesy of the artist

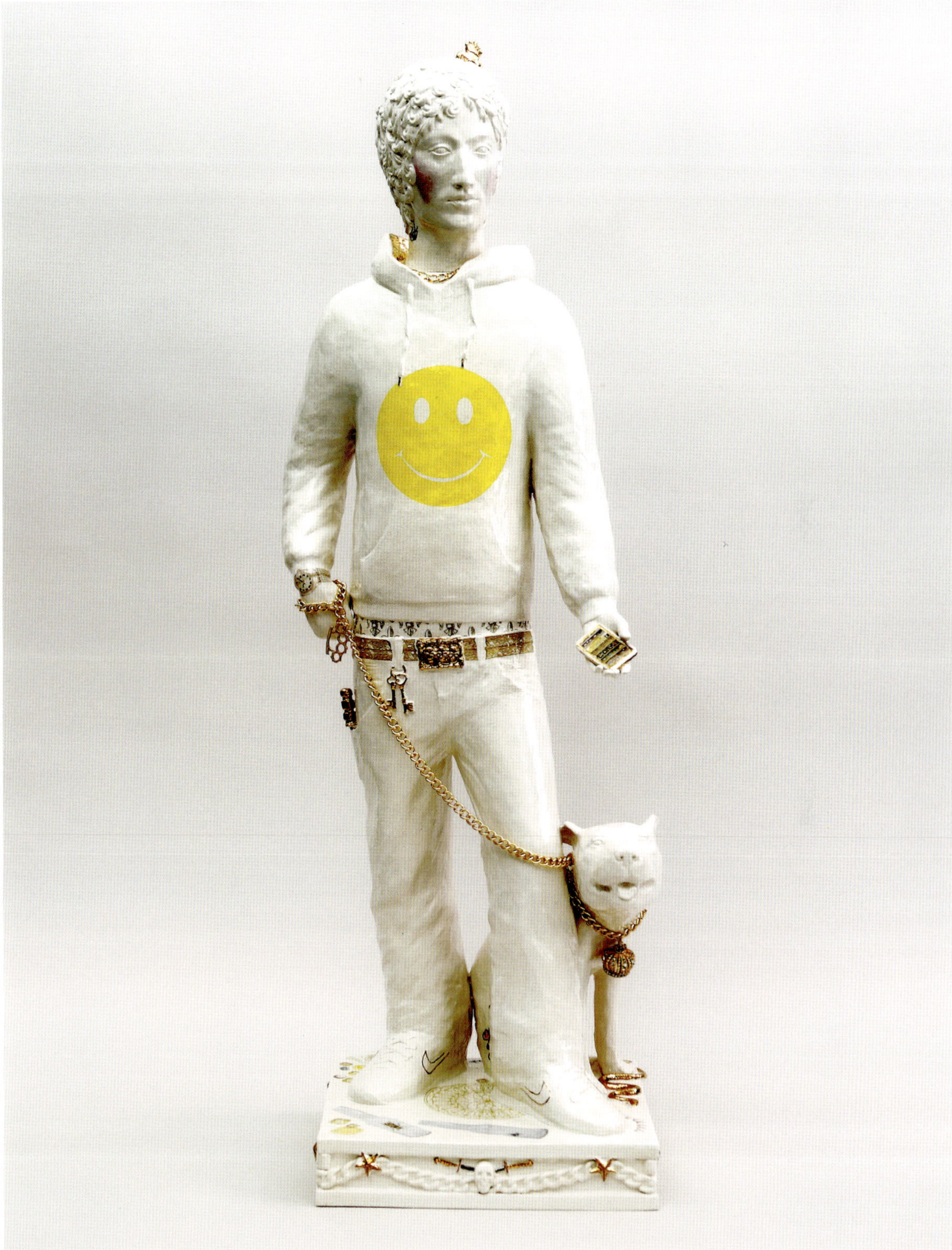

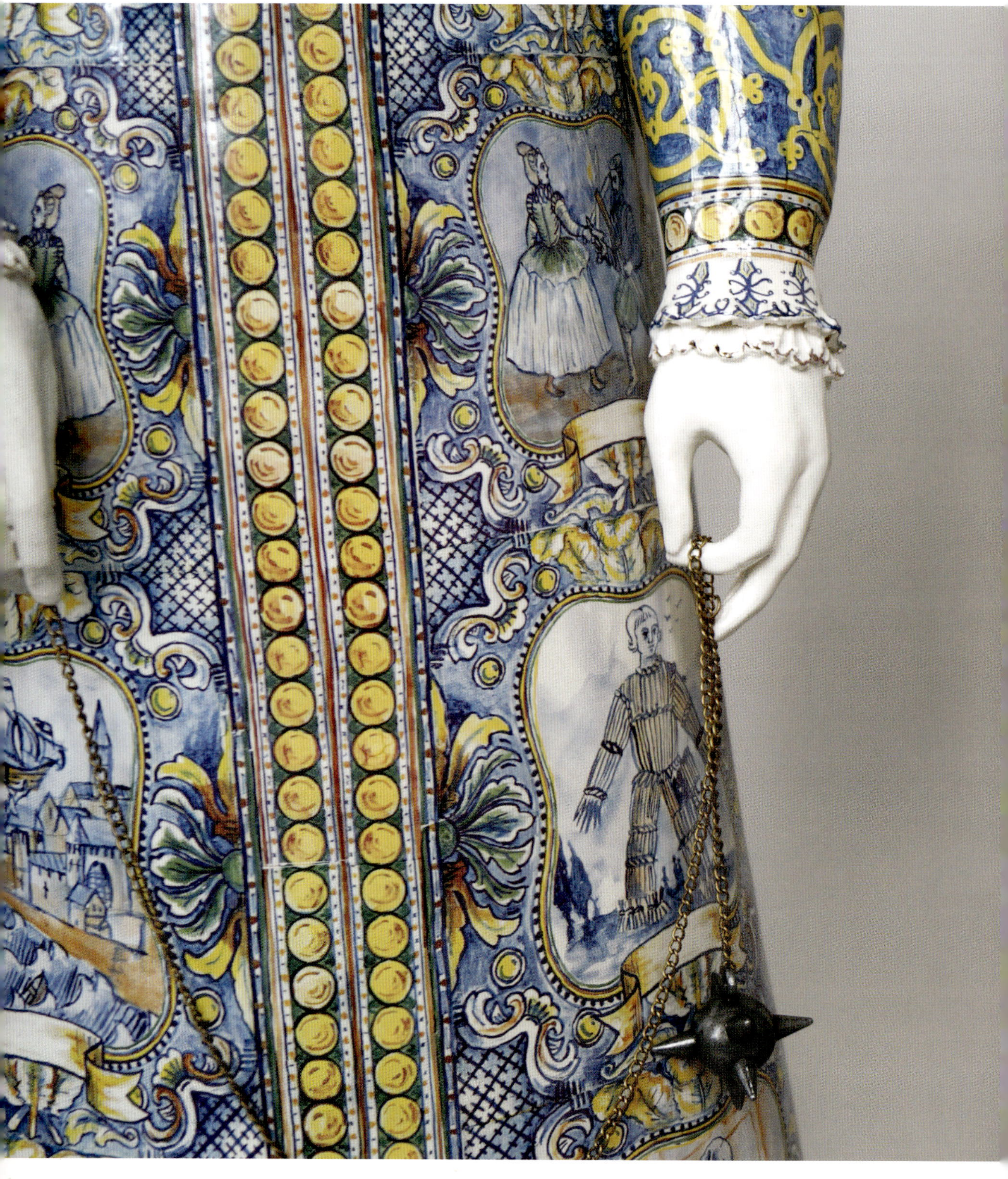

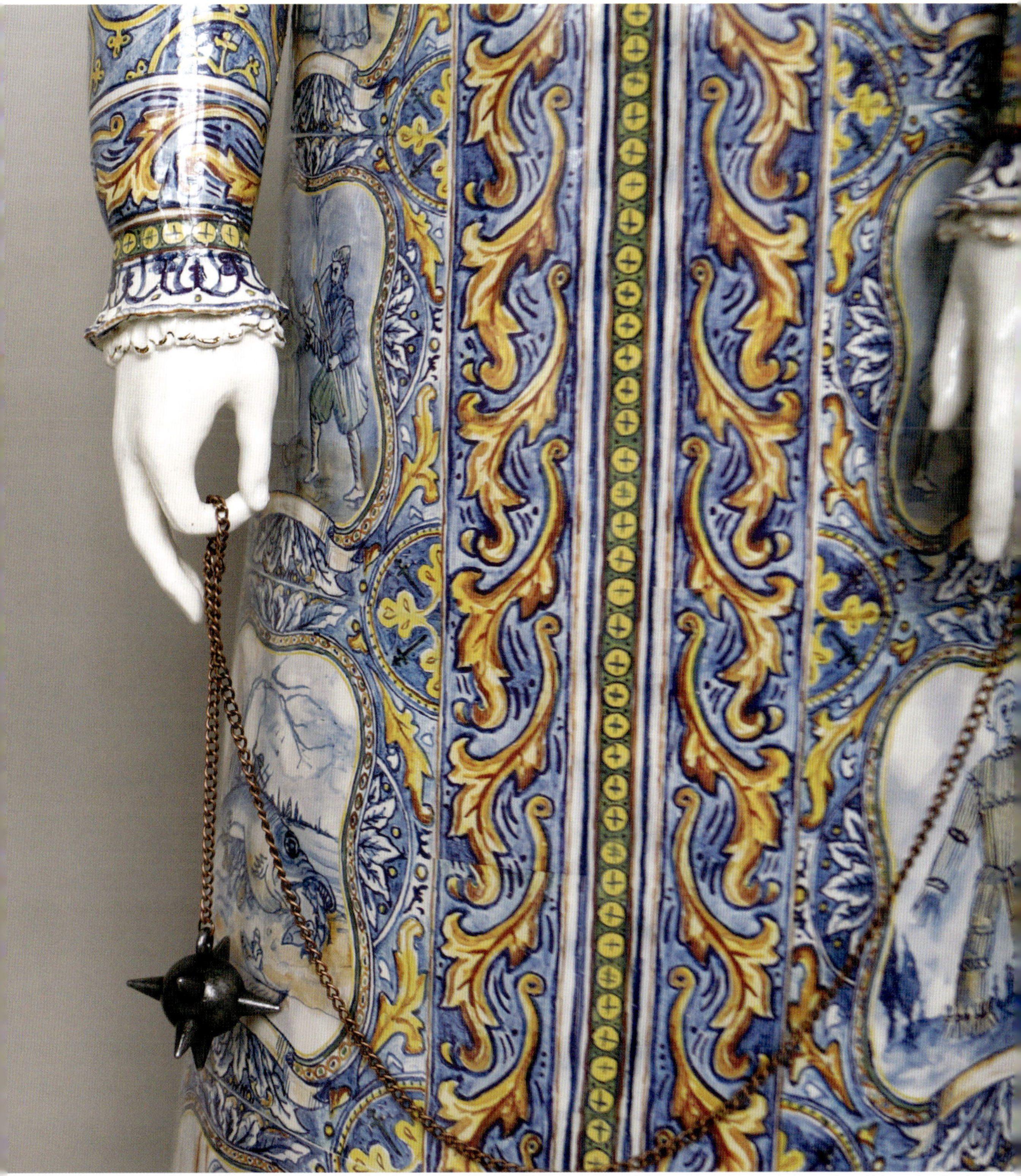

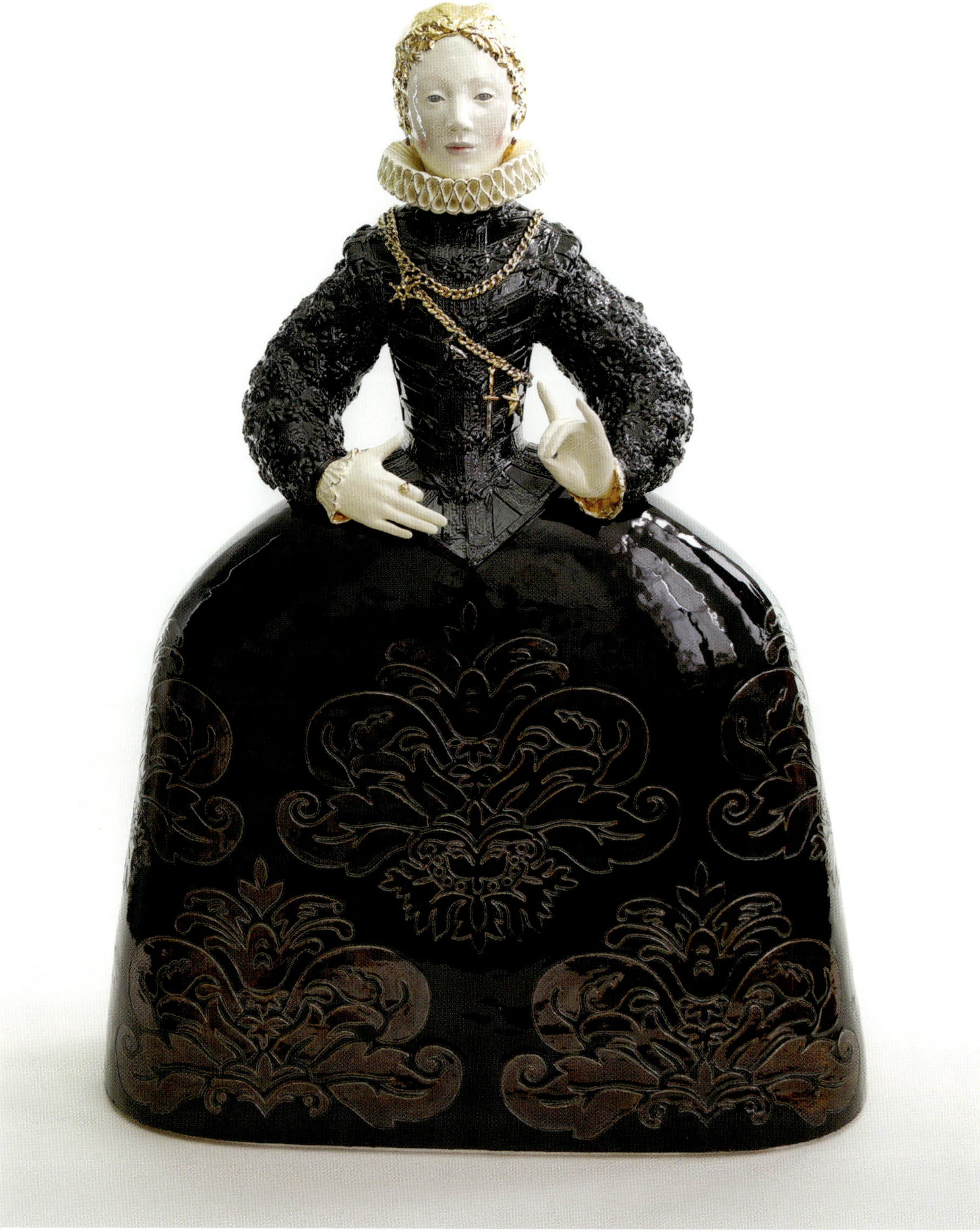

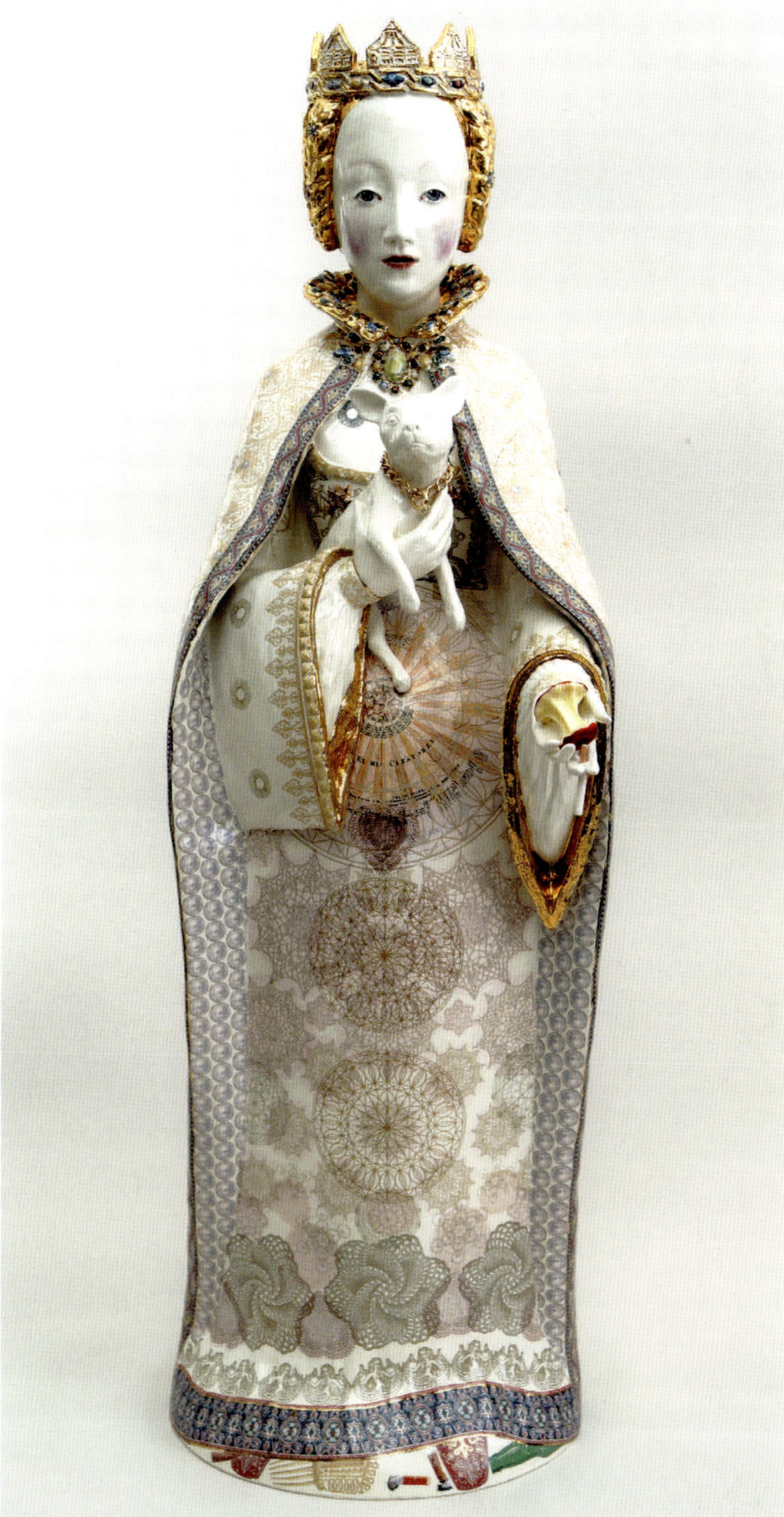

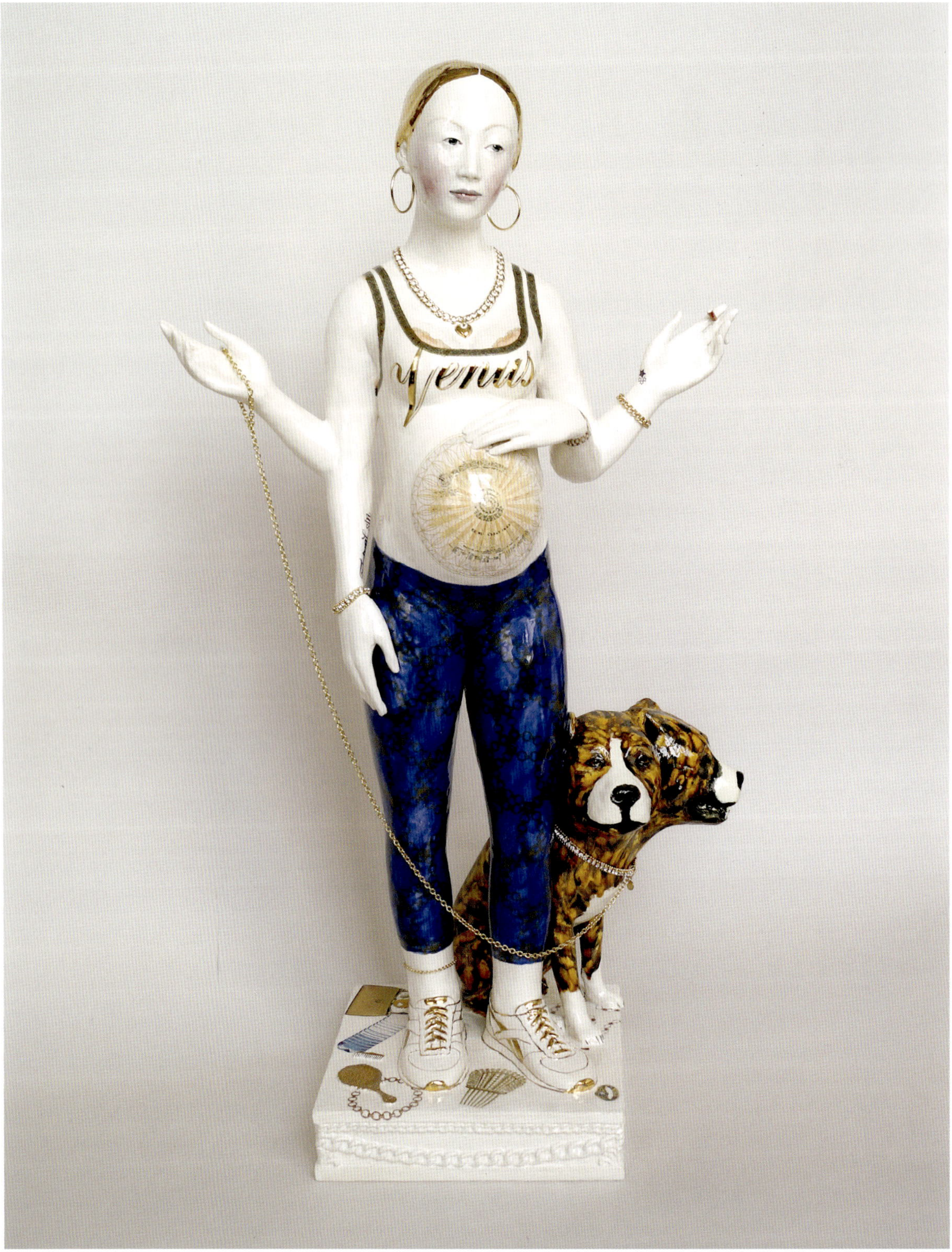
Venus

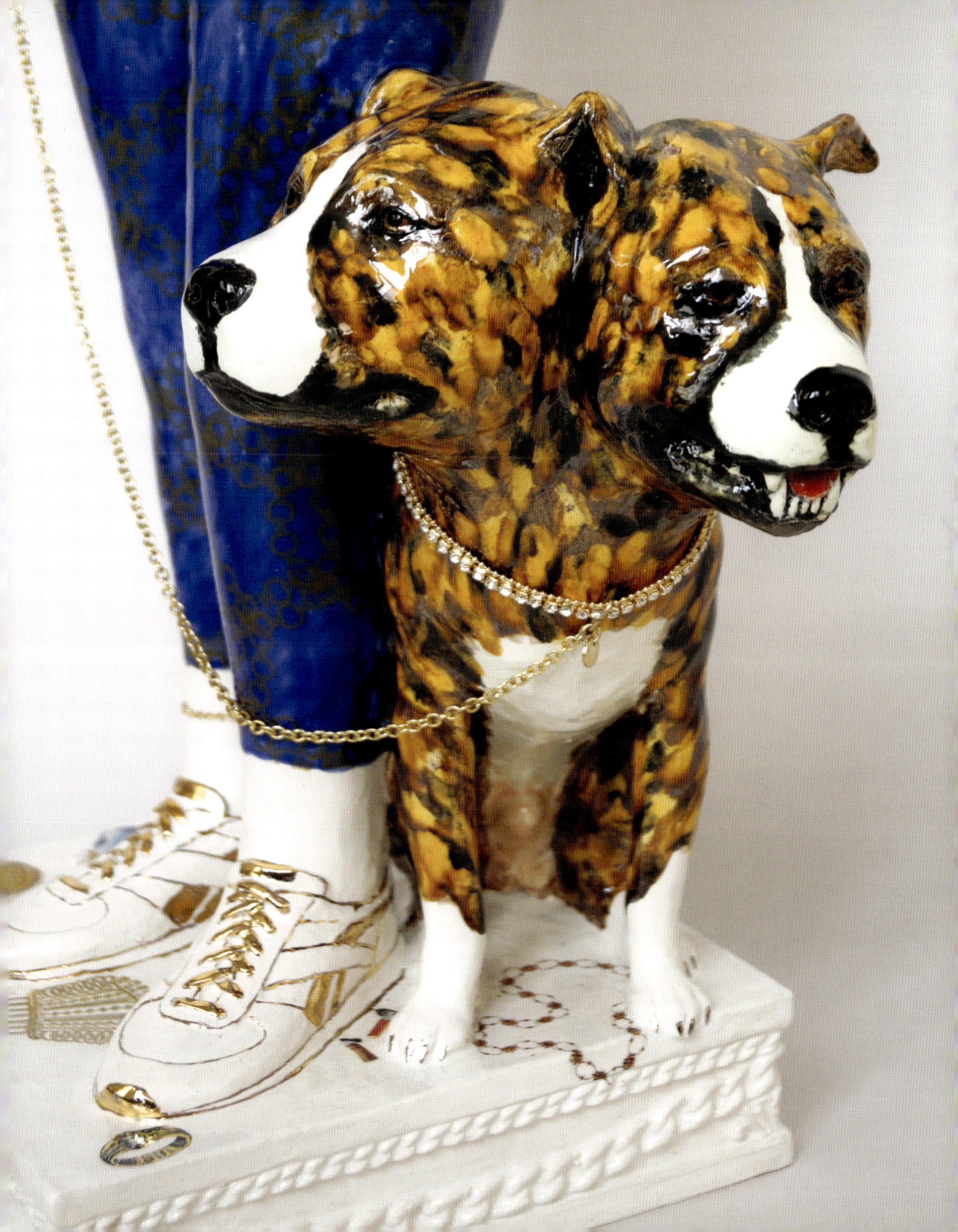

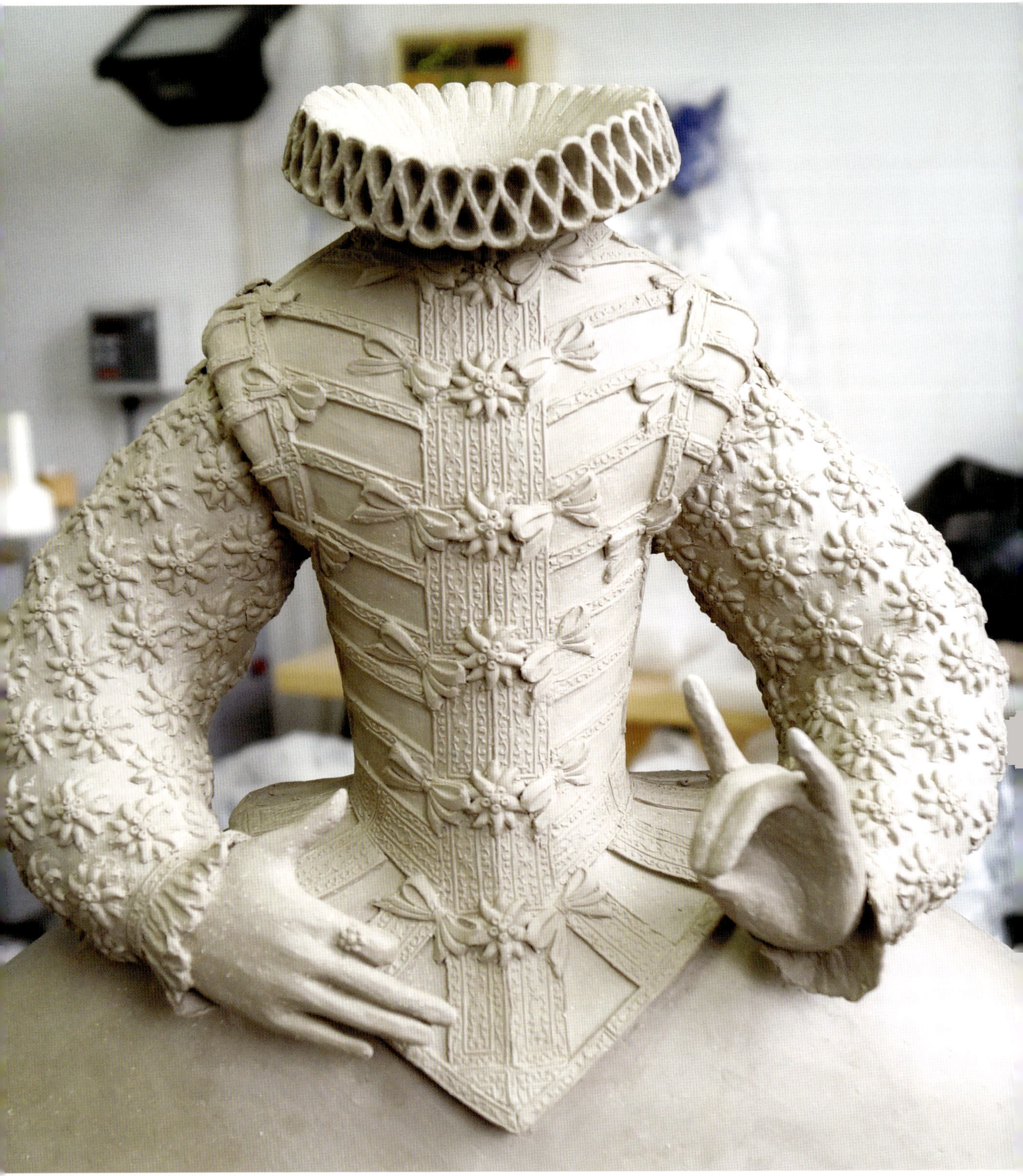

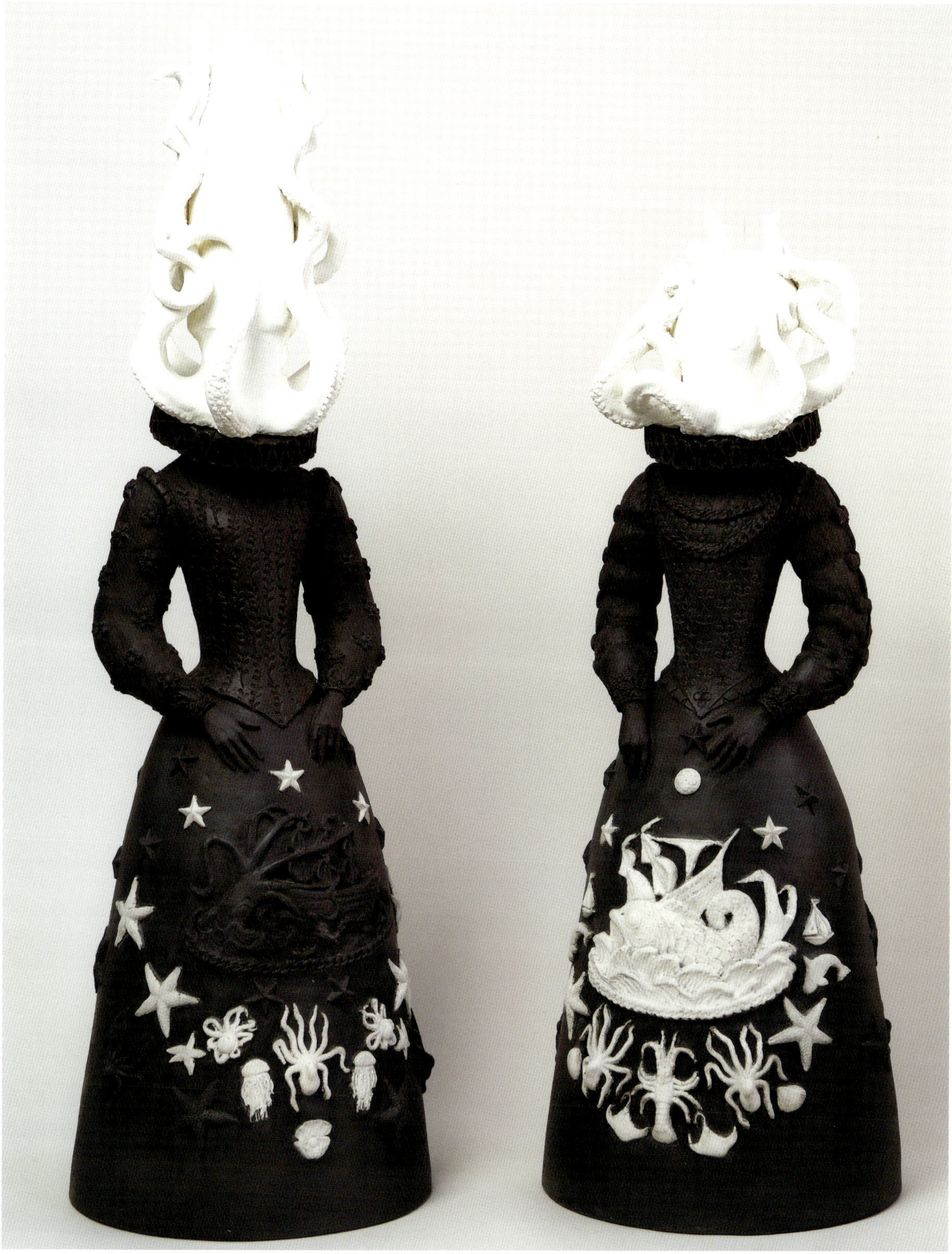

JESSICA HARRISON

LOCATION Edinburgh, Scotland
YEAR OF BIRTH 1982
MATERIALS Hand built ceramics
EXPERIENCE 7 years

Working with materials from porcelain to stone to paper, Jessica Harrison's practice explores the mechanics of perception and the fallibility of observation through an examination of the interaction between the visual and the tactile. Her work seeks to move away from traditional definitions and distinctions between bodily spaces, looking towards a mingling of skin and space, body and world, to propose alternative re-imaginings of the body, using the simplicity of materials to explore the complexity of the sensory body.

Born in St. Bees, England in 1982, Harrison moved to Scotland in 2000 to study sculpture at Edinburgh College of Art, going on to complete a practice-based PhD in sculpture.

Photos courtesy of the artist

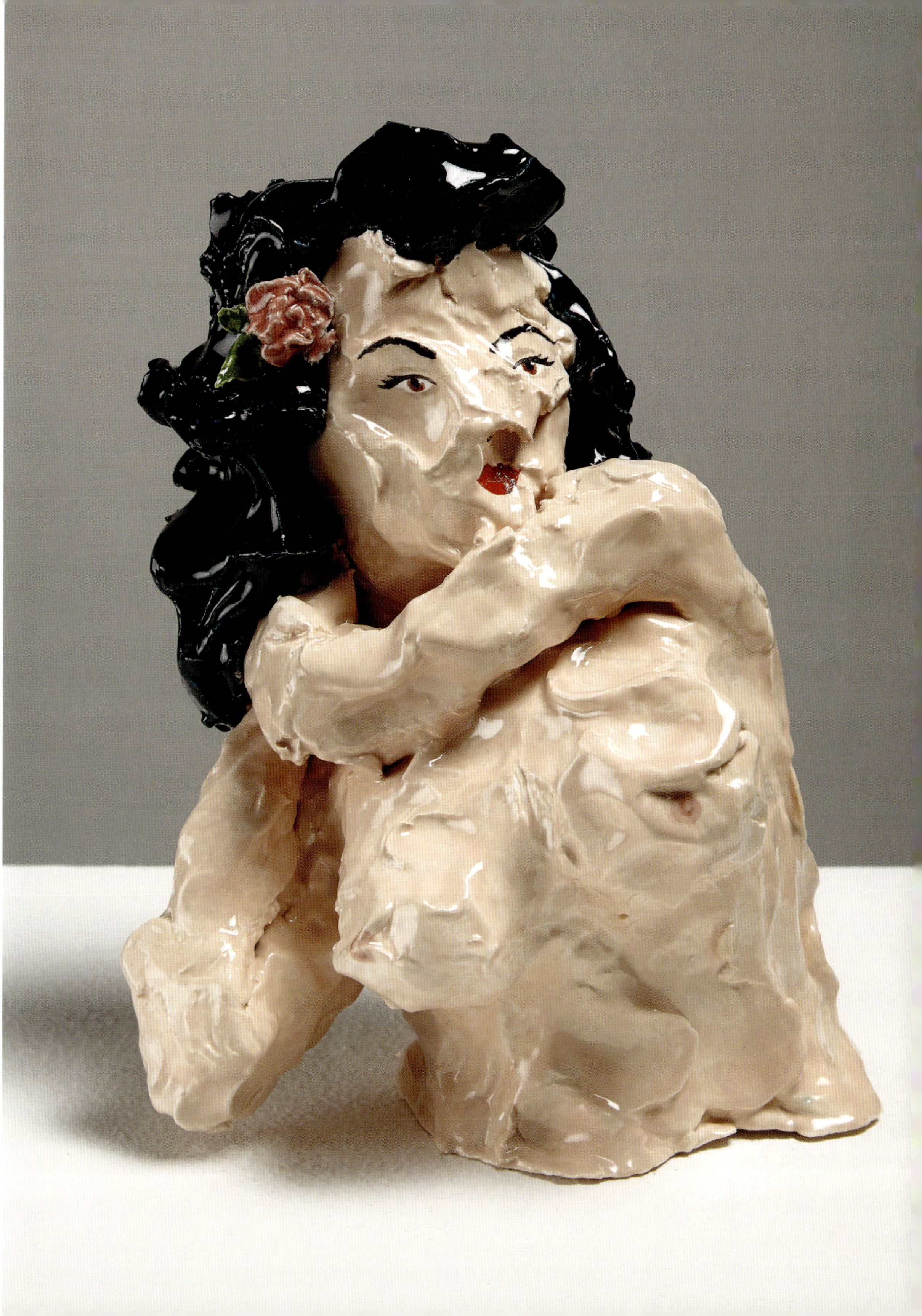

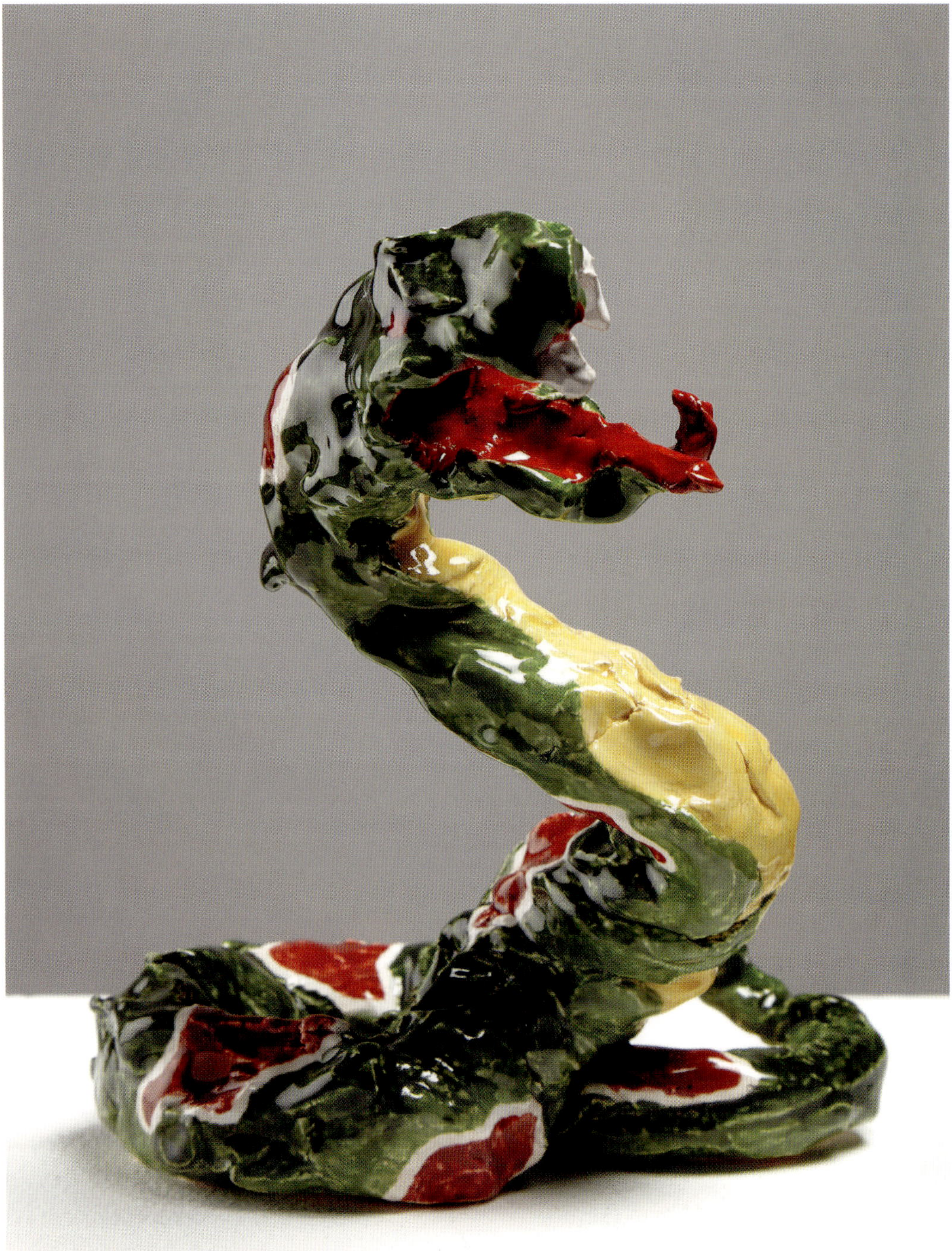

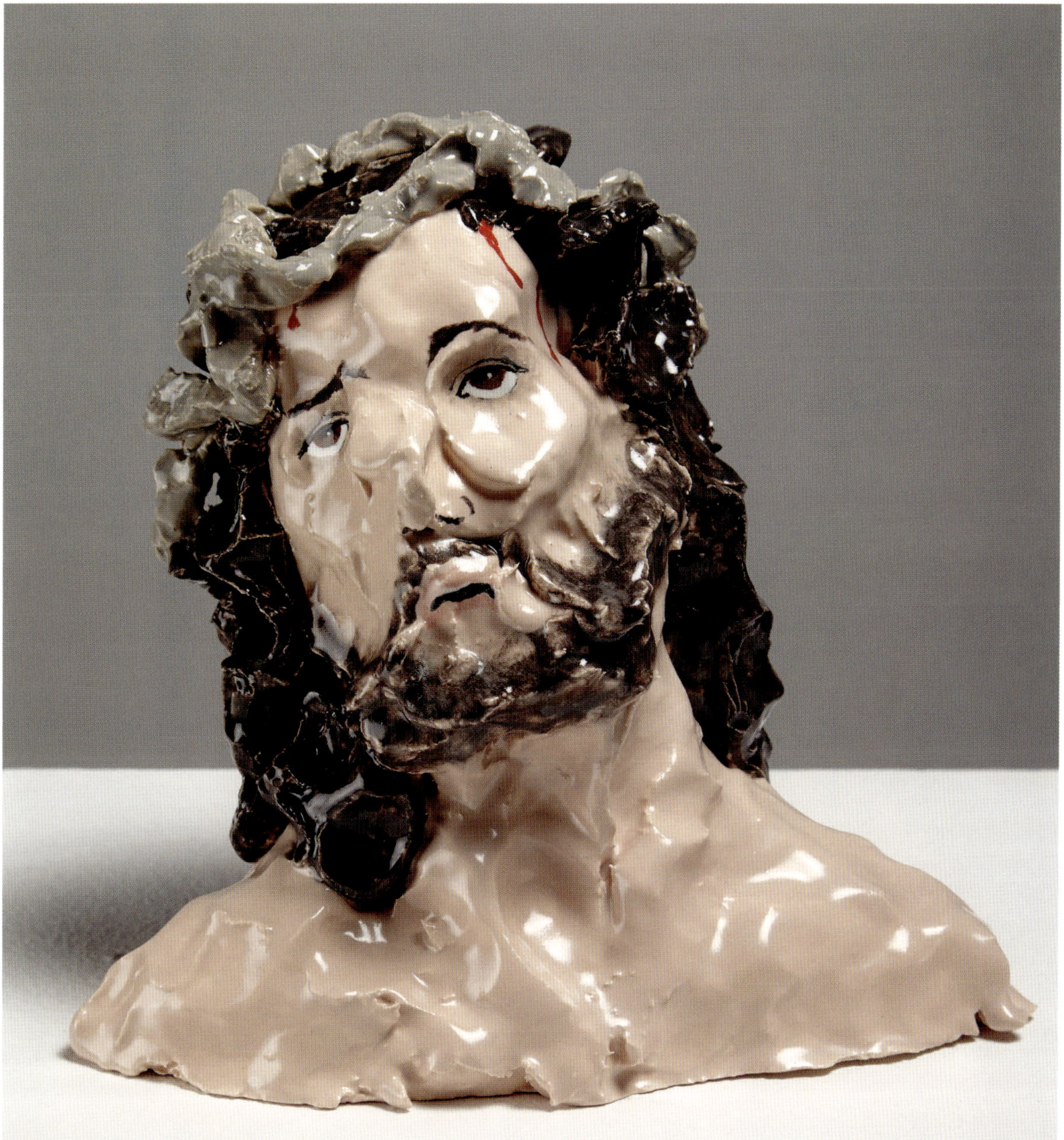

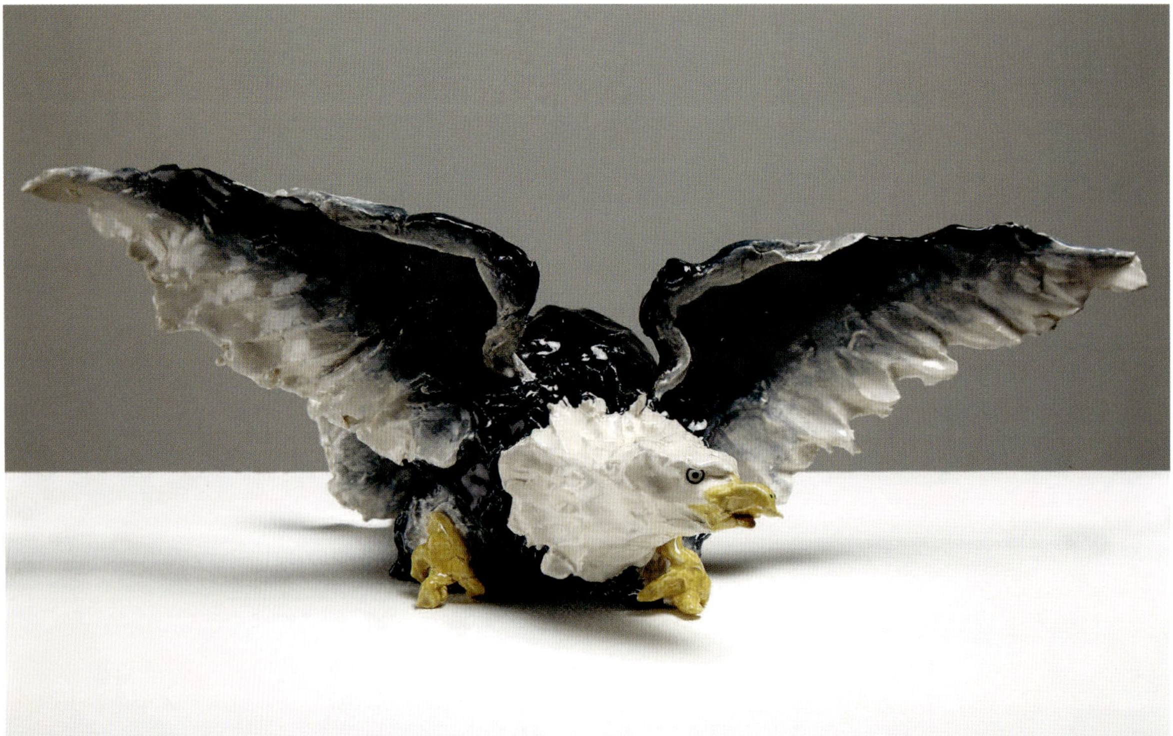

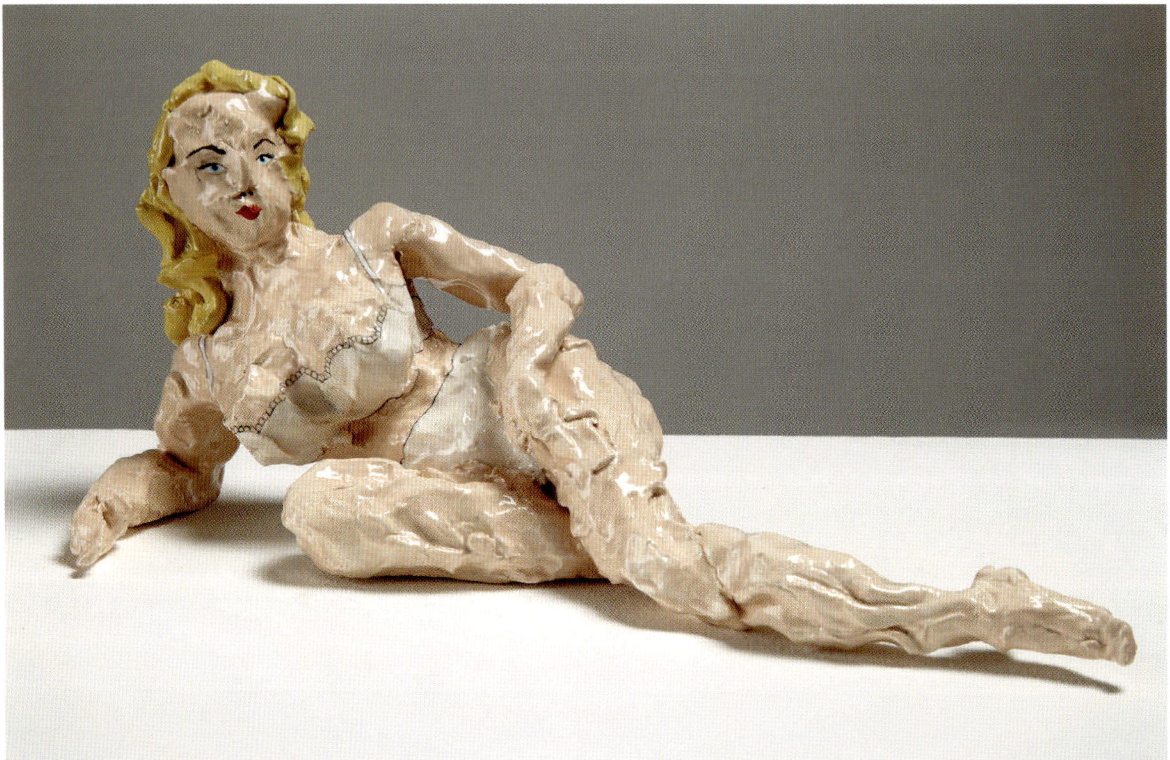

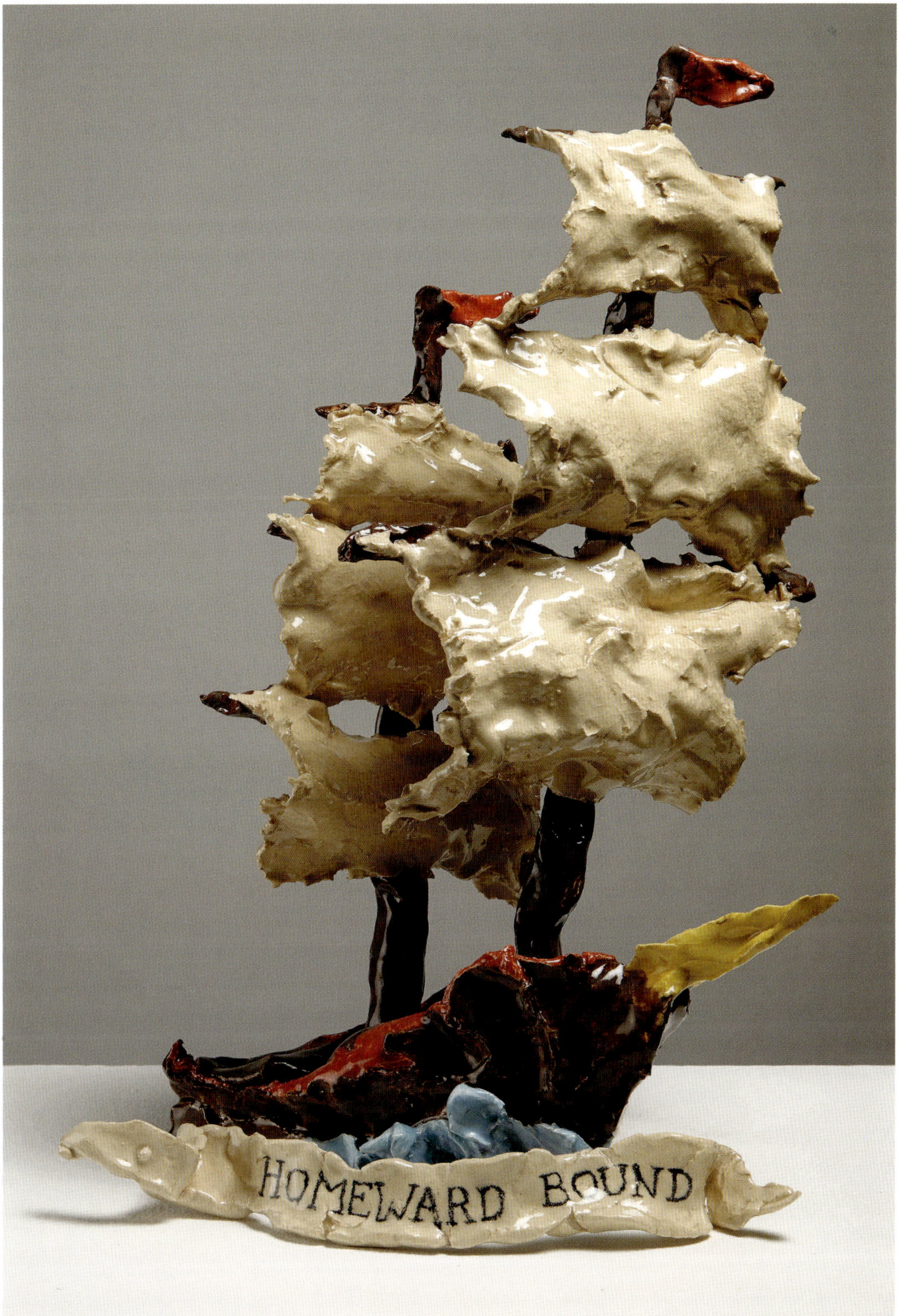
HOMEWARD BOUND

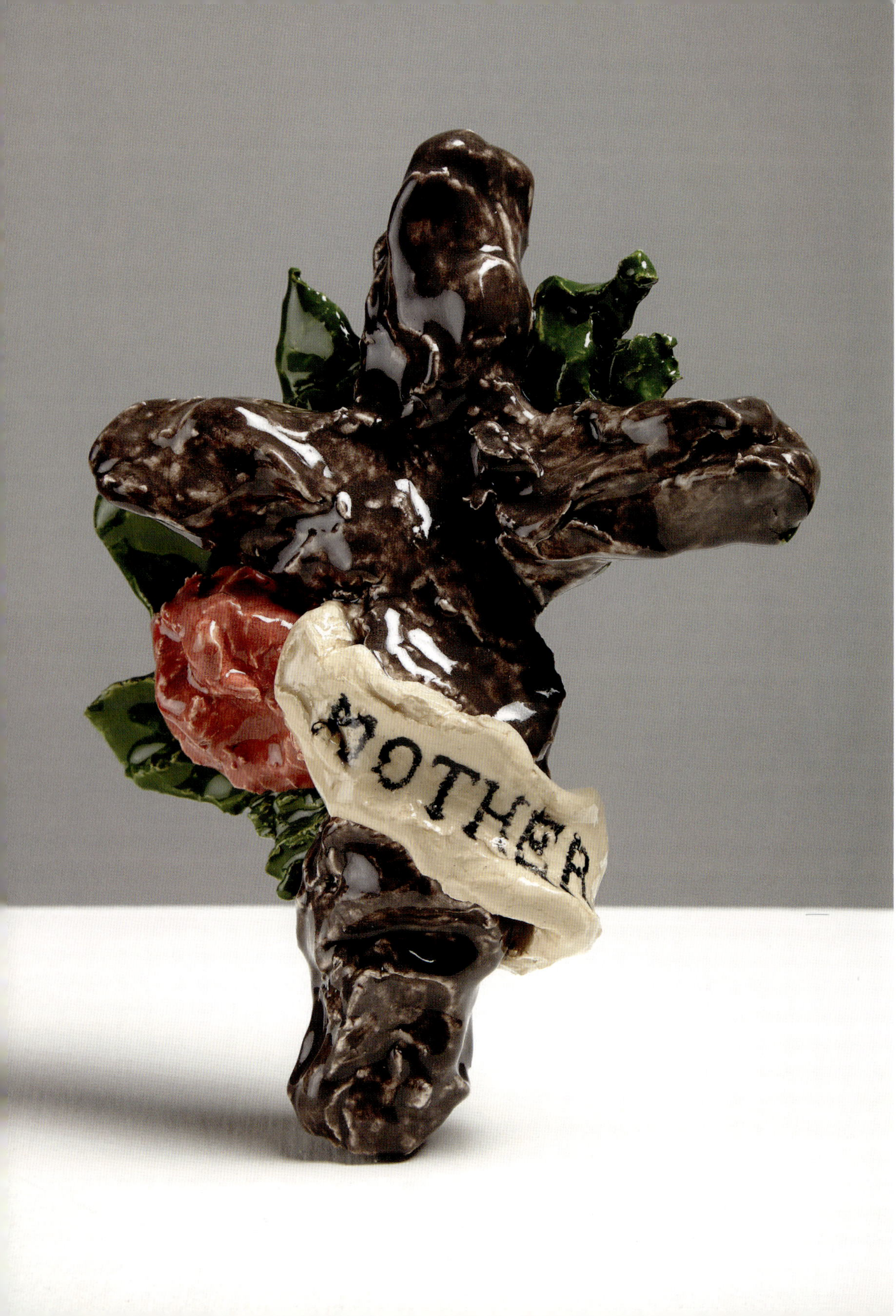
MOTHER

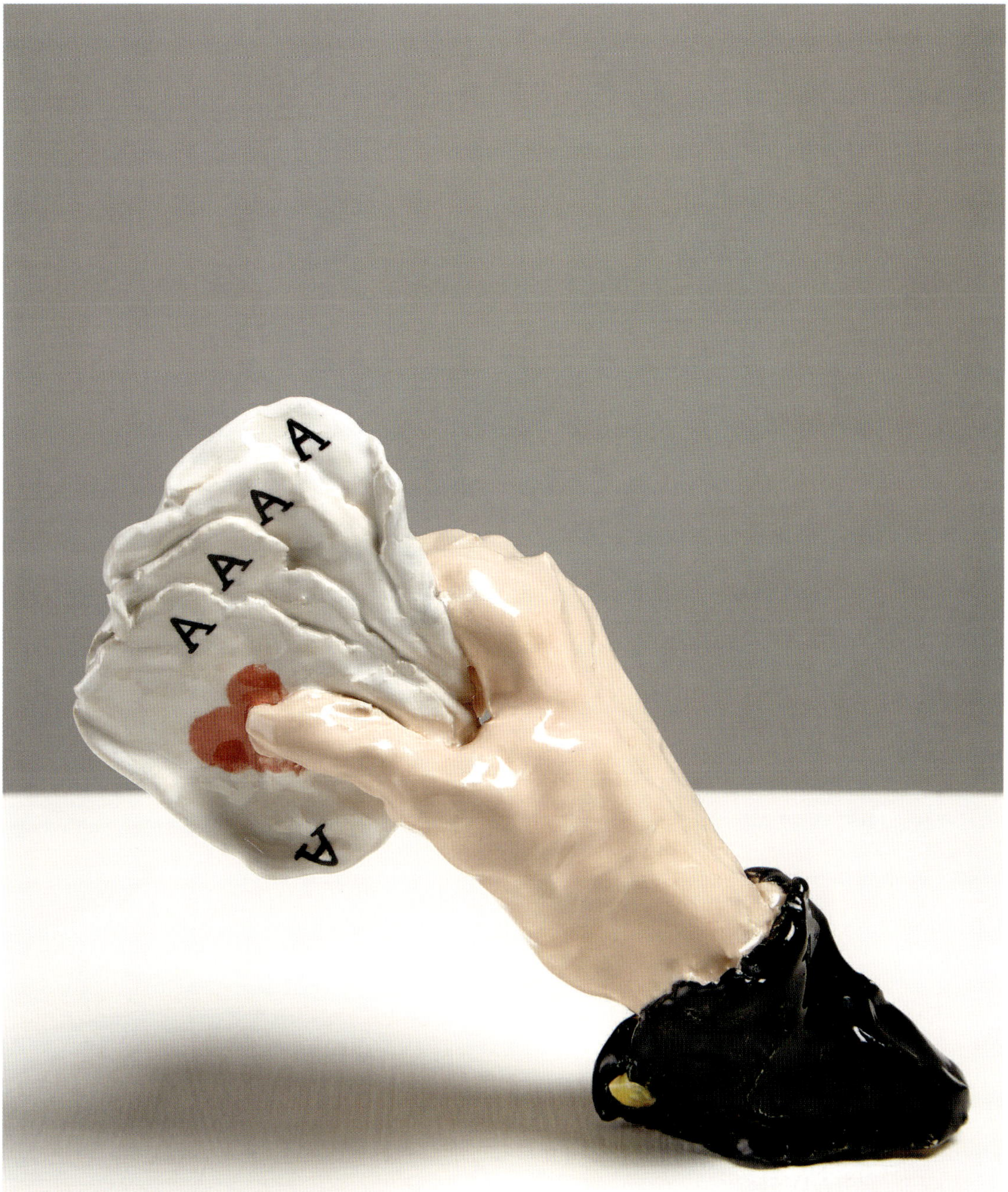

PROCESS

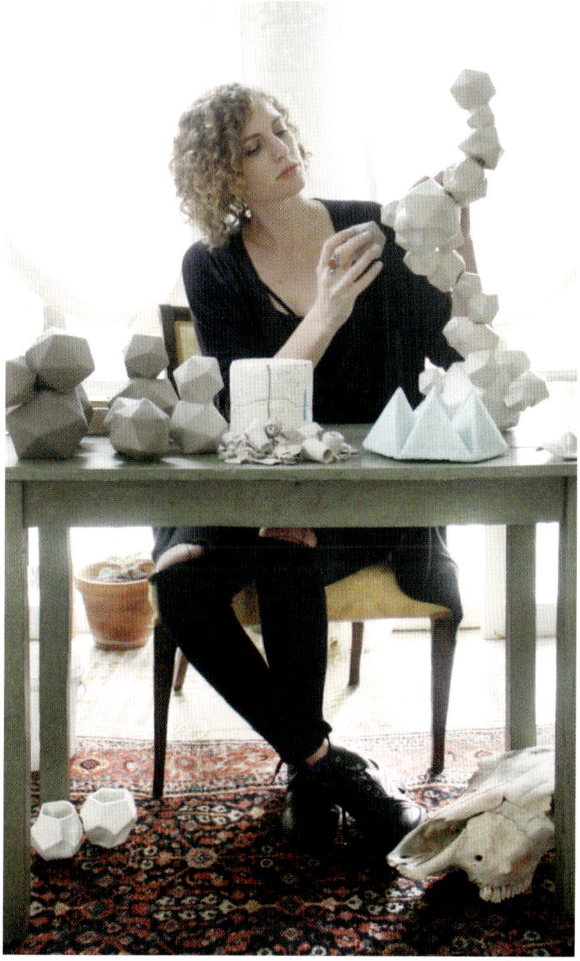
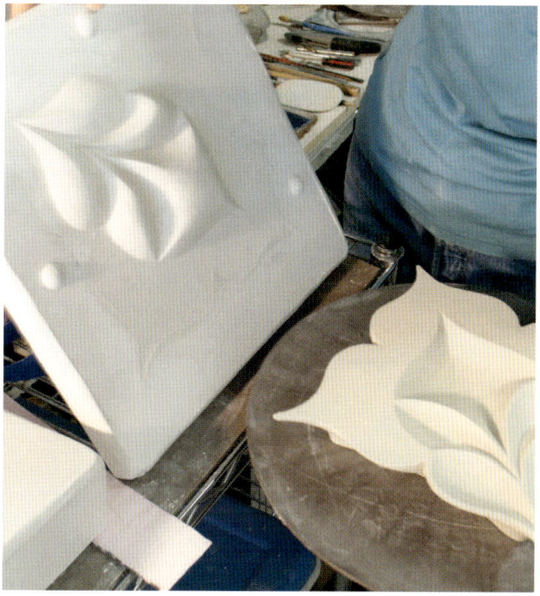
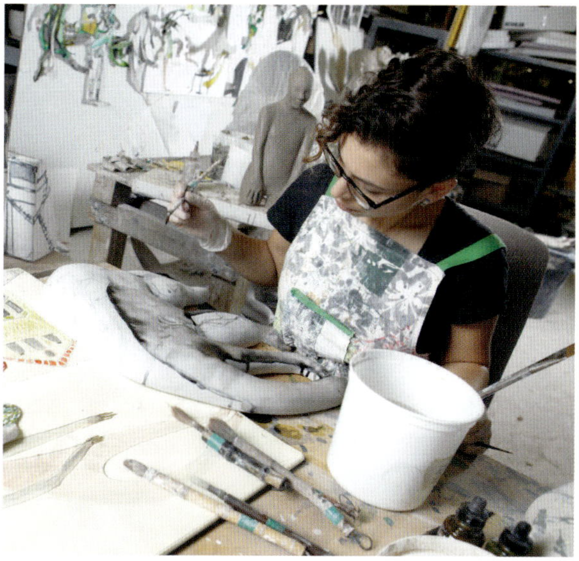
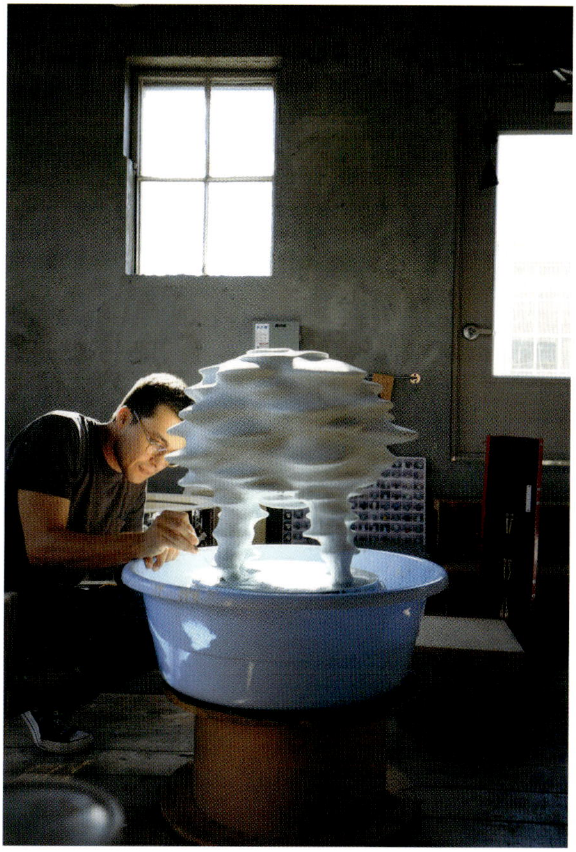

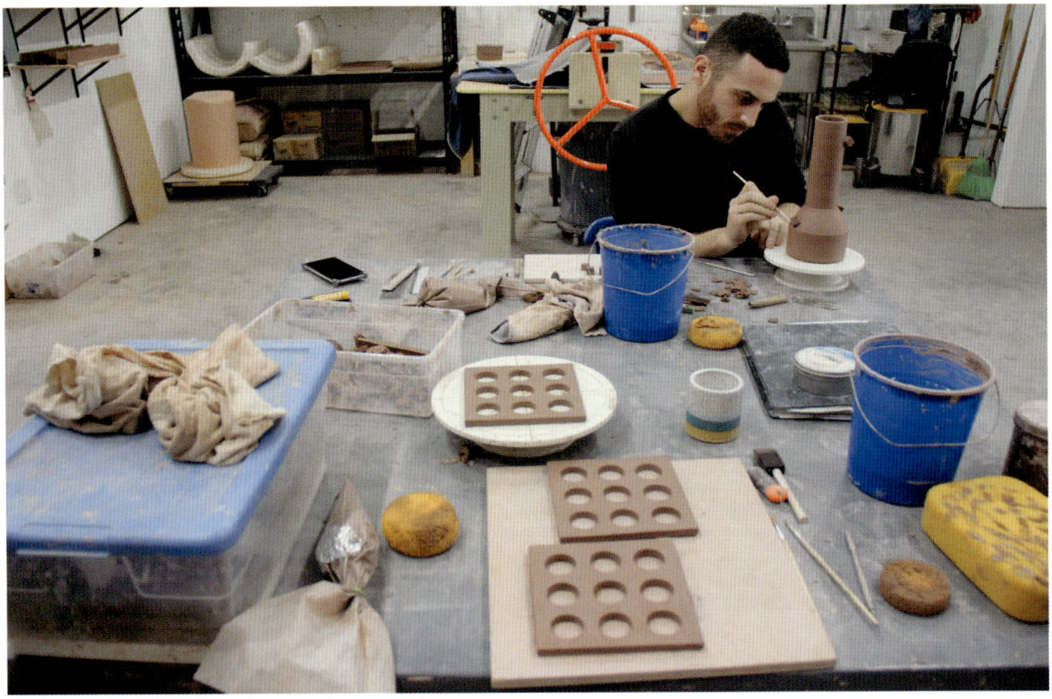

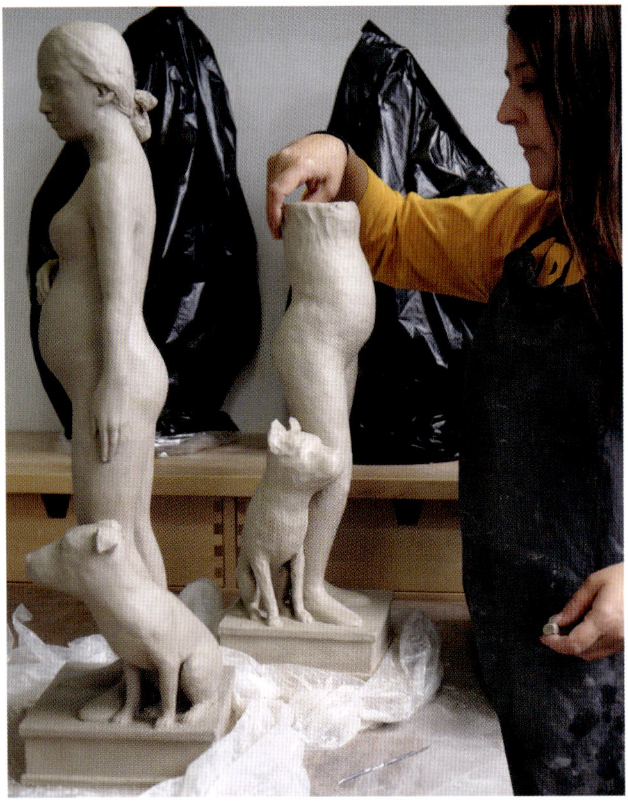

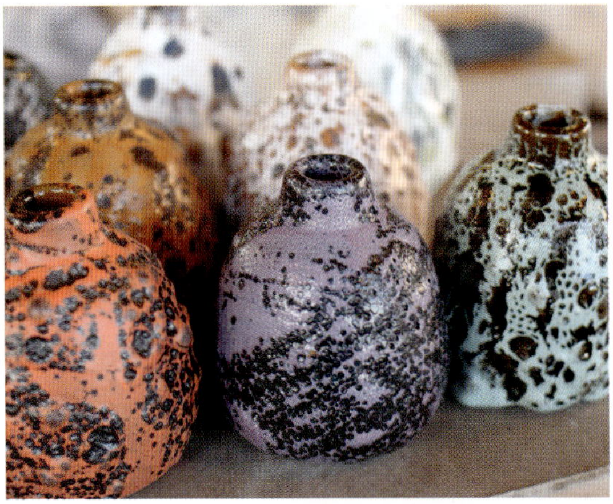

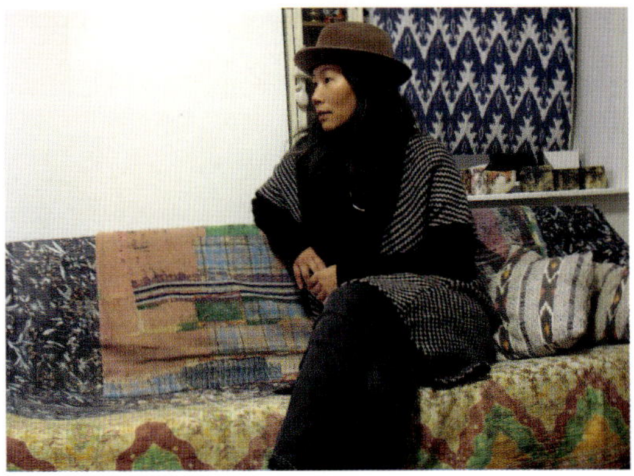

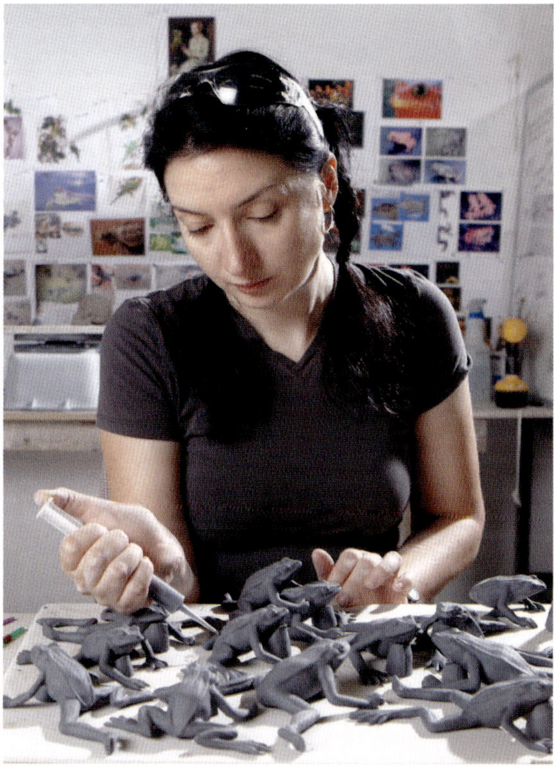

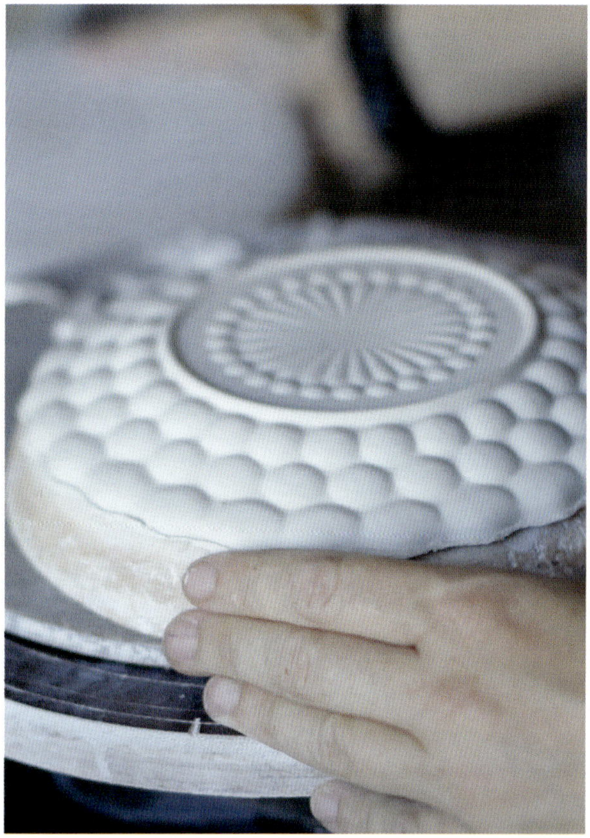

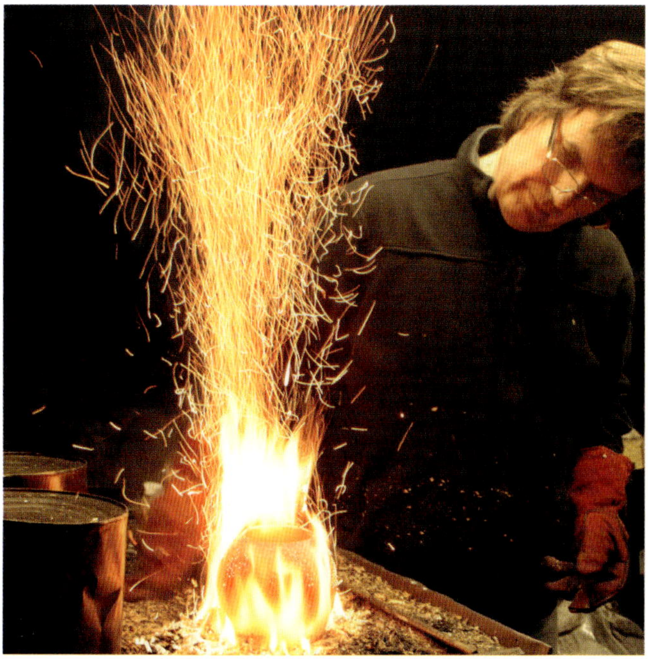

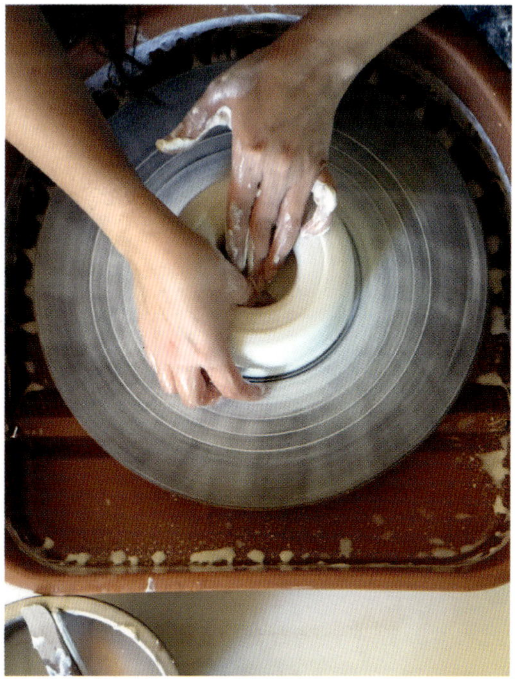

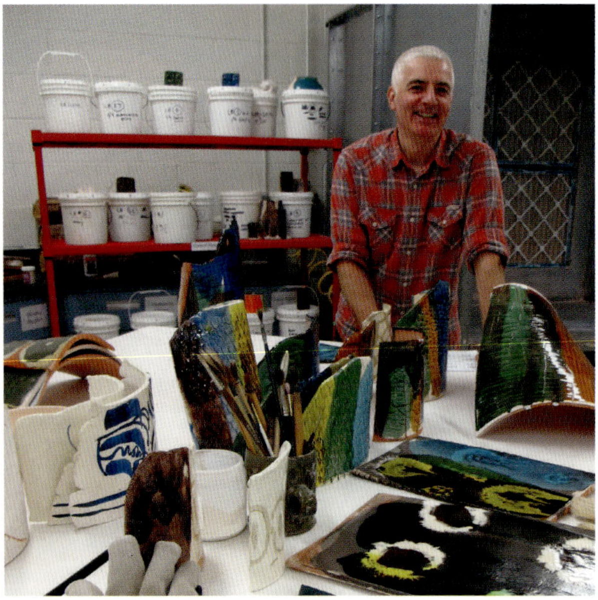
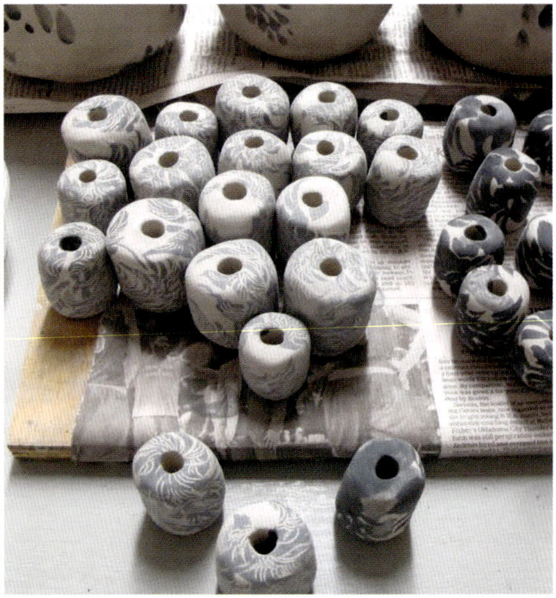
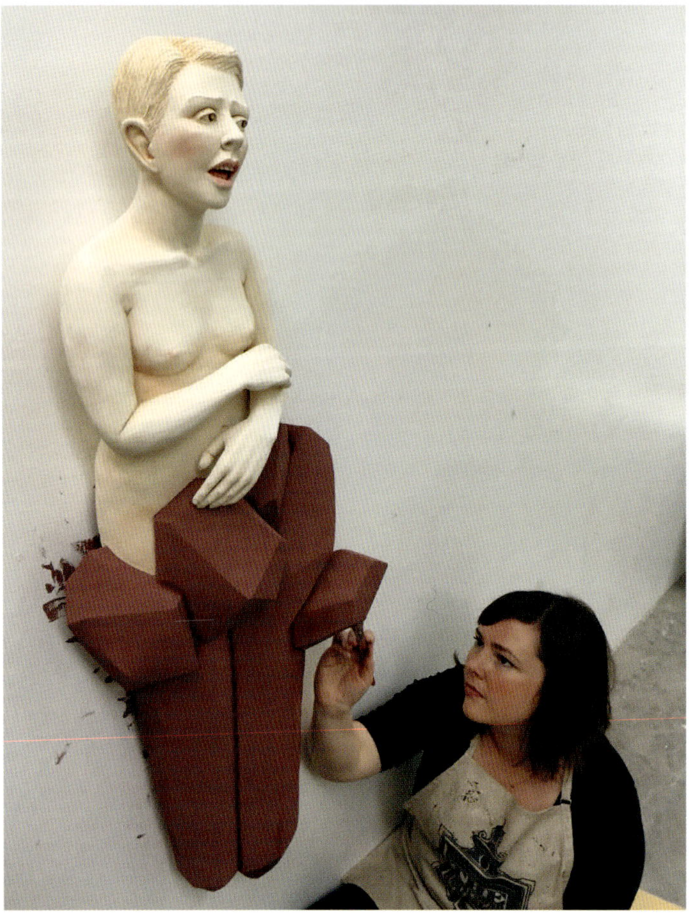
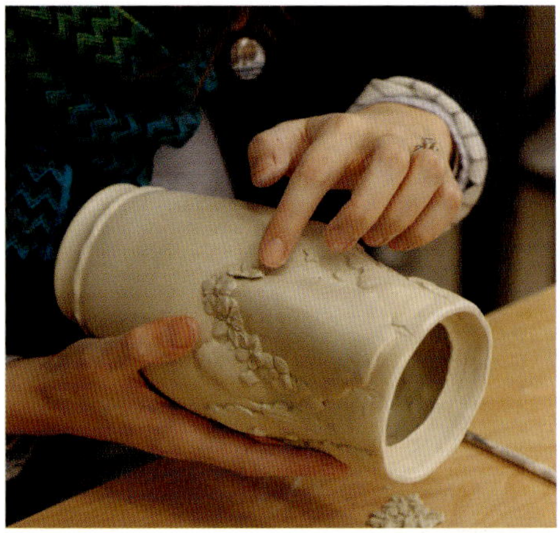
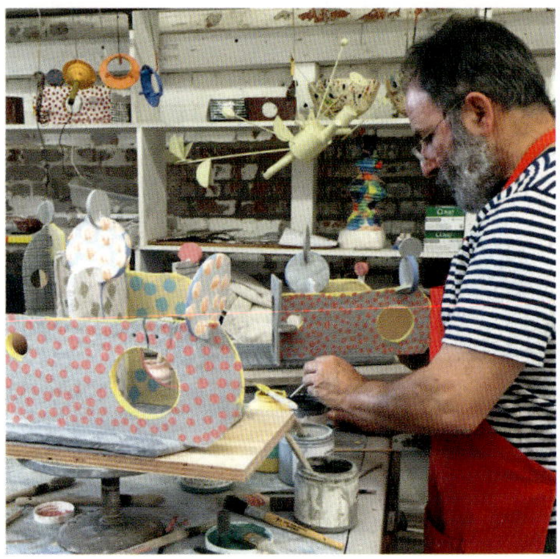

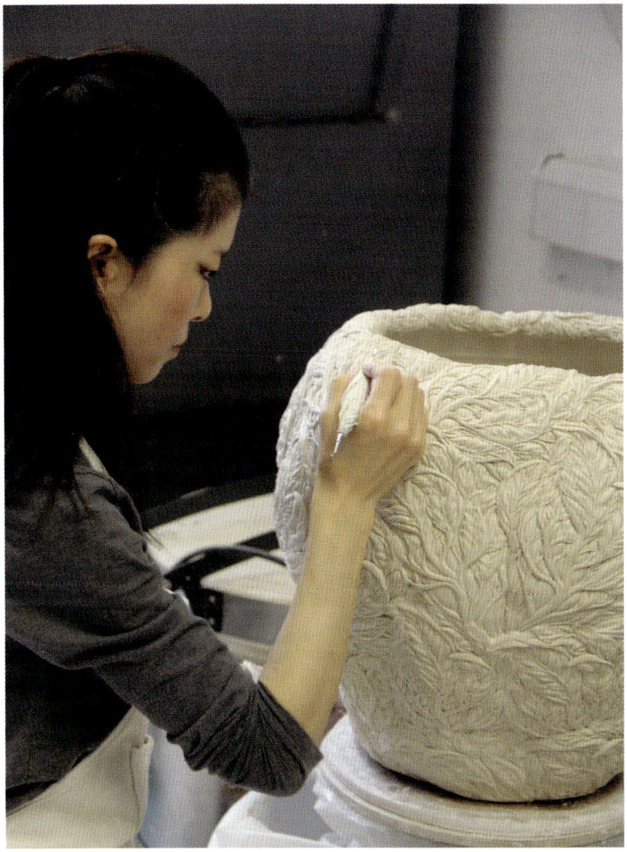
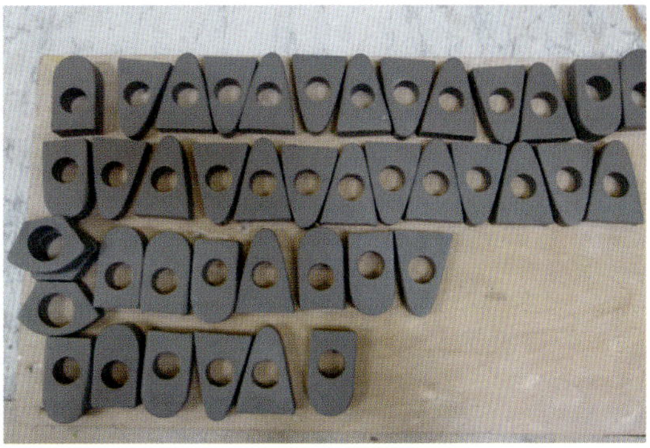
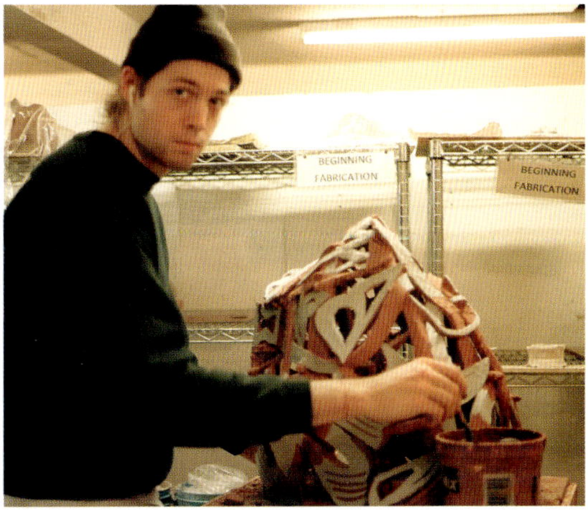
BEGINNING
FABRICATION
BEGINNING
FABRICATION
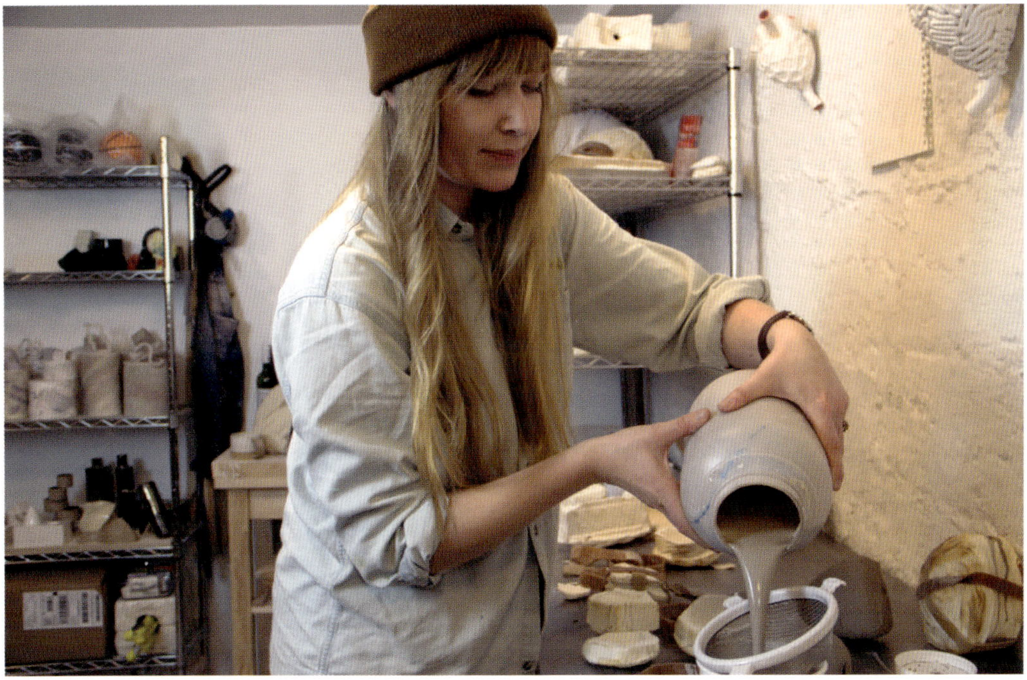

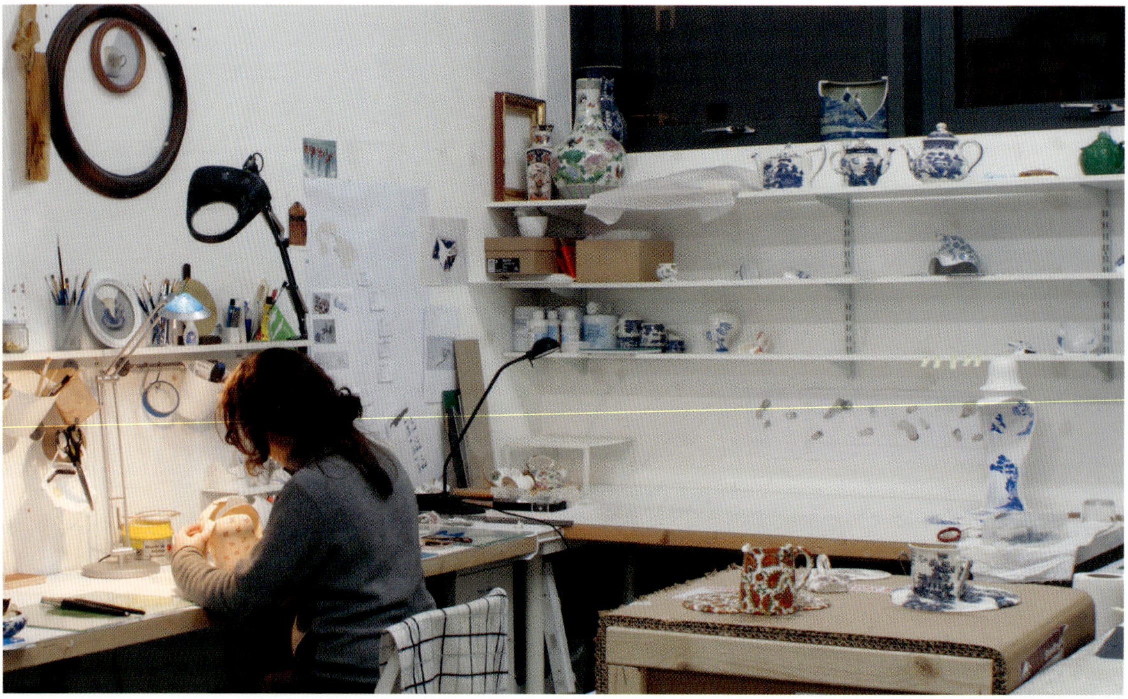

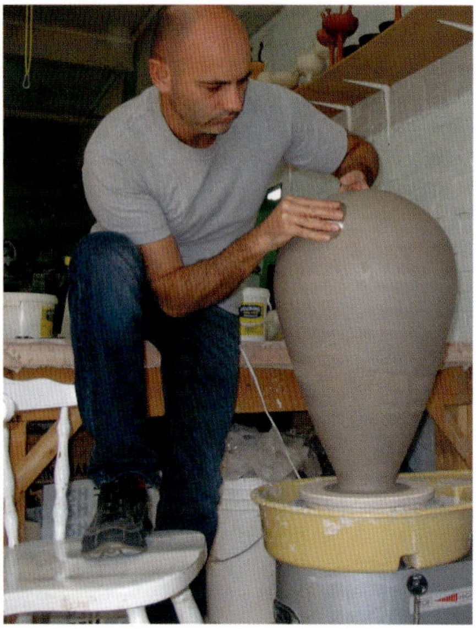

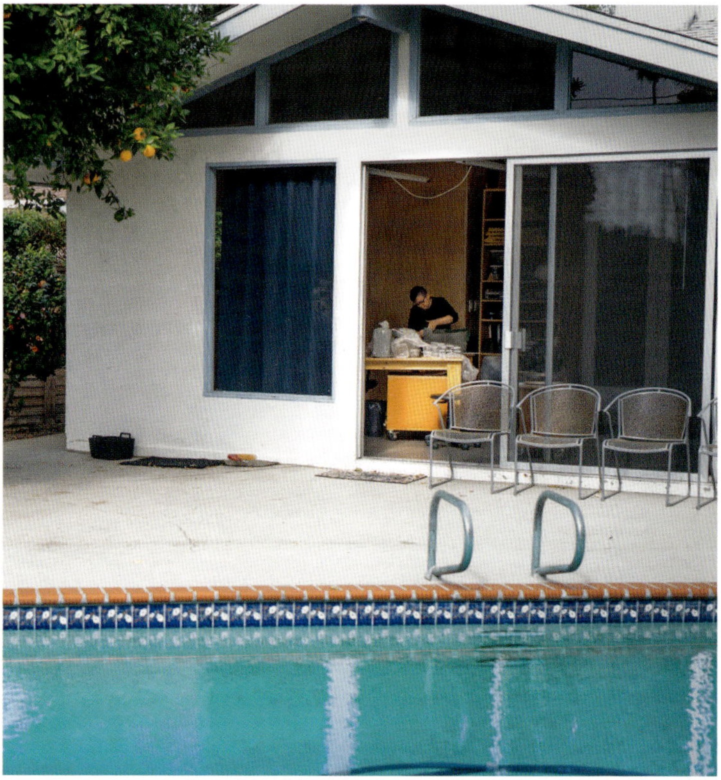

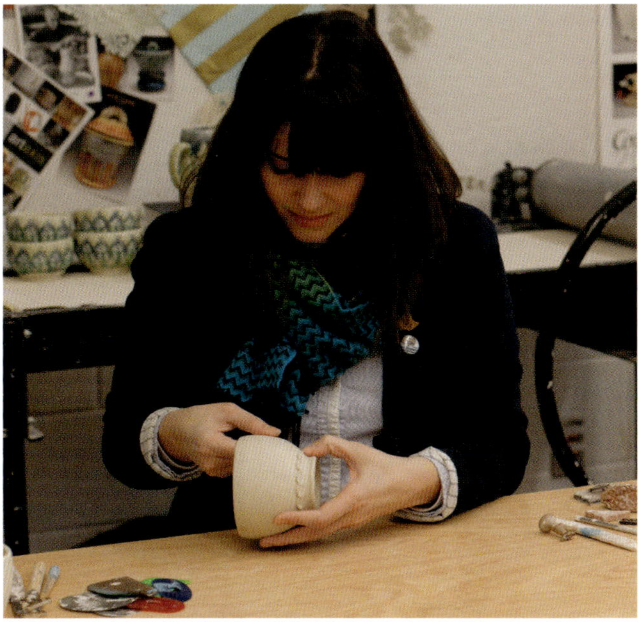
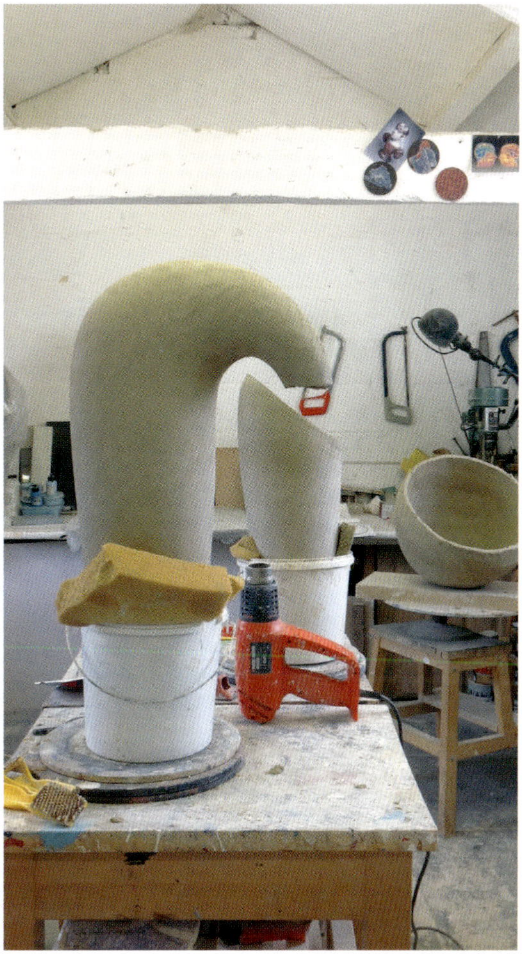
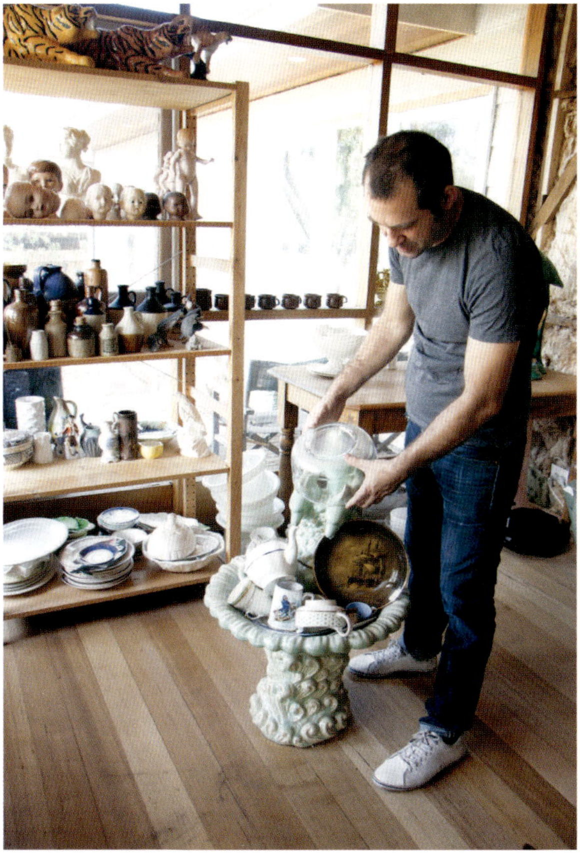
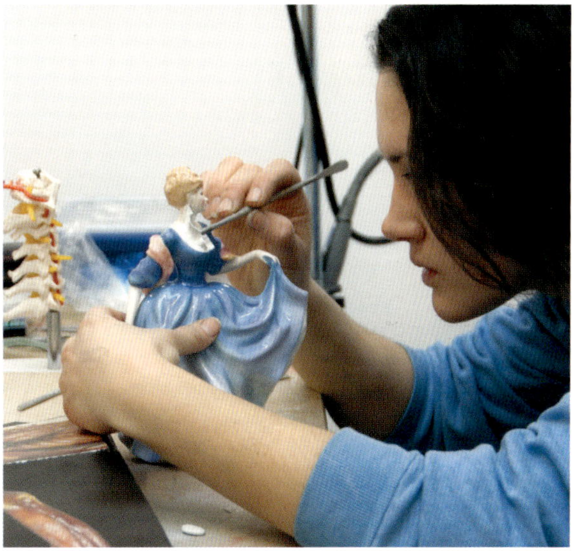

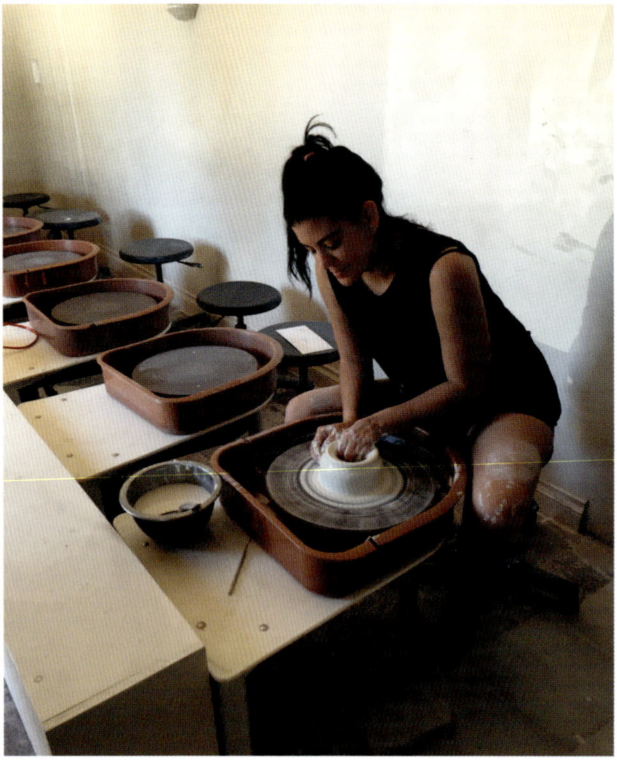

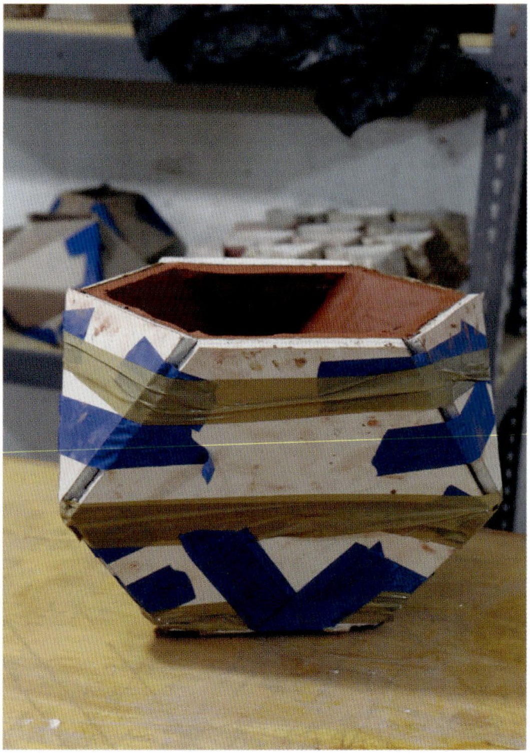

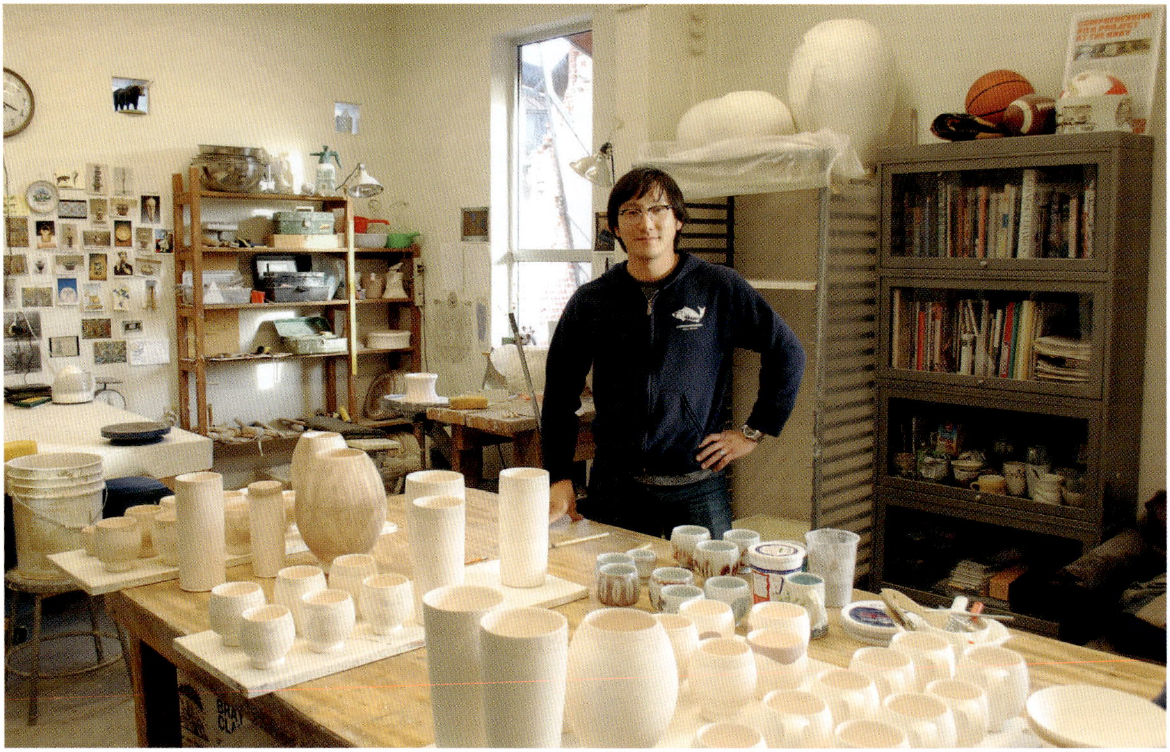

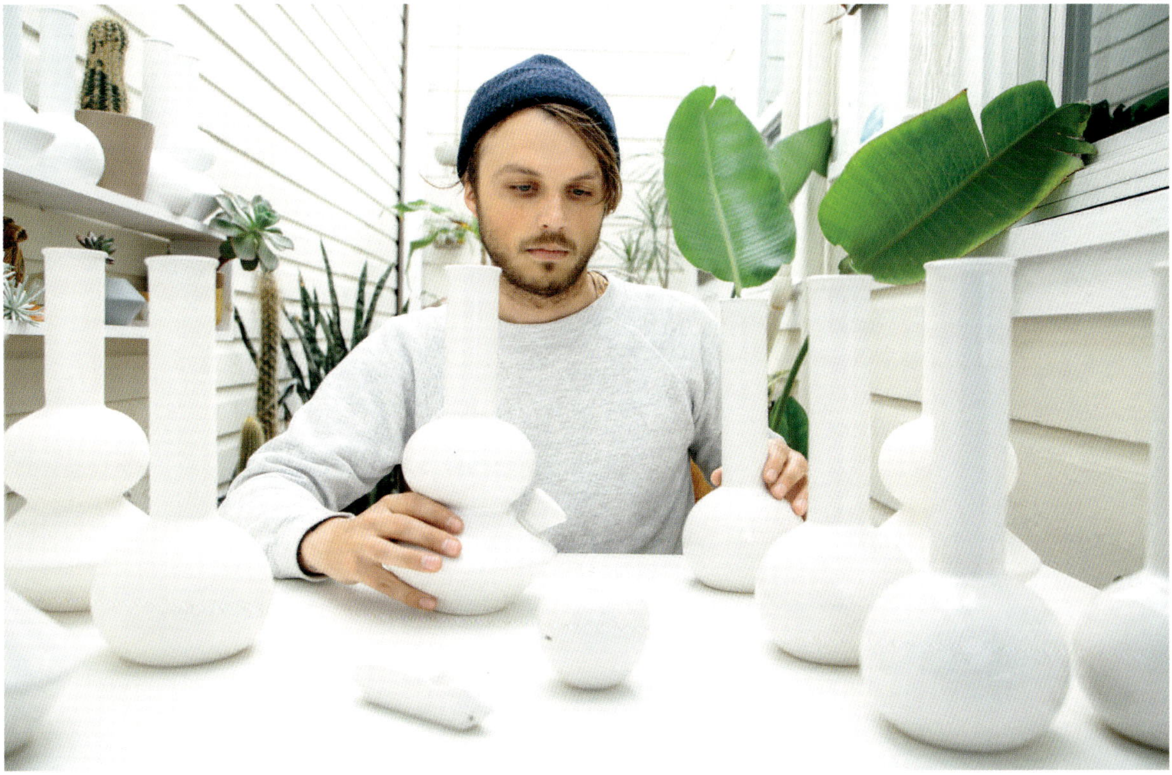

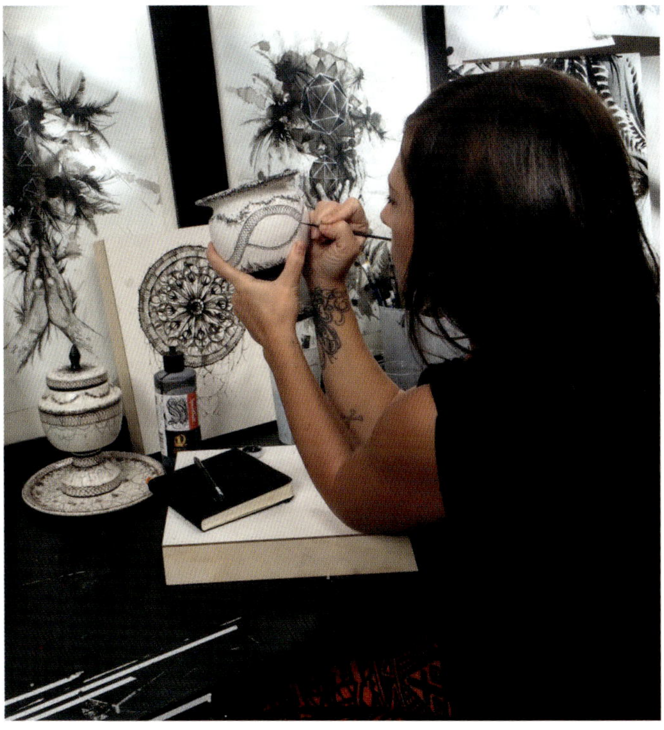

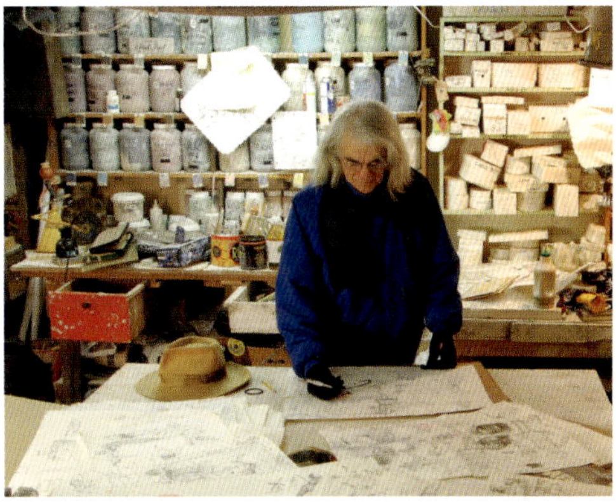

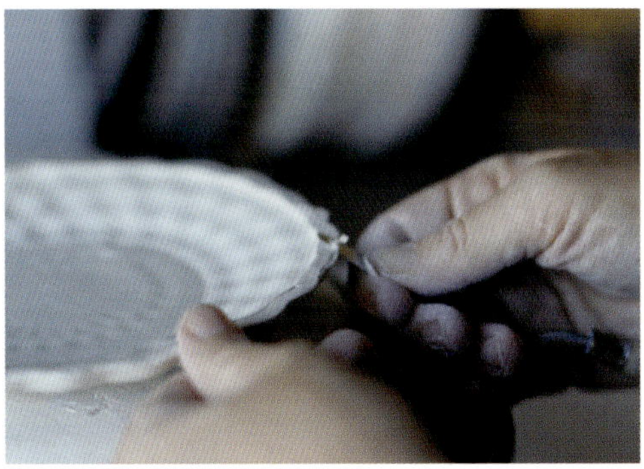

ARTISTS

BEN MEDANSKY
www.benmedansky.com

BODIL MANZ
www.bodilmanz.com

BRENDAN MONROE
www.brendanmonroe.com

CASE STUDYO
www.casestudyo.com

CHE JEN
www.chejen.com

CHRIS HAWKINS
www.chris-hawkins.co.uk

CLAIRE PARTINGTON
www.clairepartington.co.uk

CODY HOYT
www.codyhoyt.com

CRYSTAL MOREY
www.crystalmorey.com

GRIT CERAMICS
www.gritceramics.com

HEATH CERAMICS
www.heathceramics.com

HITOMI HOSONO
www.adriansassoon.com/hosono

JEN DWYER
www.jendwyerstudio.com

JENNIFER MCCURDY
www.jennifermccurdy.com

JESS RIVA COOPER
www.jessrivacooper.com

JESSICA HARRISON
www.jessicaharrison.co.uk

JON ALAMEDA
www.almedapottery.com

KATE MACDOWELL
www.katemacdowell.com

KATSUYO AOKI
www.katsuyoaoki.s1.bindsite.jp/Works.html

KIMBERLEE JOY ROTH
www.kimberleejoyroth.com

LAUREN SHAPIRO
www.laurenshapiroart.com

LAURENT CRASTE
www.laurentcraste.com

LINDSAY SCAYPTA
www.lindsayscypta.com

LIVIA MARIN
www.liviamarin.com

MARK WHALEN
www.markwhalenart.com

MATT WEDEL
www.mattwedel.com

MATTHEW GROVES
www.matthewgrovesceramics.com

NICK MAKANNA
www.nickmakanna.com

PATRICIA VOLK
www.patriciavolk.co.uk

PAUL WOOD
www.thisisnofantasy.com/artist/paul-wood/

PETER SHIRE
www.petershirestudio.com

RICHARD SHAW
www.richardshawart.com

STEVEN YOUNG LEE
www.stevenyounglee.com

SUMMERLAND CERAMICS
www.welcometosummer.land

TAKEUCHI KOUZO
www.kouzotakeuchi.com

TANYA BATURA
www.tanyabatura.com

THEONE CERAMICS
www.theoneceramics.com

ZEMER PELED
www.zemerpeled.com

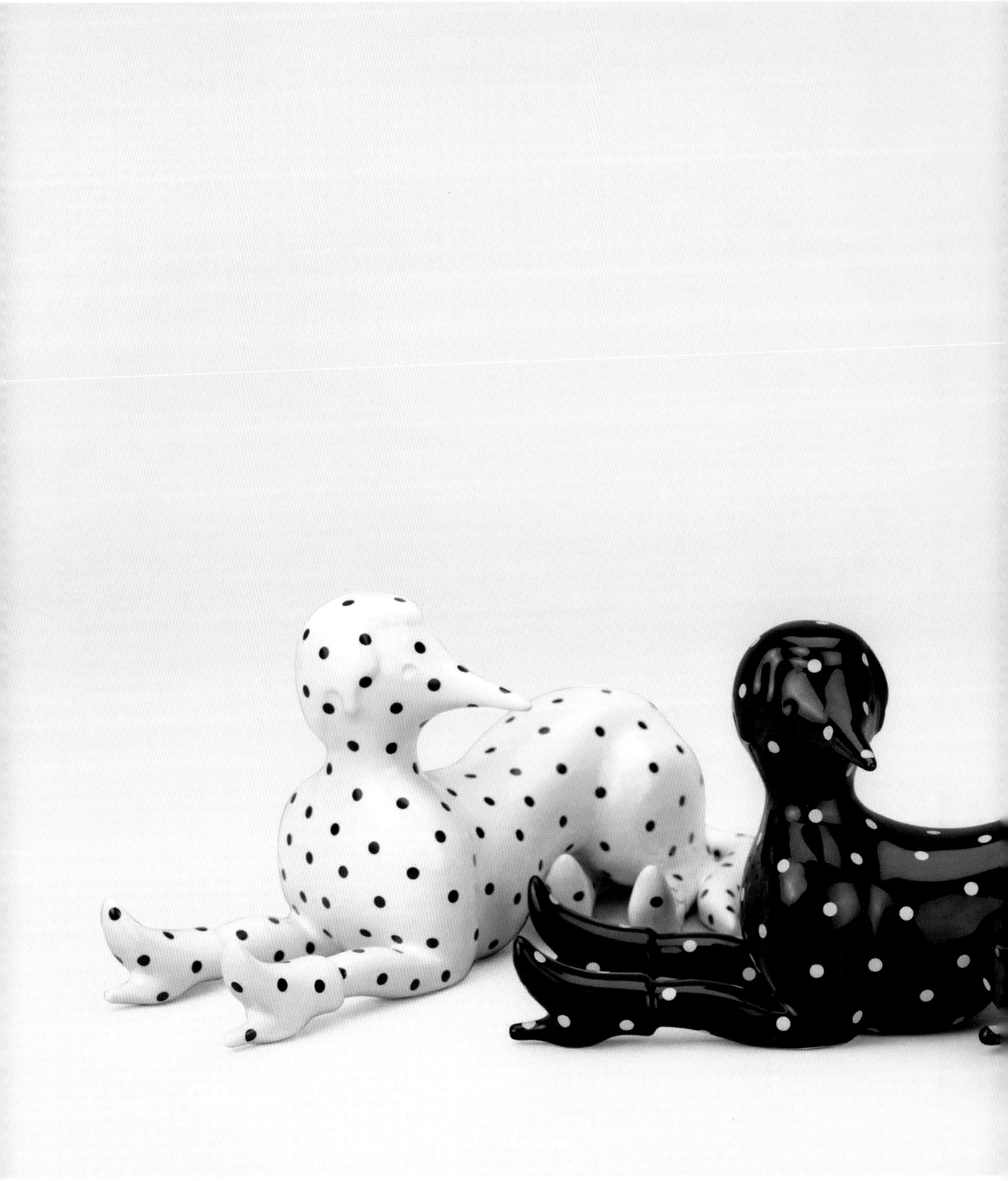

ACKNOWLEDGMENTS

I would like to thank all the artists for their generosity and permission to publish their works, as well as all the photographers who have graciously granted the rights to use their images. I am also infinitely grateful to many people whose names do not appear in the credits but who have provided support for this book. Without them, I would not have been able to share these gorgeous advances in the tradition of ceramics with readers and enthusiasts worldwide.

-Hannah Stouffer

Previous page Parra   Next page THEONE Ceramics

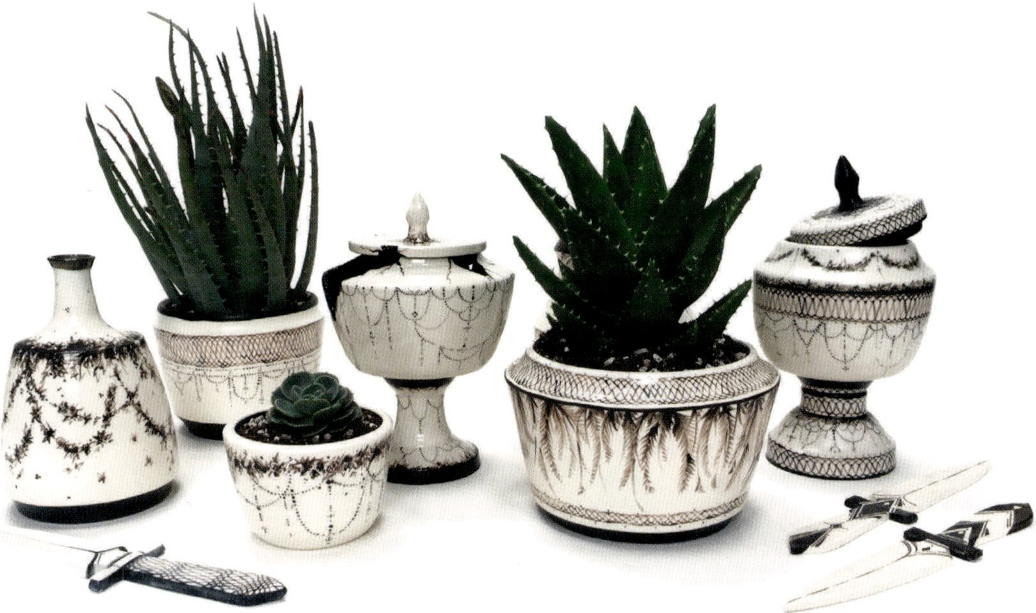